D0701014

Apollo 15

The NASA Mission Reports
Volume One

Compiled from the archives & edited
by Robert Godwin

Special thanks to:
Dave Scott
Jim Busby
Mark Kahn and NASA History Office HQ

All rights reserved under article two of the Berne Copyright Convention (1971).
We acknowledge the financial support of the Government of Canada through the
Book Publishing Industry Development Program for our publishing activities.
Published by Apogee Books an imprint of Collector's Guide Publishing Inc., Box 62034, Burlington, Ontario, Canada, L7R 4K2
Printed and bound in Canada
Apollo 15 - The NASA Mission Reports Volume One
by Robert Godwin
ISBN 1-896522-57-2
ISSN 1496-6921
Apogee Books Space Series
©2001 Apogee Books
All photos courtesy of NASA and Apogee Books

Apollo 15
The NASA Mission Reports
(from the archives of the National Aeronautics and Space Administration)

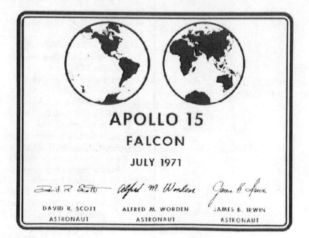

CONTENTS

PRESS KIT

ON THE MOON WITH APOLLO 15

PRE-MISSION OPERATION REPORT

POST MISSION OPERATION REPORT

TECHNICAL CREW DEBRIEFING

FOREWORD

In September 1967, NASA defined a sequence of missions of increasing complexity leading to a lunar landing mission. Complete success of all objectives of each type mission was required before proceeding to the next type. These mission types were lettered "A" through "J." Actual launches to attempt each lettered type mission would be numbered from Apollo 5 onward. Using this concept, a "D" type mission might have been launched three times before proceeding to an "E" type mission. Within this plan, "A" and "B" would be unmanned hardware test flights. "C" would be the first manned test of the Command and Service Module (CSM). "D" would be the first manned test of both the CSM and the Lunar Module (LM), including rendezvous and docking. "E" would be CSM and LM operations in a high Earth orbit, including reentry. "F" would be a full rehearsal in lunar orbit with the exception of the actual landing itself. And, of course, "G" would be the first lunar landing, but with a minimum stay time on the Moon. "H" would then be a follow-on to the lunar landing for maximum stay time using the same hardware and procedures as the "G" mission.

Finally, the "J" mission would be a major upgrade of the entire Apollo "system" including hardware, software and scientific equipment as well as significant advances in lunar surface operations and lunar orbit operations. The primary objective of the "J" missions was extended scientific exploration of the Moon, both on the surface and from orbit. This upgrade from "H" to "J" included the Lunar Roving Vehicle, double the lunar stay time, double the EVA surface excursion time, significantly more scientific equipment and experiments, and many other enhancements.

The remarkable success of Apollo was typified by the complete success of each type mission on the first attempt, leading to the first "G" mission being launched as Apollo 11. To actually go beyond the Apollo 11 "G" mission demonstrated considerable courage and confidence, especially after achieving the political objective of "landing a man on the Moon and returning him safely to Earth." But to advance beyond "G" and "H" into even one "J" mission, required a very bold and aggressive decision, especially after the near-tragic loss of the Apollo 13 crew. Just three months after that near disaster, and in the face of dwindling public support and a rapidly declining budget, NASA-of-Apollo decided to skip the final "H" type mission; press on with upgrading the "system" (hardware, software, science, and operations) to the "J" configuration, and launch not just one, but three full-up "J" missions to the most significant scientific sites on the Moon.

This commitment to the extended scientific exploration "J" missions was surely one of the most rewarding decisions of the Apollo Program. It would have been a lot easier, safer, and cheaper to finish the program with the final two "H" missions as scheduled (for if one of the final missions were to be a failure, the Program would surely end, and "Apollo" would forever have been considered a "failure"). Fortunately for the overall results and success of Apollo, NASA-of-Apollo truly made the "right" decision!

The importance of the "J" missions can perhaps best be illustrated by comparing the intellectual resources invested in the early voyages of scientific exploration of the Earth with those invested in the early voyages of scientific exploration of the Moon.

Before Captain James Cook embarked on his historic first voyage to the South Pacific and Australia, he invested three months in its preparation, using his crew of 70 to prepare the bark *Endeavour*; gather equipment, provisions, and instruments; and generally plan the expedition. After departing from Plymouth on 26 August 1768, he spent almost three years conducting the first truly scientific expedition by sea, arriving back in England on 15 July 1771. During the voyage, Captain Cook maintained a handwritten journal describing significant events and activities in several lines of narrative each day. His original *Journal* was not published until 22 years after he returned (London, Elliot Stock, 1893). A comprehensive analysis of his entries, including a detailed description of his ship, equipment, and instruments, scientific results, and anecdotes of interest, was not published until almost 185 years later (Beaglehole, *The Journals of Captain James Cook on his voyages of discovery*, Cambridge, University Press, Vol. I, 1955).

In comparison, NASA-of-Apollo invested 20 months in the preparation of the Apollo 15 mission — using more than 100,000 people to prepare the launch vehicles and spacecraft; gather equipment, provisions, and instruments; and generally plan the expedition. After departing Cape Kennedy on 26 July 1971, the crew spent 12 days conducting the first extended scientific exploration of the Moon. However, because of the enormous amount of intellectual capital invested in preparing the mission (100,000 times 20 months) as well as the added dimensions of "modern technology," the exploratory information gathered during the mission could not be comprehensively recounted in brief daily journal entries (as per Captain Cook). Only now, with the time available to compile and analyze these highly-compressed minute-by-minute events and activities do we see the emergence of comprehensive records of the mission to include technical explanations, preliminary scientific results, and anecdotes of interest.

The significance of Apollo 15, the first of the "J" missions, is perhaps best summarized by the FOREWORD to the *Apollo 15 Preliminary Science Report:*

> In richness of scientific return, the Apollo 15 voyage to the plains at Hadley compares with voyages of Darwin's *H.M.S. Beagle*, and those of the *Endeavour* and *Resolution*. Just as those epic ocean voyages set the stage for a revolution in the biological sciences and exploration generally, so also the flight of *Falcon* and *Endeavour* did the same in planetary and Earth sciences and will guide the course of future explorations.
>
> The boundary achievements of Apollo 15 cannot now be established. As the author of a following paper points out, the mission was not finished at splashdown in the Pacific, nor later with painstaking analysis in scores of laboratories of the samples and cores brought back, nor with careful study of the photographic imagery and instrument traces returned home. For the distinctive fact is that the mission is not yet over. Data still flows in daily from the isotope-powered station emplaced on the plain at Hadley, and from the Moon-encircling scientific satellite left in orbit. This data flow is of exceptional value because it now affords, for the first time, a triangulation of lunar events perceived by the three physically separated scientific stations that man has left on the Moon.
>
> This volume is the first, though assuredly not the final, effort to assemble a comprehensive accounting of the scientific knowledge so far acquired through this remarkable mission.
>
> Dr. James C. Fletcher
> *Administrator*
> *National Aeronautics and Space Administration*
> December 8, 1971

Both NASA-of-Apollo and the crew of Apollo 15 are grateful to the editors and compilers of the "NASA Mission Reports" for their contributions to the historical records of spaceflight in general, and to Apollo 15, the first "J" mission, in particular.

David R. Scott
July 2001

INTRODUCTION

By July 1971 the Apollo hardware had lived up to expectations. Now it was time for hard science.

As Commander Scott clearly describes in his Foreword, a brash decision was taken to push the Apollo/Saturn hardware to its design limitations for the last three missions. This desire to achieve as much as possible is nowhere more clearly illustrated than in the following pages. Unlike previous missions, the Apollo 15 documents stress the importance of the science to be accomplished, by concentrating less on the basics, such as diet and launch pad facilities, and zeroing in on the actual experiments.

Apollo 15 was capable of taking a full two tons of additional useful equipment into space. This was achieved by narrowing the margins in every conceivable manner. For example the launch window was deliberately placed in the summer as this reduced the inevitable winter wind factor at launch time. The Saturn V had its reserve propellant margins reduced and was launched into a slighter lower Earth Parking Orbit. Even the mighty F-1 engines were reorificed to provide more "bang-for-the-buck". The additional 4,000 pounds of payload capacity would be spoken for right down to the ounce.

The most significant addition of equipment was the first lunar car. For the first time a crew of explorers were to drive a vehicle on another world. The Lunar Rover was NASA's version of the hardy military Jeep.

In the early 1940's the United States Army asked contractors to submit proposals for an all-terrain vehicle which could meet a series of stringent requirements. It had to be tough, lightweight, high-powered and capable of enduring harsh environments. The car had to be delivered for trials within 49 days. With only hours to spare on 23rd September 1940, the Bantam motor company delivered the *Bantam* LRV (Light Reconaissance Vehicle), the first Jeep, and a legend was born. As remarkable as the original Jeep was, it barely holds a candle to the *Boeing* LRV (Lunar Roving Vehicle). The *Bantam* had a weight limit of 1275 lbs and ran on a 45 horse-power engine. Its payload was 600 lbs and it had a low crawling speed of 5 kph. It was able to navigate steep inclines and it used four wheel drive. It was also capable of low-speed cooling.

The *Boeing* LRV was an all-terrain vehicle which could be folded in half. It was delivered to NASA on March 14 1971 with two weeks to spare and 17 months after the signing of the contract. It had a "curb-weight" of 462 lbs on the Earth and yet it was capable of carrying 1080 lbs of cargo. It used four electric motors which delivered a total of one horse-power to its four-wheel drive system and had a top speed of 14 kph. It was able to maintain its balance in one-sixth gravity on slopes up to 45°, while its thermal control had to contend with temperatures ranging from -200 to +400 degrees F°. Its wheels could withstand driving over solid rock without the benefit of inflatable tires. It had double redundancy on most of its major components such as power and steering. It was powered by two electric batteries while carrying a portable television station with a range of a quarter of a million miles. It also had an onboard navigation system that would tell the driver where it was, where it was going and where it had been. Slightly more than thirty years had passed but I bet Karl Probst, the designer of the Bantam, would have been proud and amazed. Even if Boeing did have a bigger budget...

The sheer audacity of the full-up "J" missions is forgotten by most people after thirty years. Apollo 15 was the first and arguably the most seductive of these flights in part due to the astonishing landscape of Hadley-Apennine. David Scott and Jim Irwin would take their fully-loaded Lunar Module and take it straight down between mountains almost three miles high and land amidst one of the richest geologic treasure troves ever encountered by man. Not to be outdone Command Module Pilot Alfred Worden would run through a daunting schedule of science for three days in lunar orbit. The so-called "SIM" bay aboard the *Endeavour* housed a battery of scientific equipment which would photograph, catalog, scan and probe the lunar surface.

After accomplishing everything in the flight-plan, and more, *Falcon* blasted off to a perfect rendezvous with *Endeavour* before the crew fired the engines to come home. On that return voyage Worden would become the first human to perform a spacewalk in deep space, while presumably becoming the first person ever to see the Earth and the Moon in their entirety simply by turning his head.

In the following pages the reader will find an abundance of scientific information as well as details about the amazing lunar rover. However, in the interests of providing some lighter reading along the lines of earlier Apollo Press Kits I have chosen to include the short booklet "On The Moon with Apollo 15" written prior to the mission by NASA scientist Gene Simmons. It provides an easy layman's approach to the astonishing accomplishments of the Apollo team and the dauntless crew of the *Endeavour*.

Robert Godwin (Editor)
July 2001

NATIONAL AERONAUTICS AND SPACE ADMINISTRATION
WASHINGTON, D.C. 20546

TELS. WO 2-4155 WO 3-6925
FOR RELEASE: THURSDAY A.M.
July 15, 1971
RELEASE NO: 71-119K

PROJECT: APOLLO 15 - PRESS KIT

(To be launched no earlier than July 26)

NASA NEWS

NATIONAL AERONAUTICS AND SPACE ADMINISTRATION
WASHINGTON, D.C. 20546
TELS: (202) 963-6925 (202) 962-4155

Ken Atchison/Howard Allaway
(Phone 202/962-0666)

FOR RELEASE: THURSDAY, A.M.
July 15, 1971

RELEASE NO: 71-119

APOLLO 15 LAUNCH JULY 26

The 12-day Apollo 15 mission, scheduled for launch on July 26 to carry out the fourth United States manned exploration of the Moon, will:

Double the time and extend tenfold the range of lunar surface exploration as compared with earlier missions;
Deploy the third in a network of automatic scientific stations;
Conduct a new group of experiments in lunar orbit; and
Return to Earth a variety of lunar rock and soil samples.

Scientists expect the results will greatly increase man's knowledge both of the moon's history and composition and of the evolution and dynamic interaction of the Sun-Earth system.

This is so because the dry, airless, lifeless Moon still bears records of solar radiation and the early years of solar system history that have been erased from Earth. Observations of current lunar events also may increase understanding of similar processes on Earth, such as earthquakes.

The Apollo 15 lunar module will make its descent over the Apennine peaks, one of the highest mountain ranges on the Moon, to land near the rim of the canyon-like Hadley Rille. From this Hadley-Apennine lunar base, between the mountain range and the rille, Commander David R. Scott and Lunar Module Pilot James B. Irwin will explore several kilometers from the lunar module, driving an electric-powered lunar roving vehicle for the first time on the Moon.

Scott and Irwin will leave the lunar module for three exploration periods to emplace scientific experiments on the lunar surface and to make detailed geologic investigations of formations in the Apennine foothills, along the Hadley Rille rim, and to other geologic structures.

The three previous manned landings were made by Apollo 11 at Tranquillity Base, Apollo 12 in the Ocean of Storms and Apollo 14 at Fra Mauro.

The Apollo 15 mission should greatly increase the scientific return when compared to earlier exploration missions. Extensive geological sampling and survey of the Hadley-Apennine region of the Moon will be enhanced by use of the lunar roving vehicle and by the improved life support systems of the lunar module and astronaut space suit. The load-carrying capacity of the lunar module has been increased to permit landing a greater payload on the lunar surface.

Additionally, significant scientific data on the Earth-Sun-Moon system and on the Moon itself will be gathered by a series of lunar orbital experiments carried aboard the Apollo command/service modules. Most of the orbital science tasks will be accomplished by Command Module Pilot Alfred M. Worden, while his comrades are on the lunar surface. Worden is a USAF major, Scott a USAF colonel and Irwin a USAF lieutenant colonel.

During their first period of extravehicular activity (EVA) on the lunar surface, Scott and Irwin will drive the lunar roving vehicle to explore the Apennine front. After returning to the LM, they will set up the Apollo Lunar Surface Experiment Package (ALSEP) about 300 feet West of the LM.

Experiments in the Apollo 15 ALSEP are: passive seismic experiment for continuous measurement of moonquakes and meteorite impacts; lunar surface magnetometer for measuring the magnetic field at the lunar surface; solar wind spectrometer for measuring the energy and flux of solar protons and electrons reaching the Moon; suprathermal ion detector for measuring density of solar wind high and low-energy ions; cold cathode ion gauge for measuring variations in the thin lunar atmosphere; and the heat flow experiment to measure heat emanating from beneath the lunar surface.

Scott and Irwin will use for the first time a percussive drill for drilling holes in the Moon's crust for placement of the heat flow experiment sensors and for obtaining samples of the lunar crust.

Additionally, two experiments independent of the ALSEP will be set up near the LM. They are the solar wind composition experiment for determining the isotopic makeup of noble gases in the solar wind; and the laser ranging retro-reflector experiment which acts as a passive target for Earth-based lasers in measuring Earth-Moon distances over a long-term period. The solar wind composition experiment has been flown on all previous missions, and the laser reflector experiment was flown on Apollos 11 and 14. The Apollo 15 reflector has three times more reflective area than the two previous reflectors.

The second EVA will be spent in a lengthy geology traverse in which Scott and Irwin will collect documented samples and make geology investigations and photopanoramas at a series of stops along the Apennine front.

The third EVA will be a geological expedition along the Hadley Rille and northward from the LM.

At each stop in the traverses, the crew will re-aim a high-gain antenna on the lunar roving vehicle to permit a television picture of their activities to be beamed to Earth.

A suitcase-size device — called the lunar communications relay unit — for the first time will allow the crew to explore beyond the lunar horizon from the LM and still remain in contact with Earth. The communications unit relays two-way voice, biomedical telemetry and television signals from the lunar surface to Earth. Additionally, the unit permits Earth control of the television cameras during the lunar exploration.

Experiments in the Scientific Instrument Module (SIM) bay of the service module are: gamma-ray spectrometer and X-ray fluorescence which measure lunar surface chemical composition along the orbital ground track; alpha-particle spectrometer which measures alpha-particles from radioactive decay of radon gas isotopes emitted from the lunar surface; mass spectrometer which measures the composition and distribution of the lunar atmosphere; and a subsatellite carrying three experiments which is ejected into lunar orbit for relaying scientific information to Earth on the Earth's magnetosphere and its interaction with the Moon, the solar wind and the lunar gravity field.

The SIM bay also contains equipment for orbital photography including a 24-inch panoramic camera, three-inch mapping camera and a laser altimeter for accurately measuring spacecraft altitude for correlation with the mapping photos.

Worden will perform an inflight EVA to retrieve the exposed film. Selected flight experiments will be conducted during transearth coast.

Scheduled for launch at 9:34 a.m. EDT, July 26, from NASA's Kennedy Space Center, Fla., the Apollo 15 will land on the Moon on Friday July 30. The lunar module will remain on the surface about 67 hours. Splashdown will be at 26.1° North latitude by 158° West longitude in the North Central Pacific, north of Hawaii.

The prime recovery ship for Apollo 15 is the helicopter landing platform USS Okinawa.

Apollo 15 command module call sign is "Endeavour," and the lunar module is "Falcon." As in all earlier lunar landing missions, the crew will plant an American Flag on the lunar surface near the landing point. A plaque with the date of the Apollo 15 landing and signatures of the crew will be affixed to the LM front landing gear.

Apollo 15 backup crewmen are USN Capt. Richard F. Gordon, Jr., commander; Mr. Vance Brand, command module pilot; and Dr. Harrison H. Schmitt, lunar module pilot.

APOLLO 15 INCREASED OPERATIONAL CAPABILITIES

IMPROVEMENT	SYSTEM	CAPABILITY
MOBILITY	• LUNAR ROVER VEHICLE	INCREASED RANGE, CREW MOBILITY, TRAVERSE PAYLOAD CAPACITY AND EFFICIENCY OF SURFACE OPERATIONS
	• LCRU/GCTA	
	• A7LB SUIT	
EVA	• -7PLSS	IMPROVED LIFE SUPPORT SYSTEM DURATION INCREASES TOTAL EVA DURATION FROM 18 TO 40 MANHOURS
SURFACE DURATION	• LM	VEHICLE MODIFICATIONS PERMITTED NOMINAL LUNAR SURFACE STAY TIME ABOUT DOUBLE. (FROM 37 TO 67 HOURS)
ORBITAL SCIENCE	• CM/SM	ADDED SIM BAY AND EXPERIMENT CONTROLS TO PERMIT CONDUCTING ADDITIONAL ORBITAL SCIENCE
PAYLOAD CAPABILITY	• SATURN V	LAUNCH VEHICLE CAPABILITY INCREASED TO ACCOMODATE THE INCREASED WEIGHT OF THE PRIOR ITEMS

MISSION COMPARISON SUMMARY

	APOLLO 14	APOLLO 15
LAUNCH WINDOWS	1-3-3	2-2-3
LAUNCH WINDOW DURATION	3.5 HOURS	2.5 HOURS
LAUNCH AZIMUTH	72 - 96 DEGREES	80 -100 DEGREES
EARTH PARKING ORBIT	100 NM	90 NM
SPACECRAFT PAYLOAD	102,095 POUNDS	107,500 POUNDS
TRANSLUNAR TRAJECTORY	TRANSFER MANEUVER	NO TRANSFER MANEUVER
LUNAR ORBIT INCLINATION	14 DEGREES	26 DEGREES
SCIENTIFIC INSTRUMENT MODULE	NO	LUNAR ORBIT & TRANSEARTH
LUNAR DESCENT TRAJECTORY	16 DEGREES	25 DEGREES
POST LUNAR LANDING	EVA-1	SEVA AND SLEEP
EVA's	2 (4:45 AND 4:30)	3 (7-7-6)
LUNAR SURFACE STAY TIME	33.5 HOURS	67 HOURS
SUBSATELLITE DEPLOYMENT	NO	REV 74
TRANSEARTH EVA	NO	ONE HOUR
EARTH LANDING	27 DEGREES SOUTH	26 DEGREES NORTH
MISSION DURATION	9 DAYS	12 DAYS, 7 HOURS

COUNTDOWN

The Apollo 15 launch countdown will be conducted by a government-industry team of about 500 working in two control centers at the Kennedy Space Center.

Overall space vehicle operations will be controlled from Firing Room No. 1 in the Complex 39 Launch Control Center. The spacecraft countdown will be run from an Acceptance Checkout Equipment (ACE) room in the Manned Spacecraft Operations Building.

More than five months of extensive checkout of the launch vehicle and spacecraft components are completed before the space vehicle is ready for the final countdown. The prime and backup crews participate in many of these tests including mission simulations, altitude runs, a flight readiness test and a countdown demonstration test.

The space vehicle rollout - the three and one-half-mile trip from the Vehicle Assembly Building to the launch pad - took place May 11.

Apollo 15 will be the ninth Saturn V launch from Pad A (seven manned). Apollo 10 was the only launch to date from Pad B, which will be used again in 1973 for the Skylab program.

The Apollo 15 precount activities will start at T-5 days. The early tasks include electrical connections and pyrotechnic installation in the space vehicle. Mechanical buildup of the spacecraft is completed, followed by servicing of the various gases and cryogenic propellants (liquid oxygen and liquid hydrogen) to the CSM and LM. Once this is accomplished, the spacecraft batteries are placed on board and the fuel cells are activated. The final countdown begins at T-28 hours when the flight batteries are installed in the three stages and instrument unit of the launch vehicle.

At the T-9 hour mark, a built-in hold of nine hours and 34 minutes is planned to meet contingencies and provide a rest period for the launch crew. A one hour built-in hold is scheduled at T-3 hours 30 minutes.

Following are some of the highlights of the latter part of the count:

T-10 hours, 15 minutes	Start mobile service structure (MSS) move to park site
T-9 hours	Built-in hold for nine hours and 34 minutes. At end of hold, pad is cleared for LV propellant loading.
T-8 hours, 05 minutes	Launch vehicle propellant loading. Three stages (LOX in first stage, LOX and LH_2 in second and third stages). Continues thru T-3 hours 38 minutes.
T-4 hours, 15 minutes	Flight crew alerted.
T-4 hours, 00 minutes	Crew medical examination.
T-3 hours, 30 minutes	Crew breakfast.
T-3 hours, 30 minutes	One-hour built-in hold.
T-3 hours, 06 minutes	Crew departs Manned Spacecraft Operations Building for LC-39 via transfer van.
T-2 hours, 48 minutes	Crew arrival at LC-39.
T-2 hours, 40 minutes	Start flight crew ingress.
T-1 hours, 51 minutes	Space Vehicle Emergency Detection System (EDS) test (Scott participates along with launch team).
T-43 minutes	Retract Apollo access arm to stand by position (12 degrees).
T-42 minutes	Arm launch escape system. Launch vehicle power transfer test, LM switch to internal power.
T-37 minutes	Final launch vehicle range safety checks (to 35 minutes).
T-30 minutes	Launch vehicle power transfer test, LM switch over to internal power.
T-20 minutes to T-10 minutes	Shutdown LM operational instrumentation.
T-15 minutes	Spacecraft to full internal power.
T-6 minutes	Space vehicle final status checks.
T-5 minutes, 30 seconds	Arm destruct system.
T-5 minutes	Apollo access arm fully retracted.
T-3 minutes, 6 seconds	Firing command (automatic sequence).
T-50 seconds	Launch vehicle transfer to internal power.
T-8.9 seconds	Ignition start.
T-2 seconds	All engines running.
T-0	Liftoff.

NOTE: Some changes in the countdown are possible as a result of experience gained in the countdown demonstration test which occurs about two weeks before launch.

Launch Windows

Launch date	Windows (EDT) Open	Close	Sun Elevation Angle
July 26, 1971	9:34 am	12:11 pm	12.0° *
July 27, 1971 (T+24)	9:37 am	12:14 pm	23.2°
Aug. 24, 1971 (T-0)	7:59 am	10:38 am	11.3°
Aug. 25, 1971 (T+24)	8:17 am	10:55 am	22.5°
Sept. 22, 1971 (T-24)	6:37 am	9:17 am	12.0°
Sept. 23, 1971 (T-0)	7:20 am	10:00 am	12.0°
Sept. 24, 1971 (T+24)	8:33 am	11:12 am	23.0°

* Only for launch azimuth of 80°

Ground Elapsed Time Update

It is planned to update, if necessary, the actual ground elapsed time (GET) during the mission to allow the major flight plan events to occur at the pre-planned GET regardless of either a late liftoff or trajectory dispersions that would otherwise have changed the event times.

For example, if the flight plan calls for descent orbit insertion (DOI) to occur at GET 82 hours, 40 minutes and the flight time to the Moon is two minutes longer than planned due to trajectory dispersions at translunar injection, the GET clock will be turned back two minutes during the translunar coast period so that DOI occurs at the pre-planned time rather than at 82 hours, 42 minutes. It follows that the other major mission events would then also be accomplished at the preplanned times.

Updating the GET clock will accomplish in one adjustment what would otherwise require separate time adjustments for each event. By updating the GET clock, the astronauts and ground flight control personnel will be relieved of the burden of changing their checklists, flight plans, etc.

The planned times in the mission for updating GET will be kept to a minimum and will, generally, be limited to three updates. If required, they will occur at about 53, 97 and 150 hours into the mission. Both the actual GET and the update GET will be maintained in the MCC throughout the mission.

Launch and Mission Profile

The Saturn V launch vehicle (SA-510) will boost the Apollo 15 spacecraft from Launch Complex 39A at the Kennedy Space Center, Fla., at 9:34 a.m. EDT, July 26, 1971, on an azimuth of 80 degrees.

The first stage (S-1C) will lift the vehicle 38 nautical miles above the Earth. After separation the booster will fall into the Atlantic Ocean about 367 nautical miles downrange from Cape Kennedy, approximately nine minutes, 21 seconds after liftoff.

The second stage (S-II) will push the vehicle to an altitude of about 91 nautical miles. After separation, the S-II stage will follow a ballistic trajectory as it plunges into the Atlantic about 2,241 nautical miles downrange from Cape Kennedy about 19 minutes, 41 seconds into the mission.

The single engine of the third stage (S-IVB) will insert the vehicle into a 90-nautical-mile circular parking orbit before it is cut off for a coasting period. When reignited, the engine will inject the Apollo spacecraft into a translunar trajectory.

Launch Events

Time Hrs Min Sec	Event	Vehicle Wt (Pounds)	Altitude (Feet)	Velocity (Ft/Sec)	Range (Nau Mi)
00 00 00	First Motion	6,407,758	198	0	0
00 01 20	Maximum Dynamic Pressure	4,048,843	42,869	1,605	3
00 02 15.8	S-1C Center Engine Cutoff	2,388,283	155,162	5,573	26
00 02 38.7	S-1C Outboard Engines Cutoff	1,841,856	225,008	7,782	48
00 02 40.5	S-1C/S-II Separation	1,477,783	230,893	7,799	50
00 02 42.2	S-II Ignition	1,477,782	236,196	7,778	52
00 03 10.5	S-II Aft Interstage Jettison	1,406,067	320,265	8,116	86
00 03 16.2	Launch Escape Tower Jettison	1,383,533	335,636	8,210	93
00 07 38.8	S-II Center Engine Cutoff	651,648	584,545	17,362	594
00 09 9.4	S-II Outboard Engines Cutoff	476,526	576,526	21,551	876
00 09 10.4	S-II/S-IVB Separation	476,155	576,535	21,560	880
00 09 13.5	S-IV33 Ignition	377,273	576,529	21,564	890
00 11 38.8	S-IVB First Cutoff	309,898	563,570	24,233	1,422
00 11 48.8	Parking Orbit Insertion (90 nm)	309,771	563,501	24,237	1,461

Mission Events

Events	GET hrs:min	Date/EDT	Velocity change feet/sec	Purpose and resultant orbit
Translunar injection (S-IVB engine ignition)	02:56	26/12:30 pm	10,036	Injection into translunar trajectory with 68 nm pericynthion
CSM separation, docking	03:20	26/12:54 pm	—	Mating of CSM and LM
Ejection from SLA	04:15	26/01:49 pm	1	Separates CSM-LM from S-IVB/SLA
S-IVB evasive maneuver	04:39	26/2:13 pm	10	Provides separation prior to S-IVB propellant dump and thruster maneuver to cause lunar impact
Residual Propellant Dump	05:00	26/2:34 pm		
APS Impact Burn (4 min.)	05:45	26/3:19 pm		
APS Correction Burn	09:30	26/7:04 pm		
Midcourse correction 1	TLI+9 hrs	26/9:29 pm	0*	*These midcourse corrections have a nominal velocity change of 0 fps, but will be calculated in real time to correct TLI dispersions; trajectory within capability of docked DPS burn should SPS fail to ignite.
Midcourse correction 2	TLI+28 hrs	27/4:29 pm	0	
Midcourse correction 3	LOI-22 hrs	28/6:05 pm	0*	
Midcourse correction 4	LOI-5 hrs	29/11:05 am	0*	
SIM Door jettison	LOI-4.5 hrs	29/11:35 am	9	
Lunar orbit insertion	78:33 (Thurs.)	29/4:07 pm	-2,998	Inserts Apollo 15 into 58 X 170 nm elliptical lunar orbit
S-IVB impacts lunar	79:13	29/4:47 pm	—	Seismic event for Apollo 12 and 14 surface passive seismometers
Descent orbit insertion (DOI)	82:40	29/8:14 pm	-207	SPS burn places CSM/LM into 8 X 58 nm lunar orbit
CSM-LM undocking	100:14	30/1:48 pm	-	
CSM circularization	101.35	30/3:09 pm	70	Inserts CSM into 54 X 65 nm orbit (SPS burn)
LM Powered descent	104:29	30/6:03 pm	6,698	Three-phase DPS burn to brake LM out of initiation transfer orbit, vertical descent and touchdown on lunar surface
LM touchdown on lunar surface	104:41 (Friday)	30/6:15 pm	—	Lunar exploration, deploy ALSEP, collect geological samples, photography

APOLLO 15 25° APPROACH TRAJECTORY

SIGNIFICANT ENHANCEMENT OF TERRAIN CLEARANCE
SIGNIFICANT ENHANCEMENT OF VISIBILITY AND FIDELITY OF LPD
NO SIGNIFICANT INCREASE IN VERTICAL VELOCITY
MODEST INCREASE IN DELTA-V FOR REDESIGNATIONS

POWERED DESCENT PROFILE

MPAD 71-527 F

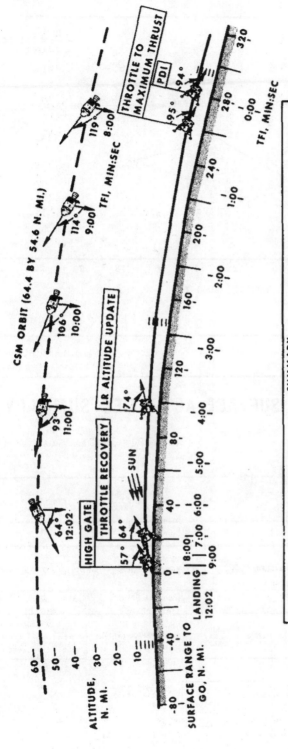

	SUMMARY					
EVENT	TFI, MIN:SEC	V_I, FPS	$\dot{H}$, FPS	$\dot{H}$, FT	ΔV, FPS	
POWERED DESCENT INITIATION	0:00	5562	-5	50,087	0	
THROTTLE TO MAXIMUM THRUST	0:26	5534	-4	49,979	28	
YAW TO VERTICAL	3:00	4111	-58	44,040	1468	
LANDING RADAR ALTITUDE UPDATE	4:06	3444	-67	39,878	2159	
LANDING RADAR VELOCITY UPDATE	5:34	2500	-85	33,623	3167	
THROTTLE RECOVERY	7:24	1163	-80	22,950	4597	
HIGH GATE	9:24	318	-162	7,029	5640	
LOW GATE	10:42	66 (76)*	-23	694	6241	
LANDING	12:02	-15 (0)*	-5	5	6698	

*(HORIZONTAL VELOCITY RELATIVE TO SURFACE)

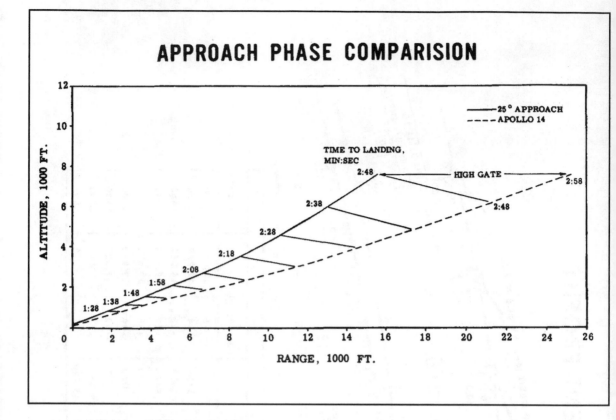

APPROACH PHASE COMPARISION

APOLLO 15 LUNAR SURFACE ACTIVITIES SUMMARY

EVA Mission Events

Events	GET hrs:min	Date/EDT
CDR starts standup EVA (SEVA) for verbal description of landing site, 360° photopanorama	106:10	Jul 30/7:44 pm
End SEVA, repressurize	106:40	8:14 pm
Depressurize LM for EVA 1	119:50	Jul 31/9:24 am
CDR steps onto surface	120:05	9:39 am
LMP steps onto surface	120:14	9:48 am
CDR places TV camera on tripod	120:16	9:50 am
LMP collects contingency sample	120:17	9:51 am
LMP climbs LM ladder to leave contingency sample on platform	120:20	9:54 am
Crew unstows LRV	120:20	9:54 am
LRV test driven	120:35	10:09 am
LRV equipment installation complete	120:58	10:32 am
Crew mounts LRV for drive to geology station No. 1—Hadley Rille rim near "elbow":	121:12	10:46 am
2—base of Apennine front between "elbow" and St. George crater:		
3—Apennine front possible debris flow area		
Start LRV traverse back to LM	123:12	12:46 pm
Arrive at LM	123:40	1:14 pm
Offload ALSEP from LM, load drill and LRRR on LRV	123:58	1:32 pm
CDR drives LRV to ALSEP site, LMP walks	124:05	1:39 pm
Crew deploys ALSEP	124:08	1:42 pm
ALSEP deploy complete, return by LRV to LM	125:49	3:23 pm
Arrive at LM	125:55	3:29 pm
LMP deploys solar wind composition experiment, CDR makes polarimetric photos	125:58	3:32 pm
Crew erects US flag	126:13	3:47 pm
Crew stows equipment at LM and on LRV	126:18	3:52 pm
Crewmen dust lunar material from each other's EMUS	126:24	3:58 pm
LMP ingresses LM, CDR sends up Sample Return Container No. 1 on transfer conveyor	126:27	4:01 pm
CDR ingresses LM	126:40	4:14 pm
Repressurize LM, end EVA 1	126:50	4:24 pm
Depressurize LM for EVA 2	141:12	Aug 1/6:46 am
CDR steps onto surface	141:23	6:57 am
LMP steps onto surface	141:37	7:11 am
Crew loads gear aboard LRV for geology traverse, begin drive to Apennine front	141:59	7:33 am
Arrive secondary crater cluster (sta.4)	142:27	8:01 am
Arrive at Front Crater, gather samples, photos of front materials on crater rim	143:16	8:50 am
Arrive at area stop 5-6 on crater rim slope, samples, photos, soil mechanics trench	144:23	9:57 am
Arrive at stop 7—secondary crater cluster near 400m crater; collect	146:11	11:45 am
documented samples, photopanorama		
Arrive at stop 8 for investigations of materials in large mare area	146:47	12:21 pm
Arrive back at LM, hoist Sample Return Container No. 2 into LM	147:10	12:44 pm
Crew ingresses LM, repressurize, End EVA 2	148:10	1:44 pm
Depressurize for EVA 3	161:50	Aug 2/3:24 am
CDR steps onto surface	162:03	3:37 am
LMP steps onto surface	162:09	3:43 am
Prepare and load LRV for geology traverse	162:11	3:45 am
Leave for stations 9-13	162:44	4:18 am
Arrive station 9—rim of Hadley Rille; photos, penetrometer, core samples,	163:08	4:42 am
documented samples		
Arrive at station 10; documented samples, photopanorama	164:01	5:35 am
Arrive at station 11—rim of Hadley Rille; documented samples,	164:17	5:51 am
photopanorama, description of near and far rille walls		
Arrive at station 12—SE rim of Chain Crater; documented samples, photopanorama,	165:00	6:34 am
seek unusual samples		
Arrive at station 13—north complex scarp between larger craters;	165:31	7:05 am
documented samples, photograph scarp, observe and describe 750m and		
390m craters, core tubes, trench, penetrometer		
Arrive station 14—fresh blocky crater in mare south of north complex;	166:43	8:17 am
photopanorama, documented samples		
Arrive back at LM	167:17	8:51 am
Load samples, film in LM; park LRV 300 feet east of LM, switch to	167:35	9:09 am
ground-controlled TV for ascent		
Crew ingress LM, end 3rd EVA	167:50	9:24 am

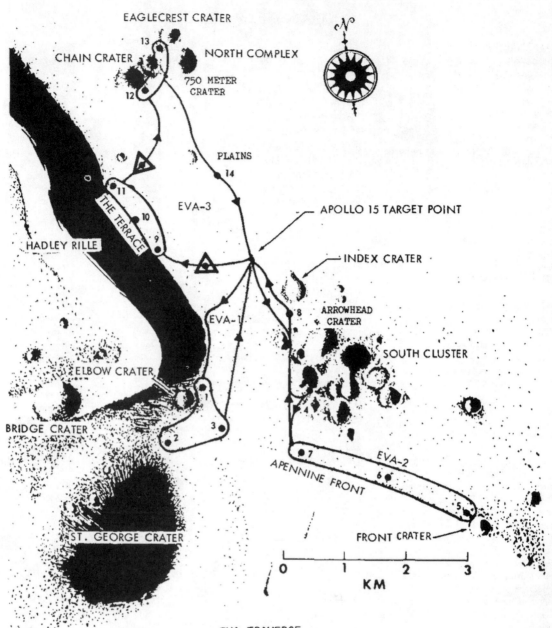

EVA TRAVERSE

TRAVERSE SUMMARY

TRAVERSE	START	END	DURATION	DISTANCE	RIDING TIME*	STATION TIME
I	1:25	3:50	2:25	7.9 KM	1:11	1:14
II	:49	6:20	5:31	16.1 KM	2:15	3:16
III	:42	5:15	4:33	12.3 KM	1:28	3:05
TOTALS				36.3 KM	4:54	7:35

• INCLUDES LRV INGRESS/EGRESS TIMES

TRAVERSE PLAN EVA-1

STATION/AREA	ACTIVITY	STATION TIME
1 (ELBOW)	RADIAL SAMPLE	
2 (ST. GEORGE)	RADIAL SAMPLE	
	COMPREHENSIVE SAMPLE	
	500mm PHOTOGRAPHY	
	STEREO PAN	
	PENETROMETER	
3	DOCUMENTED SAMPLE	
NEAR LM	ALSEP DEPLOYMENT	
	LR3 DEPLOYMENT	
	SWC DEPLOYMENT	
	MARE SAMPLING	

TRAVERSE PLAN EVA-2

STATION/AREA	ACTIVITY	STATION TIME
4 (SECONDARIES)	SOIL/RAKE SAMPLE	
	DOCUMENTED SAMPLE	
	500mm PHOTOGRAPHY	
	EXPLORATORY TRENCH	
	CORE TUBE (1)	
5 - 6 - 7 (SECONDARIES)	STATION 5: DOCUMENTED SAMPLES FROM UPSLOPE SIDE	
	DOCUMENTED SAMPLES DOWNSLOPE SIDE	
	EXPLORATORY TRENCH	
	500mm PHOTOGRAPHY	
	STATION 6 - 7: DOCUMENTED SAMPLES	
	EXPLORATORY TRENCHES	
	CORE TUBE SAMPLE	
	500mm PHOTOGRAPHY	
8 (MARE)	COMPREHENSIVE SAMPLE	
	DOUBLE CORE TUBE SAMPLE	
	DOCUMENTED SAMPLE	
	SESC	
	TRENCH	
	SOIL MECHANICS EXPERIMENT	

TRAVERSE PLAN EVA-3

STATION/AREA	ACTIVITY	STATION TIME
9 -10 (RILLE)	STATION 9: 500mm PHOTOGRAPHY	
	COMPREHENSIVE SAMPLE	
	DOUBLE CORE	
	DOCUMENTED SAMPLE	
	SESC	
	PENETROMETER	
	STATION 10: 500mm PHOTOGRAPHY	
	DOCUMENTED SAMPLE	
11 (RILLE)	500mm PHOTOGRAPHY	
	DOCUMENTED SAMPLE	
12 (N. COMPLEX/CHAIN CRATER)	DOCUMENTED SAMPLE	
	CORE TUBE	
13 (N. COMPLEX)	CRATER - DOCUMENTED SAMPLE	
	- PHOTOGRAPHY	
	SAMPLES, OBSERVATION & PHOTOGRAPHY OF:	
	EAGLE CREST	
	NORTH COMPLEX	
	SCARPS	
14 (MARE)	DOCUMENTED SAMPLE	

Mission Events (Cont'd)

Events	GET hrs:min	Date/EDT	Velocity change feet/sec	Purpose and resultant orbit
CSM plane change	165:13	2/6:47 am	309	Changes CSM orbital plane by 3.3° to coincide with LM orbital plane at time of ascent from surface
LM ascent	171:35	2/01:09 pm	6,056	Boosts ascent stage into 9 X 46 nm lunar orbit for rendezvous with CSM
Insertion into lunar orbit	171:43	2/01:17 pm		
Terminal phase initiate (TPI) (LM APB)	172:30	2/2:04 pm	52	Boosts ascent stage into 61 X 44 nm catch-up orbit; LM trails CSM by 32 nm and 15 nm below at time of TPI burn
Braking (LM RCS; 4 burns)	173:11	2/2:45 pm	31	Line-of-sight terminal phase braking to place LM in 59 X 59 nm orbit for final approach, docking
Docking	173:30	2/3:04 pm		CDR and LMP transfer back to CSM
LM jettison, separation	177:38	2/7:12 pm		Prevents recontact of CSM with LM ascent stage during remainder of lunar orbit
LM ascent stage deorbit (RCS)	179:06	2/8:40 pm	-195	ALSEP seismometers at Apollo 15, 14 and 12 landing sites record impact event
LM impact	179:31	2/9:05 am		Impact at about 5,528 fps at -4° angle, 32 nm from Apollo 15 ALSEP
CSM orbital change	221:25	4/2:59 pm	64	55 X 75 nm orbit (Rev 73)
Subsatellite ejection	222:36	4/4:10 pm		Lunar orbital science experiment
Transearth injection TEI SPS	223:44	4/5:18 pm	3,047	Inject CSM into transearth trajectory
Midcourse correction	5 TEI+17 hrs	5/10:20 am	0	Transearth midcourse corrections will be computed in real time for entry corridor control and recovery area weather avoidance
Inflight EVA	242:00	5/11:34 am		To retrieve film canisters from SM SIM bay
Midcourse correction 6	EI-22 hrs	6/6:32 pm	0	
Midcourse correction 7	EI-3 hrs	7/01:32 pm	0	
CM/SM separation	294:43	7/4:17 pm		Command module oriented for Earth atmosphere entry
Entry interface (400,000 ft)	294:58	7/4:32 pm		Command module enters atmosphere at 36,097 fps
Splashdown	295:12	7/4:46 pm		Landing 1,190 nm downrange from entry; splash at 26.1° North latitude, 158° west longitude

EVA PROCEDURES
CREWMAN PATH TO FOOT RESTRAINTS

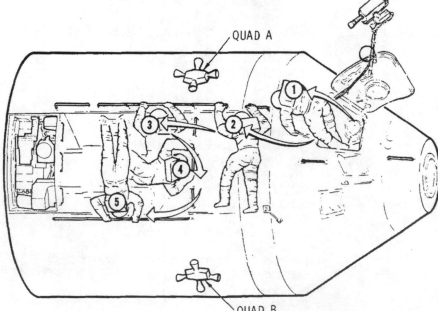

Entry Events

Event	Time from 400,000 ft. min:sec	
Entry	00:00	4:32 p.m. 7th August
Enter S-band communication blackout	00:18	
Initiate constant drag	00:54	
Maximum heating rate	01:10	
Maximum load factor (FIRST)	01:24	
Exit S-band communication blackout	03:34	
Maximum load factor (SECOND)	05:42	
Termination of CMC guidance	06:50	
Drogue parachute deployment	07:47	(altitude, 23,000 ft.)
Main parachute deployment	08:36	(altitude, 10,000 ft.)
Landing	13:26	4:45p.m. 7th August

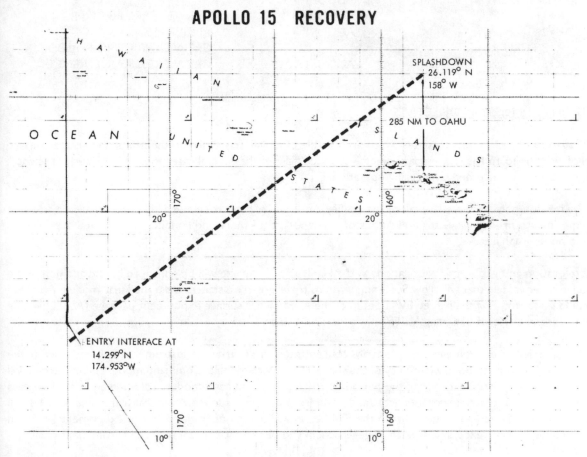

APOLLO 15 RECOVERY

Recovery Operations

Launch abort landing areas extend downrange 3,400 nautical miles from Kennedy Space Center, fanwise 50 nm above and below the limits of the variable launch azimuth (80-100 degrees) in the Atlantic Ocean.

Splashdown for a full-duration lunar landing mission launched on time July 26 will be at 4:46 p.m. EDT, August 7 at 26.1° North latitude by 158° West longitude — about 290 nm due north of Pearl Harbor, Hawaii.

The landing platform-helicopter (LPH) USS Okinawa, Apollo 15 prime recovery vessel, will be stationed near the end-of-mission aiming point prior to entry.

In addition to the primary recovery vessel located in the recovery area, HC-130 air rescue aircraft will be on standby at staging bases at Guam, Hawaii, Azores and Florida.

Apollo 15 recovery operations will be directed from the Recovery Operations Control Room in the Mission Control Center, supported by the Atlantic Recovery Control Center, Norfolk, Va., and the Pacific Recovery Control Center, Kunia, Hawaii.

The Apollo 15 crew will remain aboard the USS Okinawa until the ship reaches Pearl Harbor the day after splashdown. They will be flown from Hickam Air Force Base to Houston aboard a USAF transport aircraft. There will be no postflight quarantine of crew or spacecraft.

APOLLO 15 CREW POST-LANDING ACTIVITIES

DAYS FROM RECOVERY	DATE	ACTIVITY
SPLASHDOWN	AUGUST 7	
R+1	AUGUST 8	ARRIVE PEARL HARBOR
R+2	AUGUST 9	ARRIVE MSC
R+3 THRU R+15		CREW DEBRIEFING PERIOD
R+5	AUGUST 12	CREW PRESS CONFERENCE

APOLLO 15 MISSION OBJECTIVES

First of the Apollo J mission series which are capable of longer stay times on the Moon and greater surface mobility, Apollo 15 has four primary objectives which fall into the general categories of lunar surface science, lunar orbital science, and engineering/operational.

The mission objectives are to explore the Hadley-Apennine region, set up and activate lunar surface scientific experiments, make engineering evaluations of new Apollo equipment, and conduct lunar orbital experiments and photographic tasks.

Exploration and geological investigations at the Hadley Appenine site will be enhanced by the addition of the lunar rover vehicle that will allow Scott and Irwin to travel greater distances from the lunar module than they could on foot during their three EVAs. Setup of the Apollo lunar surface experiment package (ALSEP) will be the third in a trio of operating ALSEPs (Apollos 12, 14, and 15.)

Orbital science experiments are primarily concentrated in an array of instruments and cameras in the scientific instrument module (SIM) bay of the spacecraft service module. Command module pilot Worden will operate these instruments during the period he is flying the command module solo and again for two days following the return of astronauts Scott and Irwin from the lunar surface. After transearth injection, he will go EVA to retrieve film cassettes from the SIM bay. In addition to operating SIM bay experiments, Worden will conduct other experiments such as gegenschein and ultraviolet photography tasks from lunar orbit.

Among the engineering/operational tasks to be carried out by the Apollo 15 crew is the evaluation of the modifications to the lunar module which were made for carrying a heavier payload and for a lunar stay time of almost three days. Changes to the Apollo spacesuit and to the portable life support system (PLSS) will be evaluated. Performance of the lunar rover vehicle (LRV) and the other new J-mission equipment that goes with it—the lunar communications relay unit (LCRU) and the ground-controlled television assembly (GCTA)—also will be evaluated.

LUNAR SURFACE EXPERIMENTS

EXPERIMENT			11	12	14	15
	S-031	LUNAR PASSIVE SEISMOLOGY	X	X	X	X
	S-033	LUNAR ACTIVE SEISMOLOGY			X	
	S-034	LUNAR TRI-AXIS MAGNETOMETER		X		X
A	S-035	MEDIUM ENERGY SOLAR WIND		X		X
L	S-036	SUPRATHERMAL ION DETECTOR		X	X	X
S	S-037	LUNAR HEAT FLOW				X
E	S-038	CHARGED PARTICLE LUNAR ENVIRONMENT			X	
P	S-058	COLD CATHODE GAUGE		X	X	X
	M-515	LUNAR DUST DETECTOR		X	X	X
	S-059	LUNAR GEOLOGY INVESTIGATION	X	X	X	X
	S-078	LASER RANGING RETRO-REFLECTOR	X		X	X
	S-080	SOLAR WIND COMPOSITION			X	X
		LUNAR SURFACE CLOSE-UP CAMERA	X	X	X	
	S-198	LUNAR PORTABLE MAGNETOMETER			X	
	S-200	SOIL MECHANICS			X	X

LUNAR ORBITAL EXPERIMENTS

		11	12	14	15
SERVICE MODULE					
S-160	GAMMA-RAY SPECTROMETER				X
S-161	X-RAY FLUORESCENCE				X
S-162	ALPHA-PARTICLE SPECTROMETER				X
S-164	S-BAND TRANSPONDER			X	X
S-165	MASS SPECTROMETER				X
S-170	BISTATIC RADAR			X	X
S-173	PARTICLE MEASUREMENT (SUBSATELLITE)				X
S-174	MAGNETOMETER (SUBSATELLITE)				X
S-164	S-BAND TRANSPONDER (SUBSATELLITE)				X
	24" PANORAMIC CAMERA				X
	3" MAPPING CAMERA				X
	LASER ALTIMETER				X
COMMAND MODULE					
S-176	APOLLO WINDOW METEOROID			X	X
S-177	UV PHOTOGRAPHY - EARTH AND MOON				X
S-178	GEGENSCHEIN FROM LUNAR ORBIT			X	X

Lunar Surface Science

As in previous lunar landing missions, a contingency sample of lunar surface material will be the first scientific objective performed during the first EVA period. The Apollo 15 landing crew will devote a large portion of the first EVA to deploying experiments in the ALSEP. These instruments will remain on the Moon to transmit scientific data through the Manned Space Flight Network on long-term physical and environmental properties of the Moon. These data can be correlated with known Earth data for further knowledge on the origins of the planet and its satellite.

The ALSEP array carried on Apollo 15 has seven experiments: S-031 Passive Seismic Experiment, S-034 Lunar Surface Magnetometer Experiment, S-035 Solar Wind Spectrometer Experiment, S-036 Suprathermal Ion Detector Experiment, S-037 Heat Flow Experiment, S-058 Cold Cathode Gauge Experiment, and M-515 Lunar Dust Detector Experiment.

Two additional experiments, not part of ALSEP, will be deployed in the ALSEP area: S-078 Laser Ranging Retro-Reflector and S-080 Solar Wind Composition.

Passive Seismic Experiment: (PSE): The PSE measures seismic activity of the Moon and gathers and relays to Earth information relating to physical properties of the lunar crust and interior. The PSE reports seismic data on man-made impacts (LM ascent stage), natural impacts of meteorites, and moonquakes. Dr. Gary Latham of the Lamont-Doherty Geological Observatory (Columbia University) is responsible for PSE design and experiment data analysis.

Two similar PSEs deployed as a part of the Apollo 12 and 14 ALSEPs have transmitted to Earth data on lunar surface seismic events since deployment. The Apollo 12, 14, and 15 seismometers differ from the seismometer left at Tranquillity Base in July 1969 by the Apollo 11 crew in that the later PSEs are continuously powered by SNAP-27 radioisotope electric generators. The Apollo 11 seismometer, powered by solar cells, transmitted data only during the lunar day, and is no longer functioning.

After Apollo 15 translunar injection, an attempt will be made to impact the spent S-IVB stage and the instrument unit into the Moon. This will stimulate the passive seismometers left on the lunar surface by other Apollo crews.

Through a series of switch-selection-command and ground-commanded thrust operations, the S-IVB/IU will be directed to hit the Moon within a target area 379 nautical miles in diameter. The target point is 3.65 degrees south latitude by 7.58 degrees west longitude, near Lelande Crater about 161 nautical miles east of Apollo 14 landing site.

After the lunar module is ejected from the S-IVB, the launch vehicle will fire an auxiliary propulsion system (APS) ullage motor to separate the vehicle from the spacecraft a safe distance. Residual liquid oxygen in the almost spent S-IVB/IU will then be dumped through the engine with the vehicle positioned so the dump will slow it into an impact trajectory. Mid-course corrections will be made with the stage's APS ullage motors if necessary.

The S-IVB/IU will weigh 30,836 pounds and will be traveling 4,942 nautical-miles-an-hour at impact. It will provide an energy source at impact equivalent to about 11 tons of TNT.

After Scott and Irwin have completed their lunar surface operations and rendezvoused with the command module in lunar orbit, the lunar module ascent stage will be jettisoned and later ground-commanded to impact on the lunar surface about 25 nautical miles west of the Apollo 15 landing site at Hadley-Apennine.

Impacts of these objects of known masses and velocities will assist in calibrating the Apollo 14 PSE readouts as well as providing comparative readings between the Apollo 12 and 14 seismometers forming the first two stations of a lunar surface seismic network.

There are three major physical components of the PSE:

1. The sensor assembly consists of three long-period and one short-period vertical seismometers with orthogonally-oriented capacitance-type seismic sensors, capable of measuring along two horizontal components and one vertical component. The sensor assembly is mounted on a gimbal platform. A magnet-type sensor short-period seismometer is located on the base of the sensor assembly.

2. The leveling stool allows manual leveling of the sensor assembly by the crewman to within ±5 degrees. Final leveling to within ±3 arc seconds is accomplished by control motors.

3. The five-foot diameter hat-shaped thermal shroud covers and helps stabilize the temperature of the sensor assembly. The instrument uses thermostatically controlled heaters to protect it from the extreme cold of the lunar flight.

Background Scientific Information on the Lunar Surface Experiments

SCIENTIFIC DISCIPLINE / EXPERIMENT	GEOLOGY	GEOPHYSICS	GEOCHEMISTRY	BIOSCIENCES	GEODESY/CARTOGRAPHY	LUNAR ATMOSPHERE	PARTICLES AND FIELDS	ASTRONOMY
CONTINGENCY SAMPLE COLLECTION	AID IN DETERMINING LUNAR HISTORY BY AGING OF LUNAR SAMPLES		DETERMINE COMPOSITION OF LUNAR SURFACE BY CHEMICAL ANALYSIS OF LUNAR SAMPLES	AID IN DETERMINING POSSIBILITY OF BIOLOGICAL FORMS ON LUNAR SURFACE				
ALSEP PASSIVE SEISMIC (S-031)	AID IN DETERMINING INTERIOR STRUCTURE, TECTONISM AND VOLCANISM	AID IN DETERMINING FREE OSCILLATIONS, TIDES, SECULAR STRAINS, TILT, VELOCITY, GRAVITY CHANGES AND FREQUENCY AMPLITUDE, ATTENUATION AND DIRECTION OF SEISMIC WAVES						MEASURE METEOROID IMPACTS
HEAT FLOW (S-037)	AID IN DETERMINING LUNAR EVOLUTION FROM DATA	MEASURE VERTICAL TEMPERATURE GRADIENTS, ABSOLUTE TEMPERATURE OF THE SURFACE TO ESTABLISH VERTICAL THERMAL CONDUCTIVITY	DETERMINE BULK COMPOSITION AND CHEMICAL SORTING MAY BE INFERRED FROM DATA					DETERMINE THERMAL ENVIRONMENT
LUNAR SURFACE MAGNETOMETER (S-034)	AID IN DETERMINING MAGNETIC ANOMALIES, SUBSURFACE FEATURES AND LUNAR HISTORY	AID IN DETERMINING THERMAL STATE OF THE LUNAR INTERIOR					ESTABLISH GROSS ELECTRICAL DIFFUSIVITY; MEASURE MAGNETIC FIELD OF THE MOON	DETERMINE LUNAR RESPONSE TO FLUCTUATIONS IN THE INTERPLANETARY MAGNETIC FIELD
SOLAR WIND SPECTROMETER (S-035)		MONITOR FLUX, ENERGY STREAMING DIRECTION, AND TEMPORAL VARIATIONS IN THE SOLAR WIND PLASMA				DETERMINE PRESENCE OF ATMOSPHERE	ESTABLISH GROSS ELECTRICAL CONDUCTIVITY	
SUPRATHERMAL ION DETECTOR (S-036)		MEASURE FLUX, NUMBER DENSITY, VELOCITY AND ENERGY PER UNIT CHARGE OF POSITIVE IONS				DETERMINE IONOSPHERE/ ATMOSPHERE CHARACTERISTICS	DETERMINE AMBIENT ELECTRIC FIELD EFFECTS	
COLD CATHODE ION GAUGE (S-058)						DETERMINE DENSITY OF THE LUNAR ATMOSPHERE INCLUDING TEMPORAL VARIATIONS	PROVIDE INFORMATION ON RATE OF LOSS ON CONTAMINANTS LEFT BY ASTRONAUTS	
LUNAR DUST DETECTOR (M-515)		AID IN DETERMINING SURFACE MATERIAL TRANSPORT, PROVIDE INFORMATION ON HIGH ENERGY RADIATION, DUST ACCUMULATION AND LUNAR SURFACE TEMPERATURES				PROVIDE INFORMATION ON DUST ACCUMULATION		
LASER RANGING RETRO-REFLECTOR (S-078)		DETERMINE FACTORS ABOUT LUNAR MOTION, LUNAR LIBRATION, AND EARTH GEOPHYSICAL DATA			AID IN DETERMINING EPHEMERIS, ORIENTATION AND LIBRATION			PROVIDE INCREASED ACCURACY IN LUNAR ORBITAL DATA AND OTHER PARAMETERS
LUNAR GEOLOGY INVESTIGATION (S-059)	AID IN DETERMINING LUNAR GEOLOGICAL STRUCTURE AND HISTORY		DETERMINE CHEMICAL COMPOSITION OF LUNAR SAMPLES	LUNAR SAMPLES MAY BE TESTED FOR ABILITY TO SUPPORT LIFE FORMS AND TO DETERMINE POSSIBILITY OF BIOLOGICAL LIFE FORMS ON THE LUNAR SURFACE				
SOIL MECHANICS (S-200)	AID IN DETERMINING LUNAR HISTORY. ENABLE DETERMINATION OF COMPOSITIONAL TEXTURAL AND MECHANICAL PROPERTIES OF LUNAR SOIL		ENABLE DETERMINATION OF COMPOSITION OF LUNAR SOIL					
SOLAR WIND COMPOSITION (S-080)			DETERMINE COMPOSITION OF SOLAR WIND PLASMA			AID IN DETERMINING HISTORY OF PLANETARY ATMOSPHERE	PROVIDE INFORMATION ON THE ELEMENTAL AND ISOTOPIC COMPOSITION OF NOBLE GASES AND OTHER ELEMENTS IN THE SOLAR WIND	

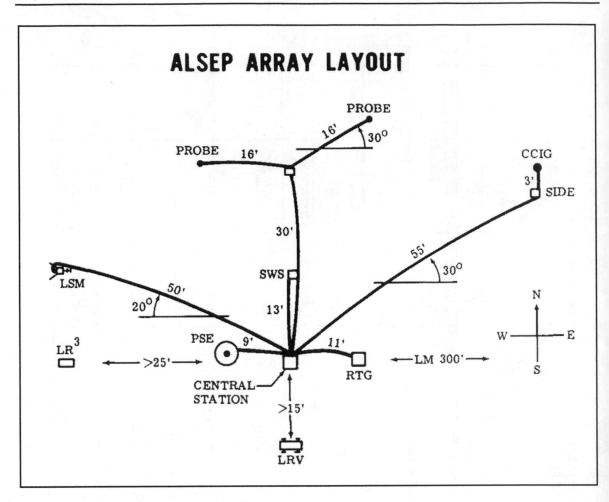

ALSEP ARRAY LAYOUT

ALSEP to Impact Distance Table		
Approximate Distance in:	Km	Statute Miles
Apollo 12 ALSEP to:		
Apollo 12 LM A/S Impact	75	45
Apollo 13 S-IVB Impact	134	85
Apollo 14 S-IVB Impact	173	105
Apollo 14 LM A/S Impact	116	70
Apollo 15 S-IVB Impact	480	300
Apollo 15 LM A/S Impact	1150	710
Apollo 14 ALSEP to:		
Apollo 14 LM A/S Impact	67	40
Apollo 15 S-IVB Impact	300	185
Apollo 15 LM A/S Impact	1070	660
Apollo 15 ALSEP to:		
Apollo 15 LM A/S Impact	50	30

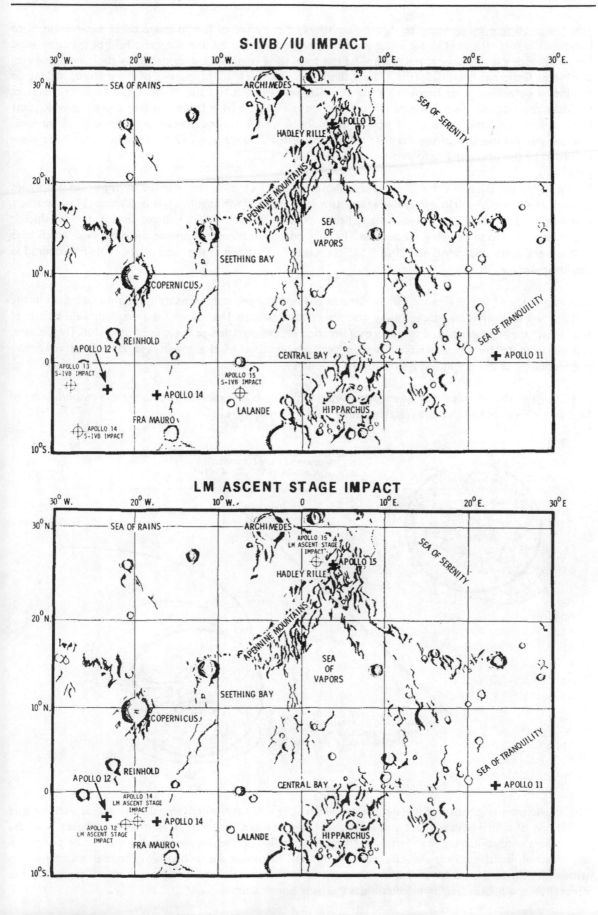

The Lunar Surface Magnetometer (LSM): The scientific objective of the magnetometer experiment is to measure the magnetic field at the lunar surface. Charged particles and the magnetic field of the solar wind impact directly on the lunar surface. Some of the solar wind particles are absorbed by the surface layer of the Moon. Others may be deflected around the Moon. The electrical properties of the material making up the Moon determine what happens to the magnetic field when it hits the Moon. If the Moon is a perfect insulator the magnetic field will pass through the Moon undisturbed. If there is material present which acts as a conductor, electric currents will flow in the Moon. A small magnetic field of approximately .35 gammas, one thousandth the size of the Earth's field was recorded at the Apollo 12 site. Similar small fields were recorded by the portable magnetometer on Apollo 14.

Two possible models are shown in the next drawing. The electric current carried by the solar wind goes through the Moon and "closes" in the space surrounding the Moon (figure a). This current (E) generates a magnetic field (M) as shown. The magnetic field carried in the solar wind will set up a system of electric currents in the Moon or along the surface. These currents will generate another magnetic field which tries to counteract the solar wind field (figure b). This results in a change in the total magnetic field measured at the lunar surface.

The magnitude of this difference can be determined by independently measuring the magnetic field in the undisturbed solar wind nearby, yet away from the Moon's surface. The value of the magnetic field change at the Moon's surface can be used to deduce information on the electrical properties of the Moon. This, in turn, can be used to better understand the internal temperature of the Moon and contribute to better understanding of the origin and history of the Moon.

The design of the tri-axis flux-gate magnetometer and analysis of experiment data are the responsibility of Dr. Palmer Dyal -NASA/Ames Research Center.

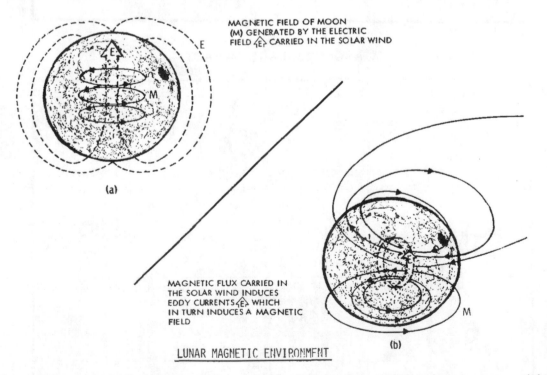

MAGNETIC FIELD OF MOON
(M) GENERATED BY THE ELECTRIC
FIELD (E) CARRIED IN THE SOLAR WIND

(a)

MAGNETIC FLUX CARRIED IN
THE SOLAR WIND INDUCES
EDDY CURRENTS (E) WHICH
IN TURN INDUCES A MAGNETIC
FIELD

M

(b)

LUNAR MAGNETIC ENVIRONMENT

The magnetometer consists of three magnetic sensors aligned in three orthogonal sensing axes, each located at the end of a fiberglass support arm extending from a central structure. This structure houses both the experiment electronics and the electro-mechanical gimbal/flip unit which allows the sensor to be pointed in any direction for site survey and calibration modes. The astronaut aligns the magnetometer experiment to within ±3 degrees east-west using a shadowgraph on the central structure, and to within ±3 degrees of the vertical using a bubble level mounted on the Y sensor boom arm.

Size, weight and power are as follows:	
Size (inches) deployed	40 high with 60 between sensor heads
Weight (pounds)	17.5
Peak Power Requirements (watts)	
Site Survey Mode	11.5
Scientific Mode	6.2
	12.3 (night)
Calibration Mode	10.8

The Magnetometer experiment operates in three modes:

Site Survey Mode — An initial site survey is performed in each of the three sensing modes for the purpose of locating and identifying any magnetic influences permanently inherent in the deployment site so that they will not affect the interpretation of the LSM sensing of magnetic flux at the lunar surface.

Scientific Mode — This is the normal operating mode wherein the strength and direction of the lunar magnetic field are measured continuously. The three magnetic sensors provide signal outputs proportional to the incidence of magnetic field components parallel to their respective axes. Each sensor will record the intensity three times per second which is faster than the magnetic field is expected to change. All sensors have the capability to sense over any one of three dynamic ranges with a resolution of 0.2 gammas.

-100 to +100 gamma*
-200 to +200 gamma
-400 to +400 gamma

*Gamma is a unit of intensity of a magnetic field. The Earth's magnetic field at the Equator, for example, is 35,000 gamma. The interplanetary magnetic field from the Sun has been recorded at 5 to 10 gamma.

Calibration Mode - This is performed automatically at 12-hour intervals to determine the absolute accuracy of the magnetometer sensors and to correct any drift from their laboratory calibration.

The Solar Wind Spectrometer: The Solar Wind Spectrometer will measure the strength, velocity and directions of the electrons and protons which emanate from the Sun and reach the lunar surface. The solar wind is the major external force working on the Moon's surface. The spectrometer measurements will help interpret the magnetic field of the Moon, the lunar atmosphere and the analysis of lunar samples.

Knowledge of the solar wind will help us understand the origin of the Sun and the physical processes at work on the Sun, i.e., the creation and acceleration of these particles and how they propagate through interplanetary space. It has been calculated that the solar wind puts one kiloton of energy into the Earth's magnetic field every second. This enormous amount of energy influences such Earth processes as the aurora, ionosphere and weather. Although it requires 20 minutes for a kiloton to strike the Moon its effects should be apparent in many ways.

In addition to the Solar Wind Spectrometer, an independent experiment (the Solar Wind Composition Experiment) will collect the gases of the solar wind for return to Earth for analysis.

The design of the spectrometer and the subsequent data analysis are the responsibility of Dr. Conway Snyder of the Jet Propulsion Laboratory.

Seven identical modified Faraday cups (an instrument that traps ionized particles) are used to detect and collect solar wind electrons and protons. One cup is to the vertical, whereas the remaining six cups surround the vertical where the angle between the normals of any two adjacent cups is approximately 60 degrees. Each cup measures the current produced by the charged particle flux entering into it. Since the cups are identical, and if particle flux is equal in each direction, equal current will be produced in each cup. If the flux is not equal in each direction, analysis of the amount of current in the seven cups will determine the variation of particle flow with direction. Also, by successively changing the voltages on the grid of the cup and measuring the corresponding current, complete energy spectra of both electrons and protons in the solar wind are produced.

Data from each cup are processed in the ALSEP data subsystem. The measurement cycle is organized into 16 sequences of 186 ten-bit words. The instrument weighs 12.5 pounds, has an input voltage of about 28.5 volts and has an average input power of about 3.2 watts. The measurement ranges are as follows:

Electrons	
High gain modulation	10.5 - 1,376 e.v. (electron volts)
Low gain modulation	6.2 - 817 e.v.
Protons	
High gain modulation	75 - 9,600 e.v.
Low gain modulation	45 - 5,700 e.v.
Field of View	6.0 Steradians
Angular Resolution	15 degrees (approximately)
Minimum Flux Detectable	106 particles/cm^2/sec

Suprathermal Ion Detector Experiment (SIDE) and Cold Cathode Gauge Experiment: The SIDE will measure flux, composition, energy and velocity of low-energy positive ions and the high-energy solar wind flux of positive ions. Combined with the SIDE is the Cold Cathode Gauge Experiment (CCGE) for measuring the density of the lunar ambient atmosphere and any variations with time or solar activity such atmosphere may have.

Data gathered by the SIDE will yield information on: (1) interaction between ions reaching the Moon from outer space and captured by lunar gravity and those that escape; (2) whether or not secondary ions are generated by ions impacting the lunar surface; (3) whether volcanic processes exist on the Moon; (4) effects of the ambient electric field; (5) loss rate of contaminants left in the landing area by the LM and the crew; and (6) ambient lunar atmosphere pressure.

Dr. John Freeman of Rice University is the SIDE principal investigator, and Dr. Francis B. Johnson of the University of Texas is the CCGE principal investigator.

The SIDE instrument consists of a velocity filter, a low-energy curved-plate analyzer ion detector and a high-energy curved-plate analyzer ion detector housed in a case measuring 15.2 by 4.5 by 13 inches, a wire mesh ground plane, and electronic circuitry to transfer data to the ALSEP central station. The SIDE case rests on folding tripod legs. Dust covers, released by ground command, protect both instruments. Total SIDE weight is 19.6 pounds.

The SIDE and the CCGE connected by a short cable, will be deployed about 55 feet northeast of the ALSEP central station, with the SIDE aligned east or west toward the subearth point and the CCGE orifice aligned along the north-south line with a clear field away from other ALSEP instruments and the LM.

The Cold Cathode Gauge on Apollo 14 is measuring a pressure of 10^{-11} to 10^{-12} torr (where one torr is equal to one millimeter of mercury and 760 millimeters of mercury equal one Earth atmosphere).

Lunar Heat Flow Experiment (HFE): The scientific objective of the Heat Flow experiment is to measure the steady-state heat flow from the lunar interior. Two predicted sources of heat are: (1) original heat at the time of the Moon's formation and (2) radioactivity. Scientists believe that heat could have been generated by the infalling of material and its subsequent compaction as the Moon was formed. Moreover, varying amounts of the radioactive elements uranium, thorium and potassium were found present in the Apollo 11 and 12 lunar samples which if present at depth, would supply significant amounts of heat. No simple way has been devised for relating the contribution of each of these sources to the present rate of heat loss. In addition to temperature, the experiment is capable of measuring the thermal conductivity of the lunar rock material.

The combined measurement of temperature and thermal conductivity gives the net heat flux from the lunar interior through the lunar surface. Similar measurements on Earth have contributed basic information to our understanding of volcanoes, earthquakes and mountain building processes. In conjunction with the seismic and magnetic data obtained on other lunar experiments the values derived from the heat flow measurements will help scientists to build more exact models of the Moon and thereby give us a better understanding of its origin and history.

The Heat Flow experiment consists of instrument probes, electronics and emplacement tool and the lunar surface drill. Each of two probes is connected by a cable to an electronics box which rests on the lunar surface. The electronics, which provide control, monitoring and data processing for the experiment, are connected to the ALSEP central station.

Each probe consists of two identical 20-inch (50 cm) long sections each of which contains a "gradient" sensor bridge, a "ring" sensor bridge and two heaters. Each bridge consists of four platinum resistors mounted in a thin-walled fiberglass cylindrical shell. Adjacent areas of the bridge are located in sensors at opposite ends of the 20-inch fiberglass probe sheath. Gradient bridges consequently measure the temperature difference between two sensor locations.

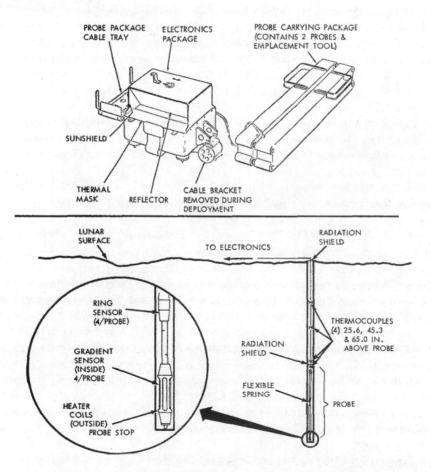

HEAT FLOW EXPERIMENT

In thermal conductivity measurements at very low values a heater surrounding the gradient sensor is energized with 0.002 watts and the gradient sensor values monitored. The rise in temperature of the gradient sensor is a function of thermal conductivity of the surrounding lunar material. For higher range of values, the heater is energized at 0.5 watts of heat and monitored by a ring sensor. The rate of temperature rise, monitored by the ring sensor is a function of the thermal conductivity of the surrounding lunar material. The ring sensor, approximately four inches from the heater, is also a platinum resistor. A total of eight thermal conductivity measurements can be made. The thermal conductivity mode of the experiment will be implemented about 20 days (500 hours) after deployment. This is to allow sufficient time for the perturbing effects of drilling and emplacing the probe in the borehole to decay; i.e., for the probe and casings to come to equilibrium with the lunar subsurface.

A 30-foot (10-meter) cable connects each probe to the electronics box. In the upper six feet of the borehole the cable contains four evenly spaced thermocouples: at the top of the probe; at 26 inches (65 cm), 45 inches (115 cm), and 66 inches (165 cm). The thermocouples will measure temperature transients propagating downward from the lunar surface. The reference junction temperature for each thermocouple is located in the electronics box. In fact, the feasibility of making a heat flow measurement depends to a large degree on the low thermal conductivity of the lunar surface layer, the regolith. Measurement of lunar surface temperature variations by Earth-based telescopes as well as the Surveyor and Apollo missions show a remarkably rapid rate of cooling. The wide fluctuations in temperature of the lunar surface (from -250 degrees F to +250 degrees) are expected to influence only the upper six feet and not the bottom three feet of the borehole.

The astronauts will use the Apollo Lunar Surface Drill (ALSD) to make a lined borehole in the lunar surface for the probes. The drilling energy will be provided by a battery-powered rotary percussive power head. The drill rod consists of fiberglass tubular sections reinforced with boron filaments (each about 20 inches or 50 cm long). A closed drill bit, placed on the first drill rod, is capable of penetrating the variety of rock including three feet of vesicular basalt (40 per cent porosity). As lunar surface penetration progresses, additional drill rod sections will be connected to the drill string. The drill string is left in place to serve as a hole casing.

An emplacement tool is used by the astronaut to insert the probe to full depth. Alignment springs position the probe within the casing and assure a well-defined radiative coupling between the probe and the borehole. Radiation shields on the hole prevent direct sunlight from reaching the bottom of the hole.

The astronaut will drill a third hole near the HFE and obtain cores of lunar material for subsequent analysis of thermal properties. Total available core length is 100 inches.

Heat flow experiments, design and data analysis are the responsibility of Dr. Marcus Langseth of the Lamont-Doherty Geological Observatory.

Lunar Dust Detector Experiment: Separates and measures high-energy radiation damage to three solar cells, measures reduction of solar cell output due to dust accumulation and measures reflected infrared energy and temperatures for computation of lunar surface temperatures. A sensor package is mounted on the ALSEP central station sunshield and a printed circuit board inside the central station monitors the data subsystem power distribution unit. Principal investigator: James R. Bates, NASA Manned Spacecraft Center.

ALSEP Central Station: The ALSEP Central Station serves as a power-distribution and data-handling point for experiments carried on each version of the ALSEP. Central Station components are the data subsystem, helical antenna, experiment electronics, power conditioning unit and dust detector. The Central Station is deployed after other experiment instruments are unstowed from the pallet.

The Central Station data subsystem receives and decodes uplink commands, times and controls experiments, collects and transmits scientific and engineering data downlink, and controls the electrical power subsystem through the power distribution and signal conditioner.

The modified axial helix S-band antenna receives and transmits a right-hand circularly-polarized signal. The antenna is manually aimed with a two-gimbal azimuth/elevation aiming mechanism. A dust detector on the Central Station, composed of three solar cells, measures the accumulation of lunar dust on ALSEP instruments.

The ALSEP electrical power subsystem draws electrical power from a SNAP-27 (Systems for Nuclear Auxiliary Power) radioisotope thermoelectric generator.

Laser Ranging Retro-Reflector (LRRR) Experiment: The LRRR will permit long-term measurements of the Earth-Moon distance by acting as a passive target for laser beams directed from observatories on Earth. Data gathered from these measurements of the round trip time for a laser beam will be used in the study of fluctuations in the Earth's rotation rate, wobbling motions of the Earth on its axis, the Moon's size and orbital shape, and the possibility of a slow decrease in the gravitational constant "G". Dr. James Faller of Wesleyan University, Middletown, CT, is LRRR principal investigator.

The LRRR is a square array of 300 fused silica reflector cubes mounted in an adjustable support structure which will be aimed toward Earth by the crew during deployment. Each cube reflects light beams back in absolute parallelism in the same direction from which they came.

By timing the round trip time for a laser pulse to reach the LRRR and return, observatories on Earth can calculate the exact distance from the observatory to the LRRR location within a tolerance of +6 cm (or one foot). A 100-cube LRRR was deployed at Tranquillity Base by the Apollo 11, and at Fra Mauro by the Apollo 14 crew. The goal is to set up LRRRs at three lunar locations to establish absolute control points in the study of Moon motion.

Solar Wind Composition Experiment:(SWC): The scientific objective of the solar wind composition experiment is to determine the elemental and isotopic composition of the noble gases in the solar wind.

As in Apollos 11, 12, and 14, the SWC detector will be deployed on the lunar surface and brought back to Earth by the crew. The detector will be exposed to the solar wind flux for 45 hours compared to 21 hours on Apollo 14, 17 hours on Apollo 12, and two hours on Apollo 11.

The solar wind detector consists of an aluminum foil four square feet in area and about 0.5 mils thick rimmed by Teflon for resistance to tearing during deployment. A staff and yard arrangement will be used to deploy the foil and to maintain the foil approximately perpendicular to the solar wind flux. Solar wind particles will penetrate into the foil, allowing cosmic rays to pass through. The particles will be firmly trapped at a depth of several hundred atomic layers. After exposure on the lunar surface, the foil is rolled up and returned to Earth. Professor Johannes Geiss, University of Berne, Switzerland, is principal investigator.

SNAP-27 — Power Source for ALSEP: A SNAP-27 unit, similar to two others on the Moon, will provide power for the ALSEP package. SNAP-27 is one of a series of radioisotope thermoelectric generators, or atomic batteries developed by the Atomic Energy Commission under its space SNAP program. The SNAP (Systems for Nuclear Auxiliary Power) program is directed at development of generators and reactors for use in space, on land, and in the sea.

While nuclear heaters were used in the seismometer package on Apollo 11, SNAP-27 on Apollo 12 marked the first use of a nuclear electrical power system on the Moon. The use of SNAP-27 on Apollo 14 marked the second use of such a unit on the Moon. The first unit has already surpassed its one-year design life by eight months, thereby allowing the simultaneous operation of two instrument stations on the Moon.

The basic SNAP-27 unit is designed to produce at least 63.5 electrical watts of power. The SNAP-27 unit is a cylindrical generator, fueled with the radioisotope plutonium-238. It is about 18 inches high and 16 inches in diameter, including the heat radiating fins. The generator, making maximum use of the lightweight material beryllium, weighs about 28 pounds unfueled.

The fuel capsule, made of a superalloy material, is 16.5 inches long and 2.5 inches in diameter. It weighs about 15.5 pounds, of which 8.36 pounds represent fuel. The plutonium-238 fuel is fully oxidized and is chemically and biologically inert.

The rugged fuel capsule is contained within a graphite fuel cask from launch through lunar landing. The cask is designed to provide reentry heating protection and added containment for the fuel capsule in the event of an aborted mission. The cylindrical cask with hemispherical ends includes a primary graphite heat shield, a secondary beryllium thermal shield, and a fuel capsule support structure. The cask is 23 inches long and eight inches in diameter and weighs about 24.5 pounds. With the fuel capsule installed, it weighs about 40 pounds. It is mounted on the lunar module descent stage.

Once the lunar module is on the Moon, an Apollo astronaut will remove the fuel capsule from the cask and insert it into the SNAP-27 generator which will have been placed on the lunar surface near the module.

The spontaneous radioactive decay of the plutonium-238 within the fuel capsule generates heat which is converted directly into electrical energy — at least 63.5 watts. There are no moving parts.

The unique properties of plutonium-238 make it an excellent isotope for use in space nuclear generators. At the end of almost 90 years, plutonium-238 is still supplying half of its original heat. In the decay process, plutonium-238 emits mainly the nuclei of helium (alpha radiation), a very mild type of radiation with a short emission range.

Before the use of the SNAP-27 system in the Apollo program was authorized, a thorough review was conducted to assure the health and safety of personnel involved in the launch and of the general public. Extensive safety analyses and tests were conducted which demonstrated that the fuel would be safely contained under almost all credible accident conditions.

Lunar Geology Investigation: The Hadley/Apennines site was selected for multiple objectives: 1) the Apennine Mountain front, 2) the sinuous Hadley Rille, 3) the dark mare material of Palus Putredinis, 4) the complex of domical hills in the mare, and 5) the arrowhead-shaped crater cluster.

The Apennine Mountain front forms the arcuate southeastern rim of Mare Imbrium. It borders Palus Putredinis and, in the area of the site, it rises 12,000 feet above the surrounding mare. The Apennine Mountain front is believed to have been exposed at the time of the excavation of the giant Imbrium basin. The cratering event must have, therefore, exposed materials which are pre-Imbrian in age. Examination and collection of this ancient material as well as deep-seated Imbrium ejecta are the prime objectives of the mission. This will be accomplished during the first and second EVA's when Scott and Irwin will select samples from the foot of the mountain scarp and from the ejecta blankets of craters which excavate mountain materials.

The second important objective of the mission is to study and sample the Hadley Rille, which runs parallel to the Apennine Mountain front and incises the Palus Putredinis mare material. The rille is a sinuous or meandering channel, much like a river gorge on Earth. It displays a V-shaped cross section, with an average slope of about 25°. The rille originates in an elongate depression near the base of the mountain some 40 miles south of the site. In the vicinity of the landing site, the rille is about one mile wide and 1,200 feet deep. The origin of the rille is not known and it is hoped that samples collected at its rim and high resolution photographs of its walls will unravel its mode of formation

The third objective of the mission is to study and sample the reasonably flat mare material of Palus Putredinis on which the LM will land. This mare material is dark and preliminary studies of crater distribution indicate that this mare surface is younger than that visited on Apollo 11, and probably is closer to the age of the Apollo 12 mare site. Systematic sampling of this surface unit will be done by visiting craters which have penetrated it to various depths. The Apollo 15 crew will use the standard lunar hand tools used on past missions for sampling. However, the hand tool carrier will be mounted on the lunar roving vehicle (LRV).

A complex of domical structures about 5 km north of the landing site constitutes another objective of the mission. The hills may be made of volcanic domes superposed on the surrounding mare or buried domical structures thinly covered by the mare-like material. Among the hills are large craters which have excavated subsurface material for sampling as well as interesting linear depressions and ridges.

The fifth sampling objective of the mission is a cluster of craters which forms an arrowhead-shaped pattern. This crater cluster is aligned along a ray from the crater Autolycus, over 100 miles northwest of the site. It is believed to be made of secondary craters from Autolycus ejecta and offers a good opportunity to study the features and perhaps sample material which originated at Autolycus.

Planned sampling sites for the mission allow therefore a thorough investigation of a variety of features. The most important of all features in the area is the Apennine Mountain front, where samples of the oldest exposed rocks on the Moon may be obtained.

In addition to planned sampling sites, Scott and Irwin will select other sites for gathering, observing and photographing geological samples. Both men will use chest-mounted Hasselblad electric data cameras for documenting the samples in their natural state. Core tube samples will also be retrieved for geological and geochemical investigation.

Soil Mechanics: Mechanical properties of the lunar soil, surface and subsurface, will be investigated through trenching at various locations, and through use of the self-recording penetrometer equipped with interchangeable cones of various sizes and a load plate. This experiment will be documented with the electric Hasselblad and the 16mm data acquisition camera.

Lunar Orbital Science

Service Module Sector 1, heretofore vacant except for a third cyrogenic oxygen tank added after the Apollo 13 incident, houses the Scientific Instrument Module (SIM) bay on the Apollo J missions.

Eight experiments are carried in the SIM bay: X-ray fluorescence detector, gamma ray spectrometer, alpha-particle spectrometer, panoramic camera, 3-inch mapping camera, laser altimeter and dual-beam mass spectrometer; a subsatellite carrying three integral experiments (particle detectors, magnetometer and S-Band transponder) comprise the eighth SIM bay experiment and will be jettisoned into lunar orbit.

Gamma-Ray Spectrometer: On a 25-foot extendable boom, measures chemical composition of lunar surface in conjunction with X-ray and alpha-particle experiments to gain a compositional "map" of the lunar surface ground track. Detects natural and cosmic rays, induced gamma radioactivity and will operate on Moon's dark and light sides. Additionally, the experiment will be extended in transearth coast to measure the radiation flux in cislunar space and record a spectrum of cosmological gamma-ray flux. The device can measure energy ranges between 0.1 to 10 million electron volts. The extendable boom is controllable from the command module cabin. Principal investigator: Dr. James R. Arnold, University of California at San Diego.

X-Ray Fluorescence Spectrometer: Second of the geochemical experiment trio for measuring the composition of the lunar surface from orbit, and detects X-ray fluorescence caused by solar X-ray interaction with the Moon. It will analyze the sunlit portion of the Moon. The experiment will measure the galactic X-ray flux during transearth coast. The device shares a compartment on the SIM bay lower shelf with the alpha-particle experiment, and the protective door may be opened and closed from the command module cabin. Principal Investigator: Dr. Isidore Adler, NASA Goddard Space Flight Center, Greenbelt, MD.

Alpha-Particle Spectrometer: Measures mono-energetic alpha particles emitted from the lunar crust and fissures as products of radon gas isotopes in the energy range of 4.7 to 9.3 million electron volts. The sensor is made up of an array of 10 silicon surface barrier detectors. The experiment will construct a "map" of lunar surface alpha-particle emissions along the orbital track and is not constrained by solar illumination. It will also measure deep-space alpha-particle background emissions in lunar orbit and in transearth coast. Protective door operation is controlled from the cabin. Principal investigator: Dr. Paul Gorenstein, American Science and Engineering, Inc., Cambridge, MA.

Mass Spectrometer: Measures composition and distribution of the ambient lunar atmosphere, identifies active lunar sources of volatiles, pinpoints contamination in the lunar atmosphere. The sunset and sunrise terminators are of special interest, since they are predicted to be regions of concentration of certain gases. Measurements over at least five lunar revolutions are desired. The mass spectrometer is on a 24-foot extendable boom. The instrument can identify species from 12 to 28 atomic mass units (AMU) with No. 1 ion counter, and 28-66 AMU with No. 2 counter. Principal investigator: Dr. John.H. Hoffman, University of Texas at Dallas.

24-inch Panoramic Camera (SM orbital photo task): Gathers stereo and high-resolution (1 meter) photographs of the lunar surface from orbit. The camera produces an image size of 15 x 180 nm with a field of view 11° downtrack and 108° cross track. The rotating lens system can be stowed face-inward to avoid contamination during effluent dumps and thruster firings. The 72-pound film cassette of 1,650 frames will be retrieved by the command module pilot during a transearth coast EVA. The 24-inch camera works in conjunction with the 3-inch mapping camera and the laser altimeter to gain data to construct a comprehensive map of the lunar surface ground track flown by this mission—about 1.16 million square miles, or 8 percent of the lunar surface.

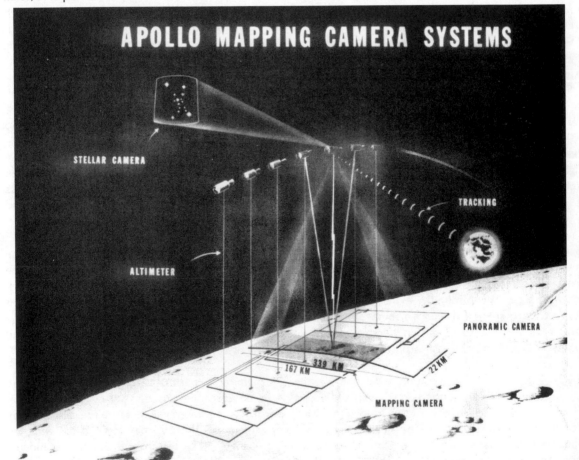

3-inch Mapping Camera: Combines 20-meter resolution terrain mapping photography on five-inch film with 3-inch focal length lens with stellar camera shooting the star field on 35mm film simultaneously at 96° from the surface camera optical axis. The stellar photos allow accurate correlation of mapping photography postflight by comparing simultaneous star field photos with lunar surface photos of the nadir (straight down). Additionally, the stellar camera provides pointing vectors for the laser altimeter during darkside passes. The 3-inch f4.5 mapping camera metric lens covers a 74° square field of view, or 92x92 nm from 60 nm altitude. The stellar camera is fitted with a 3-inch f/2.8 lens covering a 24° field with cone flats. The 23-pound film cassette containing mapping camera film (3,600) frames) and the stellar camera film will be retrieved during the same EVA described in the panorama camera discussion. The Apollo Orbital Science Photographic Team is headed by Frederick J. Doyle of the U.S. Geological Survey, McLean, VA.

Laser Altimeter: Measures spacecraft altitude above the lunar surface to within one meter. Instrument is boresighted with 3-inch mapping camera to provide altitude correlation data for the mapping camera as well as the 24-inch panoramic camera. When the mapping camera is running, the laser altimeter automatically fires a laser pulse corresponding to mid-frame ranging to the surface for each frame. The laser light source is a pulsed ruby laser operating at 6,943 angstroms, and 200-millijoule pulses of 10 nanoseconds duration. The laser has a repetition rate up to 3.75 pulses per minute. The laser altimeter working group of the Apollo Orbital Science Photographic Team is headed by Dr. William M. Kaula of the UCLA Institute of Geophysics and Planetary Physics.

Subsatellite: Ejected into lunar orbit from the SIM bay and carries three experiments: S-Band Transponder, Particle Shadows/Boundary Layer Experiment, and Subsatellite Magnetometer Experiment. The subsatellite is housed in a container resembling a rural mailbox, and when deployed is spring-ejected out-of-plane at 4 fps with a spin rate of 140 rpm. After the satellite booms are deployed, the spin rate is stabilized at about 12 rpm. The subsatellite is 31 inches long, has a 14 inch hexagonal diameter and weighs 78.5 pounds. The folded booms deploy to a length of five feet. Subsatellite electrical power is supplied by a solar cell array outputting 25 watts for dayside operation and a rechargeable silver-cadmium battery for nightside passes.

Experiments carried aboard the subsatellite are: S-Band transponder for gathering data on the lunar gravitational field, especially gravitational anomalies such as the so-called mascons; Particle Shadows/Boundary Layer for gaining knowledge of the formation and dynamics of the Earth's magnetosphere, interaction of plasmas with the Moon and the physics of solar flares using telescope particle detectors and spherical electrostatic particle detectors; and Subsatellite Magnetometer for gathering physical and electrical property data on the Moon and of plasma interaction with the Moon using a biaxial flux-gate magnetometer deployed on one of the three five-foot folding booms. Principal investigators for the subsatellite experiments are: Particle Shadows/ Boundary Layer, Dr. Kinsey A. Anderson, University of California Berkeley; Magnetometer, Dr. Paul J. Coleman, UCLA; and S-Band Transponder, Mr. William Sjogren, Jet Propulsion Laboratory.

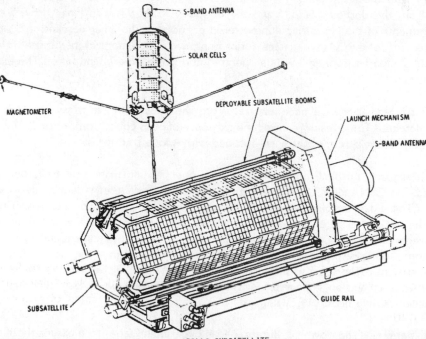

APOLLO SUBSATELLITE

Other CSM orbital science experiments and tasks not in the SIM bay include UV Photography-Earth and Moon, Gegenschein from Lunar Orbit, CSM/LM S-Band Transponder in addition to the Subsatellite, Bistatic Radar, and Apollo Window Meteoroid experiments.

UV Photography-Earth and Moon: Aimed toward gathering ultraviolet photos of the Earth and Moon for planetary atmosphere studies and investigation of lunar surface short wavelength radiation. The photos will be made with an electric Hasselblad bracket mounted in the right side window of the command module. The window is fitted with a special quartz pane that passes a large portion of the incident UV spectrum. A four-filter pack—three passing UV electromagnetic radiation and one passing visible electromagnetic radiation—is used with a 105mm lens for black and white photography; the visible spectrum filter is used with an 80mm lens for color UV photography.

Gegenschein from Lunar Orbit: This experiment is similar to the dim light photography task, and involves long exposures with a 35mm camera with 55mm f/1.2 lens camera on high speed black and white film (ASA 6,000). All photos must be made while the command module is in total darkness in lunar orbit.

Gegenschein is a faint light source covering a 20° field of view along the Earth-Sun line on the opposite side of the Earth from the Sun (anti-solar axis). One theory on the origin of Gegenschein is that particles of matter are trapped at the Moulton Point and reflect sunlight. Moulton Point is a theoretical point located 940,000 statute miles from the Earth along the anti-solar axis where the sum of all gravitational forces is zero. From lunar orbit, the Moulton Point region can be photographed from about 15 degrees off the Earth-Sun axis, and the photos should show whether Gegenschein results from the Moulton Point theory or stems from zodiacal light or from some other source. The experiment was conducted on Apollo 14.

During the same time period that photographs of the Gegenschein and the Moulton Point are taken, photographs of the same regions will be obtained from the Earth. The principal investigator is Lawrence Dunkelman of the Goddard Space Flight Center.

CSM/LM S-Band Transponder: The objective of this experiment is to detect variations in lunar gravity along the lunar surface track. These anomalies in gravity result in minute perturbations of the spacecraft motion and are indicative of magnitude and location of mass concentrations on the Moon. The Manned Space Flight Network (MSFN) and the Deep Space Network (DSN) will obtain and record S-band Doppler tracking measurements from the docked CSM/LM and the undocked CSM while in lunar orbit; S-band Doppler tracking measurements of the LM during non-powered portions of the lunar descent; and S-band Doppler tracking measurements of the LM ascent stage during non-powered portions of the descent for lunar impact. The CSM and LM S-band Transponders will be operated during the experiment period. The experiment was conducted on Apollo 14.

S-band Doppler tracking data have been analyzed from the Lunar Orbiter missions and definite gravity variations were detected. These results showed the existence of mass concentrations (mascons) in the ringed maria. Confirmation of these results has been obtained with Apollo tracking data.

With appropriate spacecraft orbital geometry much more scientific information can be gathered on the lunar gravitational field. The CSM and/or LM in low-altitude orbits can provide new detailed information on local gravity anomalies. These data can also be used in conjunction with high-altitude data to possibly provide some description on the size and shape of the perturbing masses. Correlation of these data with photographic and other scientific records will give a more complete picture of the lunar environment and support future lunar activities. Inclusion of these results is pertinent to any theory of the origin of the Moon and the study of the lunar subsurface structure. There is also the additional benefit of obtaining better navigational capabilities for future lunar missions in that an improved gravity model will be known. William Sjogren, Jet Propulsion Laboratory, Pasadena, California, is principal investigator.

Bistatic Radar Experiment: The downlink Bistatic Radar Experiment seeks to measure the electromagnetic properties of the lunar surface by monitoring that portion of the spacecraft telemetry and communications beacons which are reflected from the Moon.

The CSM S-band telemetry beacon (f = 2.2875 Gigahertz), the VHF voice communications link (f = 259.7 megahertz), and the spacecraft omni-directional and high gain antennas are used in the experiment. The spacecraft is oriented so that the radio beacon is incident on the lunar surface and is successively reoriented so that the angle at which the signal intersects the lunar surface is varied. The radio signal is reflected from the surface and is monitored on Earth. The strength of the reflected signal will vary as the angle at which it intersects the surface is varied.

By measuring the reflected signal strength as a function of angle of incidence on the lunar surface, the electromagnetic properties of the surface can be determined. The angle at which the reflected signal strength is a minimum is known as the Brewster Angle and determines the dielectric constant. The reflected signals can also be analyzed for data on lunar surface roughness and surface electrical conductivity.

The S-band signal will primarily provide data on the surface. However, the VHF signal is expected to penetrate the gardened debris layer (regolith) of the Moon and be reflected from the underlying rock strata. The reflected VHF signal will then provide information on the depth of the regolith over the Moon.

The S-band BRE signal will be monitored by the 210-foot antenna at the Goldstone, California, site and the VHF portion of the BRE signal will be monitored by the 150-foot antenna at the Stanford Research Institute in California. The experiment was flown on Apollo 14.

Lunar Bistatic Radar Experiments were also performed using the telemetry beacons from the unmanned Lunar Orbiter I in 1966 and from Explorer 35 in 1967. Taylor Howard, Stanford University, is the principal investigator.

Apollo Window Meteoroid: A passive experiment in which command module windows are scanned under high magnification pre- and postflight for evidence of meteoroid cratering flux of one-trillionth gram or larger. Such particle flux may be a factor in degradation of surfaces exposed to space environment. Principal investigator: Burton Cour-Palais, NASA Manned Spacecraft Center.

Composite Casting Demonstration

The Composite Casting technical demonstration performed on the Apollo 14 mission will be carried again on Apollo 15 to perform more tests on the effects of weightlessness on the solidification of alloys, intermetallic compounds, and reinforced composite materials. Ten samples will be processed, of which two will be directionally solidified samples of the indium-bismuth eutectic alloy, four will comprise attempts to make single crystals of the indium-bismuth intermetallic compound InBi, and four will be models of composite materials using various types of solid fibers and particles in matrices of the indium-bismuth eutectic alloy.

Hardware for the demonstration will include ten welded aluminum capsules containing the samples, a low-powered electrical resistance heater used to melt the samples, and a storage box which also serves as a heat sink for directional solidification. The entire demonstration package will weigh about ten pounds. Each of the sample capsules is 3.5 inches long and 7/8-inch in diameter. The heater unit is cylindrical, with capped openings on its top and bottom, and is operated from a 28-volt D.C. supply. The storage box is 4.25 by 5 by 3.5 inches.

To process the samples, the astronauts will insert the capsules one at a time into the heater, apply power for a prescribed time to melt the sample material, turn off the heater, and then either allow the assembly to cool without further attention or, in some cases, mount the heater on the storage box heat sink to cool. Individual samples will take from 45 to 105 minutes to process, depending on the material. All ten samples may not be processed; the deciding factor will be how much free time the astronauts have to operate the apparatus during transearth coast phase of the mission. No data will be taken on the samples in flight.

The returned samples will be evaluated on the ground by metallurgical, chemical, and physical tests. These results will be used in conjunction with those already obtained on the Apollo 14 samples to assess the prospects for further metallurgical research and eventual product manufacturing in space. The demonstration hardware was built at NASA's Marshall Space Flight Center in Huntsville, Ala.

Engineering/Operational Objectives

In addition to the lunar surface and lunar orbital experiments, there are several test objectives in the Apollo 15 mission aimed toward evaluation of new hardware from an operational or performance standpoint. These test objectives are:

*Lunar Rover Vehicle Evaluation — an assessment of the LRV's performance and handling characteristics in the lunar environment.

*EVA Communications with the lunar communications relay unit/ground commanded television assembly (LCRU/GCTA) — has the objective of demonstrating that the LCRU is capable of relaying two-way communications when the crew is beyond line-of-site from the LM, and that the GCTA can be controlled from the ground for television coverage of EVAs.

*EMU Assessment on Lunar Surface — an evaluation of the improved Apollo spacesuit (A7LB) and the -7 portable life support system (PLSS), both of which are being used for the first time on Apollo 15. The suit modifications allow greater crew mobility, and the later model PLSS allows a longer EVA stay time because of increased consumables.

*Landing Gear Performance of Modified LM — measurements of the LM landing gear stroking under a heavier load caused by J-mission modifications and additions to the basic LM structure — about 1,570 pounds over H-mission LMs.

*SIM Thermal Data — measurement of the thermal responses of the SIM Bay and the experiments stowed in the bay, and the effect of the bay upon the rest of the service module.

*SIM Bay Inspection During EVA — evaluation of the effects of SIM bay door jettison, detect any SIM bay contamination, and evaluate equipment and techniques for EVA retrieval of film cassettes.

*SIM Door Jettison Evaluation — an engineering evaluation of the SIM door jettison mechanisms and the effects of jettison on the CSM.

*LM Descent Engine Performance — evaluation of the descent engine with lengthened engine skirt, longer burn time, and new thrust chamber material.

APOLLO LUNAR HAND TOOLS

Special Environmental Container - The special environmental sample is collected in a carefully selected area and sealed in a special container which will retain a high vacuum. The container is opened in the lunar receiving laboratory (LRL) where it will provide scientists the opportunity to study lunar material in its original environment.

Extension handle - This tool is of aluminum alloy tubing with a malleable stainless steel cap designed to be used as an anvil surface. The handle is designed to be used as an extension for several other tools and to permit their use without requiring the astronaut to kneel or bend down. The handle is approximately 30 inches long and one inch in diameter. The handle contains the female half of a quick disconnect fitting designed to resist compression, tension, torsion or a combination of these loads.

Nine core tubes - These tubes are designed to be driven or augered into loose gravel, sandy material or into soft rock such as feather rock or pumice. They are about 15 inches in length and one inch in diameter and are made of aluminum tubing. Each tube is supplied with a removable non-serrated cutting edge and a screw-on cap incorporating a metal-to-metal crush seal which replaces the cutting edge. The upper end of each tube is sealed and designed to be used with the extension handle or as an anvil. Incorporated into each tube is a spring device to retain loose materials in the tube.

Adjustable Sampling Scoop - Similar to a garden scoop, the device is used for gathering sand or dust too small for the rake or tongs. The stainless steel pan is adjustable from 55 to 90 degrees. The handle is compatible with the extension handle.

Sampling hammer - This tool serves three functions, as a sampling hammer, as a pick or mattock and as a hammer to drive the core tubes or scoop. The head has a small hammer face on one end, a broad horizontal blade on the other, and large hammering flats on the sides. The handle is 14 inches long and is made of formed tubular aluminum. The hammer has on its lower end a quick-disconnect to allow attachment to the extension handle for use as a hoe. The head weight has been increased to provide more impact force.

Collection Bags - Two types of bags are provided for collecting lunar surface samples: the sample collection bag with pockets for holding core tubes, the special environmental sample and magnetic shield sample containers, and capable of holding large surface samples; and the 7 ½ X 8 inch Teflon documented sample bags in a 20-bag dispenser mounted on the lunar hand tool carrier. Both types of bags are stowed in the Apollo lunar sample return containers (ALSRC).

Tongs - The tongs are designed to allow the astronaut to retrieve small samples from the lunar surface while in a standing position. The tines are of such angles, length and number to allow samples of from 3/8 up to 2½-inch diameter to be picked up. The tool is 32 inches long overall.

Spring scale - To weigh two rock boxes and other bags containing lunar material samples, to maintain weight budget for return to Earth.

Gnomon - This tool consists of a weighted staff suspended on a two ring gimbal and supported by a tripod. The staff extends 12 inches above the gimbal and is painted with a gray scale and a color scale. The gnomon is used as a photographic reference to indicate local vertical, Sun angle and scale. The gnomon has a required accuracy of vertical indication of 20 minutes of arc. Magnetic damping is incorporated to reduce oscillations.

Color chart - The color chart is painted with three primary colors and a gray scale. It is used in calibration for lunar photography. The scale is mounted on the tool carrier but may easily be removed and returned to Earth for reference. The color chart is six inches in size.

Self-recording Penetrometer - Used in the soil mechanics experiment to measure the characteristics and mechanical properties of the lunar surface material. The penetrometer consists of a 30-inch penetration shaft and recording drum. Three interchangeable penetration cones (0.2, 0.5 and 1.0 square-inch cross sections) and a 1 X 5-inch pressure plate may be attached to the shaft. The crewman forces the penetrometer into the surface and a stylus scribes a force vs. depth plot on the recording drum. The drum can record up to 24 force-depth plots. The upper housing containing the recording drum is detached at the conclusion of the experiment for return to Earth and analysis by the principal investigator.

Lunar rake - Used by the crew for gathering samples ranging from one-half inch to one inch in size. The rake is adjustable and is fitted with stainless steel tines. A ten-inch rake handle adapts to the tool extension handle.

Apollo Lunar Hand Tool Carrier - An aluminum rack upon which the tools described above are stowed for lunar surface EVAs. The carrier differs from the folding carriers used on previous missions in that it mounts on the rear pallet of the lunar roving vehicle. The carrier may be hand carried during treks away from the LRV and is fitted with folding legs.

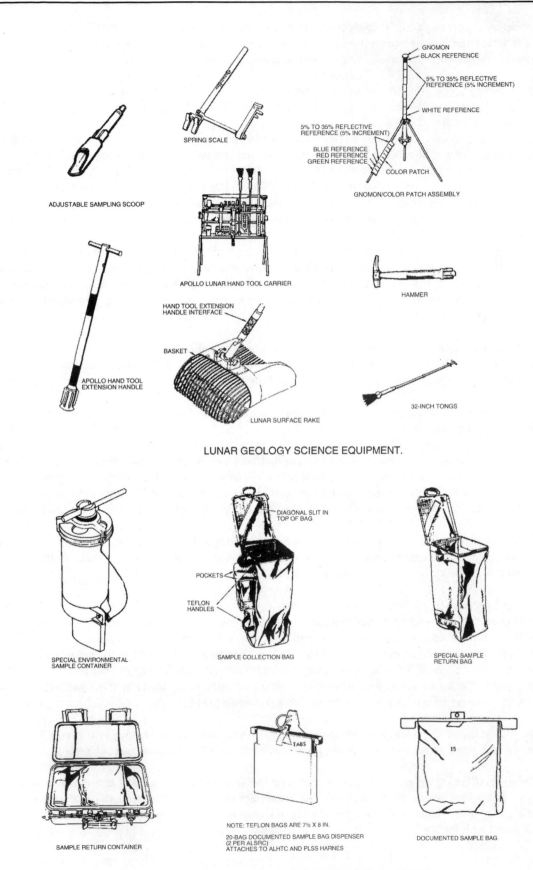

LUNAR GEOLOGY SCIENCE EQUIPMENT.

LUNAR GEOLOGY SAMPLE CONTAINERS

HADLEY-APENNINE LANDING SITE

The Apollo 15 landing site is located at 26° 04' 54" North latitude by 3° 39' 30" East longitude at the foot of the Apennine mountain range. The Apennines rise up to more than 15,000 feet along the southeastern edge of the Mare Imbrium (Sea of Rains).

The Apennine escarpment — highest on the Moon — is higher above the flatlands than the east face of the Sierra Nevadas in California and the Himalayan front rising above the plains of India and Nepal. The landing site has been selected to allow astronauts Scott and Irwin to drive from the LM to the Apennine front during two of the EVAs.

A meandering canyon, Rima Hadley (Hadley Rille), approaches the Apennine front near the landing site and the combination of lurain provides an opportunity for the crew to explore and sample a mare basin, a mountain front and a lunar rille in a single mission.

Hadley Rille is a V-shaped gorge paralleling the Apennines along the eastern edge of Mare Imbrium. The rille meanders down from an elongated depression in the mountains and across the Palus Putredenis (Swamp of Decay), merging with a second rille about 62 miles (100 kilometers) to the north. Hadley rille averages about a kilometer and a half in width and about 1,300 feet (400 meters) in depth throughout most of its length.

Large rocks have rolled down to the rille floor from fresh exposures of what are thought to be stratified mare beds along the tops of the rille walls. Selenographers are curious about the origin of the Moon's sinuous rilles, and some scientists believe the rilles were caused by some sort of fluid flow mechanism—possibly volcanic.

Material sampled from the Apennines may yield specimens of ancient rocks predating the formation and filling of the major mare basins, while the rille may provide samples of material dredged up by the impact of forming the 1.4-mile-wide (2.2 km) Hadley C crater to the south of the landing site and on the west side of Hadley rille. Secondary crater clusters in the landing site vicinity are believed to have been formed by ejecta from the Copernican-age craters Aristillus and Autolycus which lie to the north of the landing site.

Mount Hadley, Hadley Rille and the various Hadley craters in the region of the landing site are named for British scientist-mathematician John Hadley (1682-1744) who made improvements in reflector telescope design and invented the reflecting quadrant — an ancestor of the mariner's sextant.

SITE SCIENCE RATIONALE

	APOLLO 11	APOLLO 12	APOLLO 14	APOLLO 15
TYPE	MARE	MARE	HILLY UPLAND	MOUNTAIN FRONT I RILLE/ MARE
MATERIAL	BASALTIC LAVA	BASALTIC LAVA	DEEP-SEATED CRUSTAL MATERIAL	• DEEPER - SEATED CRUSTAL MATERIAL • BASALTIC LAVA
PROCESS	BASIN FILLING	BASIN FILLING	EJECTA BLANKET FORMATION	• MOUNTAIN SCARP • BASIN FILLING • RILLE FORMATION
AGE	OLDER MARE FILLING	YOUNGER MARE FILLING	EARLY HISTORY OF MOON o PRE-MARE MATERIAL o IMBRIUM BASIN FORMATION	• COMPOSITION AND AGE OF APENNINE FRONT MATERIAL • RILLE ORIGIN AND AGE • AGE OF IMBRIUM MARE FILL

LUNAR ROVING VEHICLE

The lunar roving vehicle (LRV) will transport two astronauts on three exploration traverses of the Hadley-Apennine area of the Moon during the Apollo 15 mission. The LRV will also carry tools, scientific equipment, communications gear, and lunar samples.

The four-wheel, lightweight vehicle will greatly extend the lunar area that can be explored by man. The LRV can be operated by either astronaut.

The lunar roving vehicle will be the first manned surface transportation system designed to operate on the Moon. It marks the beginning of a new technology and represents an ambitious experiment to overcome many new and challenging problems for which there is no precedent in terrestrial vehicle design and operations.

First, the LRV must be folded up into a very small package in order to fit within the tight, pie-shaped confines of Quad 1 of the lunar module which will transport it to the Moon. After landing, the LRV must unfold itself from its stowed configuration and deploy itself to the lunar surface in its operational configuration, all with minimum assistance from the astronauts.

The lack of an atmosphere on the Moon, the extremes of surface temperature, the very small gravity, and the many unknowns associated with the lunar soil and topography impose requirements on the LRV which have no counterpart in Earth vehicles and for which no terrestrial experience exists. The fact that the LRV must be able to operate on a surface which can reach 250 degrees Fahrenheit and in a vacuum which rules out air cooling required the development of new concepts of thermal control.

The one-sixth gravity introduces a host of entirely new problems in vehicle dynamics, stability, and control. It makes much more uncertain such operations as turning, braking, and accelerating which will be totally different experiences than on Earth. The reduced gravity will also lead to large pitching, bouncing, and swaying motions as the vehicle travels over craters rocks, undulations, and other roughnesses of the lunar surface.

Many uncertainties also exist in the mechanical properties of the lunar soil involved in wheel/soil interaction. The interaction of lunar-soil mechanical properties, terrain roughness and vehicle controllability in one-sixth gravity will determine the performance of the LRV on the Moon.

Thus the LRV, while it is being used to increase the effectiveness of lunar exploration, will be exploring entirely new regimes of vehicle operational conditions in a new and hostile environment, markedly different from Earth conditions. The new knowledge to be gained from this mission should play an important role in shaping the course of future lunar and planetary exploration systems.

The LRV is built by the Boeing Co., Aerospace Group, at its Kent Space Center near Seattle, Wash., under contract to the NASA-Marshall Space Flight Center. Boeing's major subcontractor is the Delco Electronics Division of the General Motors Corp. Three flight vehicles have been built, plus seven test and training units, spare components, and related equipment.

General Description

The lunar roving vehicle is ten feet, two inches long; has a six-foot tread width; is 44.8 inches high; and has a 7.5-foot wheelbase. Each wheel is individually powered by a quarter-horse-power electric motor (providing a total of one horsepower) and the vehicle's top speed will be about eight miles an hour on a relatively smooth surface.

Two 36-volt batteries provide the vehicle's power, although either battery can power all vehicle systems if required. The front and rear wheels have separate steering systems, but if one steering system fails, it can be disconnected and the vehicle will operate with the other system.

Weighing approximately 460 pounds (Earth weight) when deployed on the Moon, the LRV will carry a total payload weight of about 1,080 pounds, more than twice its own weight. This cargo includes two astronauts and their portable life support systems (about 800 pounds), 100 pounds of communications equipment, 120 pounds of scientific equipment and photographic gear, and 60 pounds of lunar samples.

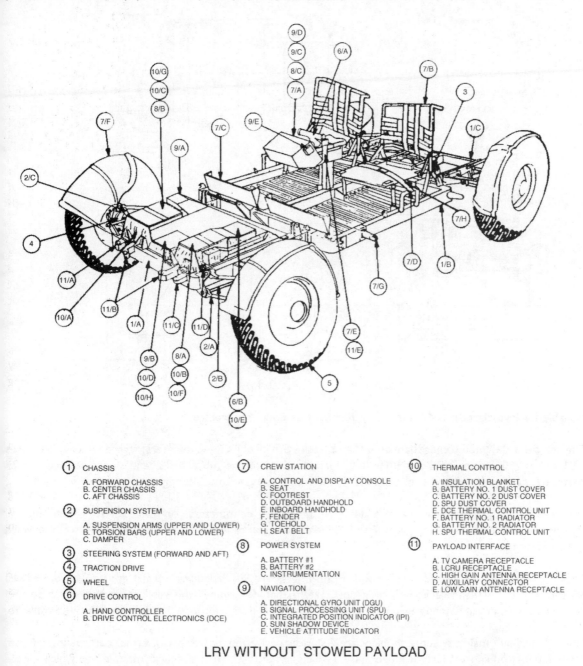

① CHASSIS

    A. FORWARD CHASSIS
    B. CENTER CHASSIS
    C. AFT CHASSIS

② SUSPENSION SYSTEM

    A. SUSPENSION ARMS (UPPER AND LOWER)
    B. TORSION BARS (UPPER AND LOWER)
    C. DAMPER

③ STEERING SYSTEM (FORWARD AND AFT)

④ TRACTION DRIVE

⑤ WHEEL

⑥ DRIVE CONTROL

    A. HAND CONTROLLER
    B. DRIVE CONTROL ELECTRONICS (DCE)

⑦ CREW STATION

    A. CONTROL AND DISPLAY CONSOLE
    B. SEAT
    C. FOOTREST
    D. OUTBOARD HANDHOLD
    E. INBOARD HANDHOLD
    F. FENDER
    G. TOEHOLD
    H. SEAT BELT

⑧ POWER SYSTEM

    A. BATTERY #1
    B. BATTERY #2
    C. INSTRUMENTATION

⑨ NAVIGATION

    A. DIRECTIONAL GYRO UNIT (DGU)
    B. SIGNAL PROCESSING UNIT (SPU)
    C. INTEGRATED POSITION INDICATOR (IPI)
    D. SUN SHADOW DEVICE
    E. VEHICLE ATTITUDE INDICATOR

⑩ THERMAL CONTROL

    A. INSULATION BLANKET
    B. BATTERY NO. 1 DUST COVER
    C. BATTERY NO. 2 DUST COVER
    D. SPU DUST COVER
    E. DCE THERMAL CONTROL UNIT
    F. BATTERY NO. 1 RADIATOR
    G. BATTERY NO. 2 RADIATOR
    H. SPU THERMAL CONTROL UNIT

⑪ PAYLOAD INTERFACE

    A. TV CAMERA RECEPTACLE
    B. LCRU RECEPTACLE
    C. HIGH GAIN ANTENNA RECEPTACLE
    D. AUXILIARY CONNECTOR
    E. LOW GAIN ANTENNA RECEPTACLE

LRV WITHOUT STOWED PAYLOAD

The LRV will travel to the Moon folded inside stowage Quadrant 1 of the lunar module's descent stage. During the first lunar surface EVA period the astronauts will manually deploy the vehicle and prepare it for cargo loading and operation.

The LRV is designed to operate for 78 hours during the lunar day. It can make several exploration sorties up to a cumulative distance of 40 miles (65 kilometers). Because of limitations in the astronauts' portable life support systems (PLSS), however, the vehicle's range will be restricted to a radius of about six miles (9.5 kilometers) from the lunar module. This provides a walk-back capability to the LM should the LRV become immobile at the maximum radius from the LM. This six-mile radius contains about 113 square miles which is

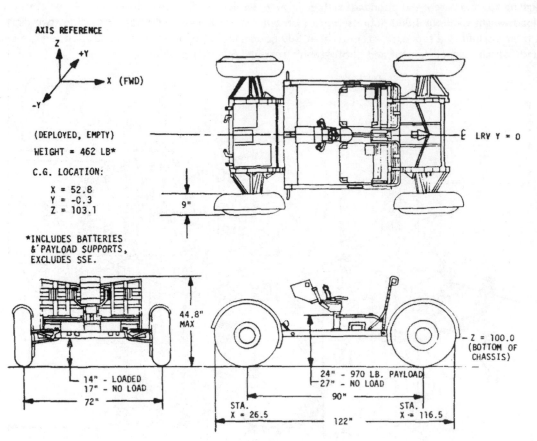

LRV COMPONENTS AND DIMENSIONS

available for investigation, some ten times the area that could be explored on foot.

The vehicle is designed to negotiate step-like obstacles 9.8 inches (25 centimeters) high, and cross crevasses 22.4 inches (50 centimeters) wide. The fully loaded vehicle can climb and descend slopes as steep as 25 degrees. A parking brake can stop and hold the LRV on slopes of up to 30 degrees.

The vehicle has ground clearance of at least 14 inches (35 centimeters) on a flat surface. Pitch and roll stability angles are at least 45 degrees with a full load. The turn radius is approximately 10 feet with forward and aft steering.

Both crewmen will be seated so that both front wheels are visible during normal driving. The driver will navigate through a dead reckoning navigation system that determines the vehicle heading, direction and distance between the LRV and the lunar module, and the total distance traveled at any point during a traverse.

The LRV has five major systems: mobility, crew station, navigation, power, and thermal control. In addition, space support equipment includes mechanisms which attach the LRV to the lunar module and which enable deployment of the LRV to the lunar surface.

Auxiliary equipment (also called stowed equipment) will be provided to the LRV by the Manned Spacecraft Center, Houston. This equipment includes the lunar communications relay unit (LCRU) and its high and low gain antennas, the ground control television assembly (GCTA), a motion picture camera, scientific equipment, astronaut tools, and sample stowage bags.

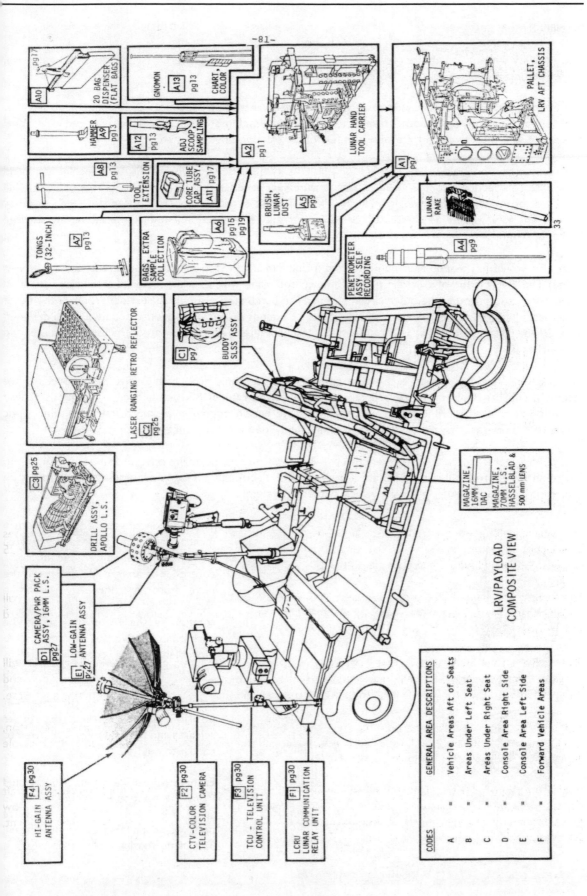

-81-

A10 pg17 — 20 BAG DISPENSER (FLAT BAGS)

GNOMON — A13 pg13 — CHART, COLOR

HAMMER A9 pg13

A12 pg13 — ADJ SCOOP, SAMPLING

A2 pg11

LUNAR HAND TOOL CARRIER

A1 pg7 — PALLET, LRV AFT CHASSIS

A8 pg13 — TOOL EXTENSION

CORE TUBE CAP ASSY. A11 pg17

BRUSH, LUNAR DUST A5 pg9

LUNAR RAKE

33

TONGS (32-INCH) A7 pg13

BAGS, EXTRA SAMPLE COLLECTION A6 pg15 pg19

PENETROMETER ASSY, SELF RECORDING A4 pg9

LASER RANGING RETRO REFLECTOR C2 pg25

BUDDY SLSS ASSY C1 pg7

DRILL ASSY, APOLLO L.S. C3 pg25

MAGAZINE, 16MM DAC

MAGAZINE, 70MM L.S. HASSELBLAD & 500 mm LENS

CAMERA/PWR PACK ASSY, 16MM L.S. D1 pg27

LOW-GAIN ANTENNA ASSY E1 pg27

LRV/PAYLOAD COMPOSITE VIEW

HI-GAIN ANTENNA ASSY F4 pg30

CTV-COLOR TELEVISION CAMERA F2 pg30

TCU - TELEVISION CONTROL UNIT F3 pg30

LCRU LUNAR COMMUNICATION RELAY UNIT F1 pg30

CODES		GENERAL AREA DESCRIPTIONS
A	=	Vehicle Areas Aft of Seats
B	"	Areas Under Left Seat
C	"	Areas Under Right Seat
D	"	Console Area Right Side
E	"	Console Area Left Side
F	"	Forward Vehicle Areas

Mobility System

The mobility system has the largest number of subsystems, including the chassis, wheels, traction drive, suspension, steering, and drive control electronics.

The aluminum chassis is divided into forward, center and aft sections that support all equipment and systems. The forward section holds both batteries, the navigation system's signal processing unit and directional gyroscope, and the drive control electronics (DCE) for traction drive and steering.

The center section holds the crew station, with its two seats, control and display console, and hand controller. This section's floor is made of beaded aluminum panels, structurally capable of supporting the full weight of both astronauts standing in lunar gravity. The aft section is a platform for the LRV's scientific payload. The forward and aft chassis sections fold over the center section and lock in place during stowage in the lunar module.

Each LRV wheel has a spun aluminum hub and a titanium bump stop (inner frame) inside the tire (outer frame). The tire is made of a woven mesh of zinc-coated piano wire to which titanium treads arc riveted in a chevron pattern around the outer circumference. The bump stop prevents excess deflection of the outer wire mesh during heavy impact. Each wheel weighs 12 pounds on Earth (two lunar pounds) and is designed for a driving distance of at least 112 statute miles (180 kilometers). The wheels are 32 inches in diameter and nine inches wide.

The traction drive attached to each wheel consists of a harmonic drive unit, a drive motor, and a brake assembly. The harmonic drives reduce motor speed at the rate of 80-to-1, allowing continuous vehicle operation at all speeds without gear shifting. Each drive has an odometer pickup that transmits magnetic pulses to the navigation system's signal processing unit. (Odometers measure distance travelled.)

The quarter-horsepower, direct current, brush-type drive motors normally operate from a 36-volt input. Motor speed control is furnished from the drive control electronics package. Suspension system fittings on each motor form the king-pin for the vehicle's steering system.

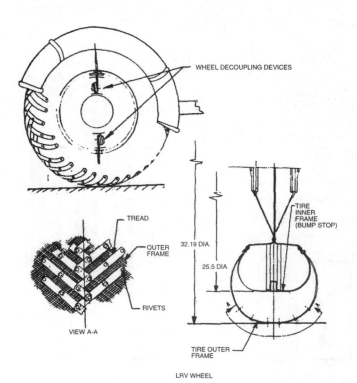

VIEW A-A

LRV WHEEL

The traction drive is equipped with a mechanical brake, cable-connected to the hand controller. Moving the controller rearward de-energizes the drive motor and forces hinged brake shoes against a brake drum, stopping rotation of the wheel hub about the harmonic drive. Full rearward movement of the controller engages and locks the parking brakes. To disengage the parking brake, the controller is moved to the steer left position at which time the brake releases and the controller is allowed to return to neutral.

Each wheel can be manually uncoupled from the traction drive and brake to allow "free-wheeling" about the drive housing, independent of the drive train. The same mechanism will re-engage a wheel.

The chassis is suspended from each wheel by a pair of parallel arms mounted on torsion bars and connected to each traction drive. A damper (shock absorber) is a part of each suspension system. Deflection of the system and the tires allows a 14-inch ground clearance when the vehicle is fully loaded, and 17 inches when unloaded. The suspension systems can be folded about 135 degrees over the center chassis for stowage in the lunar module.

Both the front and rear wheels have independent steering systems that allow a "wall-to-wall" turning radius of 122 inches (exactly the vehicle's length). Each system has a small, 1/10th-horsepower, 5,000-rpm motor driving through a 257-to-1 reduction into a gear that connects with the traction drive motor by steering arms and a tie rod. A steering vane, attached between the chassis and the steering arms, allows the extreme steering angles required for the short turn radius.

If a steering malfunction occurs on either the front or rear steering assembly, the steering linkage to that set of wheels can be disengaged and the mission can continue with the remaining active steering assembly. A crewman can reconnect the rear steering assembly if desired.

The vehicle is driven by a T-shaped hand controller located on the control and display console post between the two crewmen. The controller maneuvers the vehicle forward, reverse, left and right, and controls speed and braking.

A knob that determines whether the vehicle moves forward or reverse is located on the T-handle's vertical stem. With the knob pushed down, the hand controller can only be moved forward. When the knob is pushed up and the controller moved rearward, the LRV can be operated in reverse.

Drive control electronics accept forward and reverse speed control signals from the hand controller, and electronic circuitry will switch drive power off and on. The electronics also provide magnetic pulses from the wheels to the navigation system for odometer and speedometer readouts.

Crew Station

The LRV crew station consists of the control and display console, seats and seat belts, an armrest, footrests, inboard and outboard handholds, toeholds, floor panels, and fenders.

The control and display console is separated into two main parts: the upper portion holds navigation system gauges and the lower portion holds vehicle monitors and controls.

Attached to the upper left side of the console is an attitude indicator that shows vehicle pitch and roll. Pitch is indicated upslope or downslope within a range of ± 25 degrees; roll is indicated as 25 degrees left or right. Readings, normally made with the vehicle stopped, are transmitted verbally to Houston's Mission Control Center for periodic navigation computation.

At the console's top left is an integrated position indicator (IPI). The indicator's outer circumference is a large dial that shows the vehicle's heading (direction) with respect to lunar north. Inside the circular dial are three indicators that display readings of bearing, distance and range. The bearing indicator shows direction to the Lunar Module, the distance indicator records distance traveled by the LRV, and the range indicator displays distance to the Lunar Module.

The distance and range indicators have total scale capacities of 99.9 kilometers (62 statute miles). If the navigation system loses power, the bearing and range readings will remain displayed.

In the center of the console's upper half is a Sun shadow device (Sun compass) that can determine the LRV's heading with respect to the Sun. The device casts a shadow on a graduated scale when it is pulled up at right angles from the console. The point where the Sun's shadow intersects the scale will be read by the crew to Mission Control, which will tell the crew what heading to set into the navigation system. The device can be

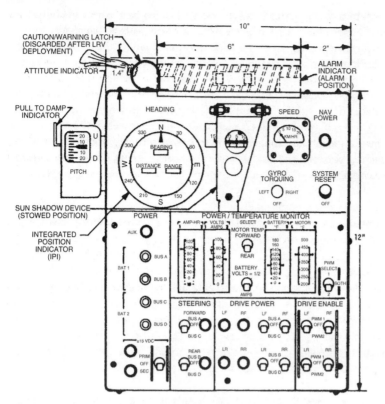

LRV CREW STATION COMPONENTS - CONTROL AND DISPLAY CONSOLE

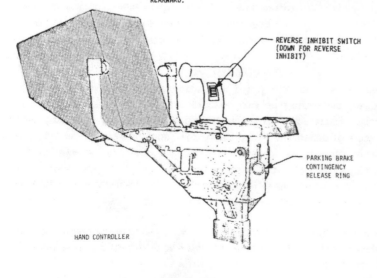

HAND CONTROLLER OPERATION:

T-HANDLE PIVOT FORWARD - INCREASED DEFLECTION FROM NEUTRAL INCREASES FORWARD SPEED.

T-HANDLE PIVOT REARWARD - INCREASED DEFLECTION FROM NEUTRAL INCREASES REVERSE SPEED.

T-HANDLE PIVOT LEFT - INCREASED DEFLECTION FROM NEUTRAL INCREASES LEFT STEERING ANGLE.

T-HANDLE PIVOT RIGHT - INCREASED DEFLECTION FROM NEUTRAL INCREASES RIGHT STEERING ANGLE.

T-HANDLE DISPLACED REARWARD - REARWARD MOVEMENT INCREASES BRAKING FORCE. FULL 3 INCH
REARWARD APPLIES PARKING BRAKE. MOVING INTO BRAKE
POSITION DISABLES THROTTLE CONTROL AT 15° MOVEMENT
REARWARD.

HAND CONTROLLER

used at Sun elevation angles up to 75 degrees.

A speed indicator shows LRV velocity from 0 to 20 kilometers an hour (0-12 statute mph). This display is driven by odometer pulses from the right rear wheel through the navigation system's signal processing unit.

A gyro torquing switch adjusts the heading indicator during navigation system resettings, and a system reset switch returns the bearing, distance, and range indicators to zero.

Down the left side of the console's lower half are switches that allow power from either battery to feed a dual bus system. Next to these switches are two power monitors that give readings of ampere hours remaining in the batteries, and either volts or amperes from each battery. To the right of these are two temperature monitors that show readouts from the batteries and the drive motors. Below these monitors are switches that control the steering motors and drive motors.

An alarm indicator (caution and warning flag) atop the console pops up if a temperature goes above limits in either battery or in any of the drive motors. The indicator can be reset.

The LRV's seats are tubular aluminum frames spanned by nylon strips. They are folded flat onto the center chassis during launch and are erected by the crewmen after the LRV is deployed. The seat backs support and restrain the astronauts' portable life support systems (PLSS) from moving sideways when crewmen are sitting on the LRV. The seat bottoms have cutouts for access to PLSS flow control valves and provisions for vertical support of the PLSS. The seat belts are made of nylon webbing. They consist of an adjustable web section and a metal hook that is snapped over the outboard handhold.

The armrest, located directly behind the hand controller, supports the arm of the crewman who is using the controller. The footrests are attached to the center floor section and may be adjusted prior to launch if required to fit each crewman. They are stowed against the center chassis floor and secured by pads until deployment by the crewmen.

The inboard handholds are made of one-inch aluminum tubing and help the crewmen get in and out of the LRV. The handholds also have attachment receptacles for the 16mm camera and the low gain antenna (auxiliary equipment). The outboard handholds are integral parts of the chassis and provide crew comfort and stability when seated on the LRV.

Toeholds are provided to help crewmen leave the LRV. They are made by dismantling the LRV support tripods and inserting the tripod center member legs into chassis receptacles on each side of the vehicle to form the toeholds. The toeholds also can be used as a tool to engage and disengage the wheel decoupling mechanism.

The vehicle's fenders, made of lightweight fiberglass, are designed to prevent lunar dust from being thrown on the astronauts, their scientific payload, and sensitive vehicle parts, or from obstructing astronaut vision while driving. The fender front and rear sections are retracted during flight and extended by the crewmen after deployment.

Navigation System

The dead reckoning navigation system is based on the principle of starting a sortie from a known point, recording direction relative to the LM and distance traveled, and periodically calculating vehicle position relative to the LM from these data.

The system contains three major components: a directional gyroscope that provides the vehicle's heading; odometers on each wheel's traction drive unit that provide distance information; and a signal processing unit (essentially a small, solid-state computer) that determines bearing and range to the LM, distance traveled, and velocity.

All navigation system readings are displayed on the control and display console. Components are activated by pressing the system reset button, which moves all digital displays and internal registers to zero. The system will be reset at the beginning of each LRV traverse.

The directional gyroscope is aligned by measuring the inclination of the LRV (using the attitude indicator) and measuring vehicle orientation with respect to the Sun (using the Sun shadow device). This information is relayed to Mission Control, where a heading angle is calculated and read back to the crew. The gyro is then adjusted until the heading indicator reads the same as the calculated value.

Nine odometer magnetic pulses are generated for each wheel revolution, and these signals enter logic in the signal processing unit (SPU). The SPU selects pulses from the third fastest wheel to insure that the pulses are not based on a wheel that has inoperative odometer pulses or has excessive slip. (Because the SPU cannot distinguish between forward and reverse wheel rotation, reverse operation of the vehicle will add to the odometer reading.) The SPU sends outputs directly to the distance indicator and to the range and bearing indicators through its digital computer. Odometer pulses from the right rear wheel are sent to the speed indicator.

The Sun shadow device is a kind of compass that can determine the LRV's heading in relation to the Sun. It will be used at the beginning of each sortie to establish the initial heading, and then be used periodically during sorties to check for slight drift in the gyro unit.

Power System

The power system consists of two 36-volt, non-rechargeable batteries, distribution wiring, connectors, switches, circuit breakers, and meters to control and monitor electrical power.

The batteries are encased in magnesium and are of plexiglass monoblock (common cell walls) construction, with silver-zinc plates in potassium hydroxide electrolyte. Each battery has 23 cells and a 115-ampere-hour capacity.

Both batteries are used simultaneously with an approximately equal load during LRV operation. Each battery can carry the entire LRV electrical load, however, and the circuitry is designed so that, if one battery fails, the load can be switched to the other battery.

The batteries are located on the forward chassis section, enclosed by a thermal blanket and dust covers. Battery No. 1 (left side) is connected thermally to the navigation system's signal processing unit and is a partial heat sink for that unit. Battery No. 2 (right side) is thermally tied to the navigation system's directional gyro and serves it as a heat sink.

The batteries are activated when installed on the LRV at the launch pad about five days before launch. They are monitored for voltage and temperature on the ground until about T-20.5 hours in the countdown. On the Moon the batteries are monitored for temperature, voltage, output current, and remaining ampere hours through displays on the control console.

During normal LRV operation, all mobility power will be turned off if a stop is to exceed five minutes, but the navigation system's power will stay on during each complete sortie.

For battery survival their temperature must remain between 40 and 125 degrees F. When either battery reaches 125 degrees, or when any motor reaches 400 degrees, temperature switches actuate to flip up the caution and warning flag atop the control console.

An auxiliary connector, located at the front of the vehicle, provides 150 watts of 36-volt power for the lunar communications relay unit (LCRU), whose power cable is attached to the connector before launch.

Thermal Control

Thermal control is used on the LRV to protect temperature-sensitive components during all phases of the mission. Thermal controls include special surface finishes, multi-layer insulation, space radiators, second-surface mirrors, thermal straps, and fusible mass heat sinks.

The basic concept of thermal control is to store heat while the vehicle is running and to cool by radiation between sorties.

During operation, heat is stored in several thermal fusible mass tank heat sinks and in the two batteries. Space radiators are located atop the signal processing unit, the drive control electronics, and the batteries. Fused silica second-surface mirrors are bonded to the radiators to lessen solar energy absorbed by the exposed radiators. The radiators are only exposed while the LRV is parked between sorties.

During sorties, the radiators are protected from lunar surface dust by three dust covers. The radiators are manually opened at the end of each sortie and held by a latch that holds them open until battery temperatures cool down to 45 degrees F (±5 degrees), at which time the covers automatically close.

A multi-layer insulation blanket protects components from harsh environments. The blanket's exterior and some parts of its interior are covered with a layer of Beta cloth to protect against wear.

All instruments on the control and display console are mounted to an aluminum plate isolated by radiation shields and fiberglass mounts. Console external surfaces are coated with thermal control paint and the face plate is anodized, as are all handholds, footrests, seat tubular sections, and center and aft floor panels.

Stowage and Deployment

Certain LRV equipment (called space support equipment) is required to attach the folded vehicle to the Lunar Module during transit to the Moon and during deployment on the surface.

The LRV's forward and aft chassis sections, and the four suspension systems, are folded inward over the center chassis inside the LM's Quadrant 1. The center chassis' aft end is pointing up, and the LRV is attached to the LM at three points.

The upper point is attached to the aft end of the center chassis and the LM through a strut that extends horizontally from the LM quadrant's apex. The lower points are attached between the forward sides of the center chassis, through the LRV tripods, to supports in the LM quadrant.

LRV DEPLOYMENT SEQUENCE

- LRV STOWED IN QUADRANT
- ASTRONAUT REMOTELY INITIATES AND EXECUTES DEPLOYMENT

- ASTRONAUT LOWERS LRV FROM STORAGE BAY WITH FIRST REEL

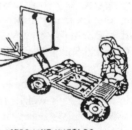

- AFT CHASSIS UNFOLDS
- REAR WHEELS UNFOLD
- AFT CHASSIS LOCKS IN POSITION

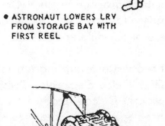

- FORWARD CHASSIS UNFOLDS
- FRONT WHEELS UNFOLD

- FORWARD CHASSIS LOCKS IN POSITION. ASTRONAUT LOWERS LRV TO SURFACE WITH SECOND REEL.

- ASTRONAUT UNFOLDS SEATS, FOOTRESTS, ETC. (FINAL STOP)

The vehicle's deployment mechanism consists of the cables, shock absorbers, pin retract mechanisms, telescoping tubes, pushoff rod, and other gear.

LRV deployment is essentially manual. A crewman first releases a mylar deployment cable, attached to the center rear edge of the LRV's aft chassis. He hands the cable to the second crewman who stands by during the entire deployment operation ready to help the first crewman.

The first crewman then ascends the LM ladder part-way and pulls a D-ring on the side of the descent stage. This deploys the LRV out at the top about five inches (4 degrees) until it is stopped by two steel deployment cables, attached to the upper corners of the vehicle. The crewman then descends the ladder, walks around to the LRV's right side and pulls the end of a mylar deployment tape from a stowage bag in the LM. The crewman unreels this tape, hand-over-hand, to deploy the vehicle.

As the tape is pulled, two support cables are unreeled, causing a pushoff tube to push the vehicle's center of gravity over-center so it will swivel outward from the top. When the chassis reaches a 45-degree angle from the LM, release pins on the forward and aft chassis are pulled, the aft chassis unfolds, the aft wheels are unfolded by the upper torsion bars and deployed, and all latches are engaged.

As the crewman unwinds the tape, the LRV continues lowering to the surface. At a 73-degree angle from the LM, the forward chassis and wheels are sprung open and into place. The crewman continues to pull the deployment tape until the aft wheels are on the surface and the support cables are slack. He then removes the two slack cables from the LRV and walks around the vehicle to its left side. There he unstows a second mylar deployment tape. Pulling this tape completes the lowering of the vehicle to the surface and causes telescoping tubes attached between the LM and the LRV's forward end to guide the vehicle away from the LM. The crewman then pulls a release lanyard on the forward chassis' right side that allows the telescoping tubes to fall away.

The two crewmen then deploy the fender extensions on each wheel, insert the toeholds, deploy the handholds and footrests, set the control and display console in its upright position, release the seat belts, unfold the seats, and remove locking pins and latches from several places on the vehicle.

One crewman will then board the LRV and make sure that all controls are working. He will back the vehicle away from the Lunar Module and drive it to a position near the LM quadrant where the auxiliary equipment is stored, verifying as he drives that all LRV controls and displays are operating. At the new parking spot, the LRV will be powered down while the two astronauts load the auxiliary equipment aboard the vehicle.

Development Background

The manned Lunar Roving Vehicle development program began in October 1969 when the Boeing Co. was awarded a contract to build four (later changed to three) flight model LRVs. The Apollo 15 LRV was delivered to NASA March 15, 1971, two weeks ahead of schedule, and less than 17 months after contract award.

During this extremely short development and test program, more than 70 major tests have been conducted by Boeing and its major subcontractor, GM's Delco Electronics Division. Tests and technical reviews have been held at Boeing's Kent, Washington plant; at Delco's laboratories near Santa Barbara, California, and at NASA's Marshall Space Flight Center, Manned Spacecraft Center, and Kennedy Space Center.

Seven LRV test units have been built to aid development of the three flight vehicles: an LRV mass unit to determine if the LRV's weight might cause stresses or strains in the LM's structure; two one-sixth-weight units to test the LRV's deployment mechanism; a mobility unit to test the mobility system which was later converted to an Earth trainer (one-G trainer) unit for astronaut training; a vibration unit to verify the strength of the LRV structure; and a qualification unit to test vibrations, temperature extremes and vacuums to prove that the LRV will withstand all operating conditions.

Boeing produces the vehicle's chassis, crew station, navigation system, power system, deployment system, ground support equipment, and vehicle integration and assembly. Delco produces the mobility system and built the one-G astronaut training vehicle. Eagle-Picher Industries, Inc., Joplin, Missouri builds the LRV batteries, and the United Shoe Machinery Corp., Wakefield, Massachusetts provides the harmonic drive unit.

LUNAR COMMUNICATIONS RELAY UNIT (LCRU)

The range from which the Apollo 15 crew can operate from the lunar module during EVAs is extended over the lunar horizon by a suitcase-size device called the lunar communications relay unit (LCRU). The LCRU acts as a portable relay station for voice, TV, and telemetry directly between the crew and Mission Control Center instead of through the lunar module communications system.

Completely self-contained with its own power supply and folding S-Band antenna, the LCRU may be mounted on a rack at the front of the lunar roving vehicle (LRV) or hand-carried by a crewman. In addition to providing communications relay, the LCRU relays ground-command signals to the ground commanded television assembly (GCTA) for remote aiming and focussing the lunar surface color television camera. The GCTA is described in another section of this press kit.

Between stops with the lunar roving vehicle, crew voice is beamed Earthward by a low-gain helical S-Band antenna. At each traverse stop, the crew must boresight the high-gain parabolic antenna toward Earth before television signals can be transmitted. VHF signals from the crew portable life support system (PLSS) transceivers are converted to S-Band by the LCRU for relay to the ground, and conversely, from S-Band to VHF on the uplink to the EVA crewmen.

The LCRU measures 22 x 16 x 6 inches not including antennas, and weighs 55 Earth pounds (9.2 lunar pounds). A protective thermal blanket around the LCRU can be peeled back to vary the amount of radiation surface which consists of 196 square inches of radiating mirrors to reflect solar heat. Additionally, wax packages on top of the LCRU enclosure stabilize the LCRU temperature by a melt-freeze cycle. The LCRU interior is pressurized to 7.5 psia differential (one-half atmosphere).

Internal power is provided to the LCRU by a 19-cell silver-zinc battery with a potassium hydroxide electrolyte. The battery weighs nine Earth pounds (1.5 lunar pounds) and measures 4.7 x 9.4 x 4.65 inches. The battery is rated at 400 watt hours, and delivers 29 volts at a 3.1-ampere current load. The LCRU may also draw power from the LRV batteries.

Three types of antennas are fitted to the LCRU system: a low-gain helical antenna for relaying voice and data when the LRV is moving and in other instances when the high-gain antenna is not deployed: a three-foot diameter parabolic rib-mesh high-gain antenna for relaying a television signal: and a VHF omni antenna for receiving crew voice and data from the PLSS transceivers. The high-gain antenna has an optical sight which allows the crewman to boresight on Earth for optimum signal strength. The Earth subtends one-half degree angle when viewed from the lunar surface.

The LCRU can operate in several modes: mobile on the LRV, fixed base such as when the LRV is parked, hand-carried in contingency situations such as LRV failure, and remote by ground control for tilting the television camera to picture LM ascent.

Detailed technical and performance data on the LCRU is available at the Houston News Center query desk.

TELEVISION AND GROUND CONTROLLED TELEVISION ASSEMBLY

Two different color television cameras will be used during the Apollo 15 mission. One, manufactured by Westinghouse, will be used in the command module. It will be fitted with a two-inch black and white monitor to aid the crew in focus and exposure adjustment.

The other camera, manufactured by RCA, is for lunar surface use and will be operated from three different positions, mounted on the LM MESA, mounted on a tripod and connected to the LM by a 100-foot cable, and installed in the LRV with signal transmission through the lunar communication relay unit rather than through the LM communications system as in the other two models.

While on the LRV, the camera will be mounted on the ground controlled television assembly (GCTA). The camera can be aimed and controlled by astronauts or it can be remotely controlled by personnel located in the Mission Control Center. Remote command capability includes camera "on" and "off", pan, tilt, zoom, iris open/closed (f2.2 to f22) and peak or average automatic light control.

The GCTA is capable of tilting the TV camera upward 85 degrees, downward 45 degrees, and panning the camera 340 degrees between mechanical stops. Pan and tilt rates are three degrees per second.

The TV lens can be zoomed from a focal length of 12.5mm to 75mm corresponding to a field of view from three to nine degrees.

At the end of the third EVA, the crew will park the LRV about 300 feet east of the LM so that the color TV camera can cover the LM ascent from the lunar surface. Because of a time delay in a signal going the quarter million miles out to the Moon, Mission Control must anticipate ascent engine ignition by about two seconds with the tilt command.

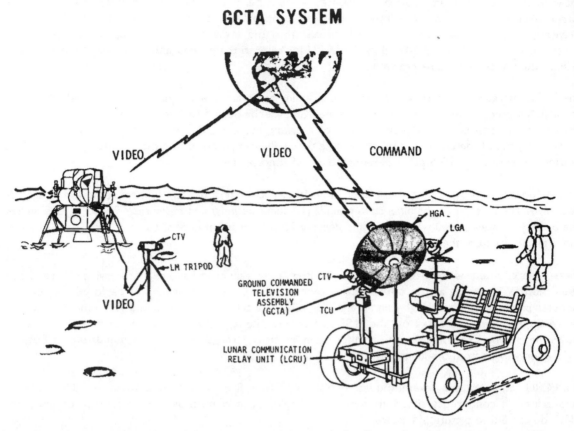

It is planned to view the solar eclipse occurring on August 6, if sufficient battery power remains. The total eclipse extends from 2:24 p.m. EDT to 5:06 p.m. EDT, bracketed by periods of partial eclipse. During the solar eclipse, the camera will be used to make several lunar surface, solar and astronomical observations. Of particular importance will be the observations of the lunar surface under changing lighting conditions. Observations planned during this period include views of the LM, the crescent Sun, the corona edge, the Apennine front, the zodiacal light, the Milky Way, Saturn, Mercury, the eclipse ring, foreground rocks, the lunar horizon and other lunar surface features.

The GCTA and camera each weigh approximately 13 pounds. The overall length of the camera is 18.1 inches, its width is 6.7 inches, and its height is 10.13 inches. The GCTA, and LCRU are built by RCA.

APOLLO 15 TV SCHEDULE

Day	Date	CDT	GET	Duration	Activity	Vehicle	Station
Monday	July 26	11:59 am	3:25	25 min.	Transposition and docking	CSM	Goldstone
Tuesday	July 27	6:20 pm	33:46	45 min.	IVA to LM	CSM	Goldstone
Friday	July 30	9:22 am	96:48	14 min	Landing Site out window	SCM	Madrid
Saturday	July 31	8:34 am	120:00	6 hr. 40 min.	EVA 1	LM/LRV	Honeysuckle/ Madrid
Sunday	Aug. 1	6:09 am	141:35	6 hr. 25 min.	EVA 2	LM/LRV	Parkes/Honeysuckle/Madrid
Monday	Aug. 2	2:49 am	162:15	5 hr. 45 min.	EVA 3	LM/LRV	Parkes/Honeysuckle/Madrid
Monday	Aug. 2	12:04 am	171:30	30 min.	LM liftoff	LRV	Madrid
Monday	Aug. 2	1:37 pm	173:03	6 min.	Rendezvous	CSM	Madrid
Monday	Aug. 2	2:00 pm	173:26	5 min.	Docking	CSM	Madrid
Thursday	Aug. 5	10:44 am	242:10	30 min.	Trans-Earth EVA	CSM	Honeysuckle
Friday	Aug. 6	3:00 pm	270:26	30 min.	Trans-Earth Coast	CSM	Madrid

PROBABLE AREAS FOR NEAR LM LUNAR SURFACE ACTIVITIES

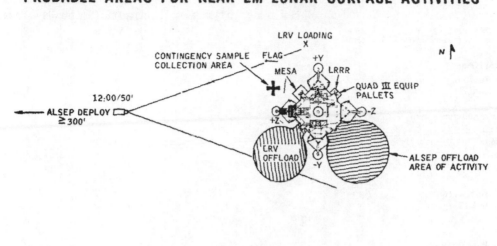

PHOTOGRAPHIC EQUIPMENT

Still and motion pictures will be made of most spacecraft maneuvers and crew lunar surface activities. During lunar surface operations, emphasis will be on documenting placement of lunar surface experiments and on recording in their natural state the lunar surface features.

Command Module lunar orbit photographic tasks and experiments include high-resolution photography to support future landing missions, photography of surface features of special scientific interest and astronomical phenomena such as Gegenschein, zodiacal light, libration points, galactic poles and the Earth's dark side.

Camera equipment stowed in the Apollo 15 command module consists of one 70mm Hasselblad electric camera, a 16mm Maurer motion picture camera, and a 35mm Nikon F single-lens reflex camera. The command module Hasselblad electric camera is normally fitted with an 80mm f/2.8 Zeiss Planar lens, but a bayonet-mount 250mm lens can be fitted for long-distance Earth/ Moon photos. A 105mm f/4.3 Zeiss W Sonnar is provided for the ultraviolet photography experiment.

The 35mm Nikon F is fitted with a 55mm f/1.2 lens for the Gegenschein and dim-light photographic experiments.

The Maurer 16mm motion picture camera in the command module has lenses of 10, 18 and 75mm focal length available. Accessories include a right-angle mirror, a power cable and a sextant adapter which allows the camera to film through the navigation sextant optical system.

Cameras stowed in the lunar module are two 70mm Hasselblad data cameras fitted with 60mm Zeiss Metric lenses, an electric Hasselblad with 500mm lens and two 16mm Maurer motion picture cameras with 10mm lenses. One of the Hasselblads and one of the motion picture cameras are stowed in the modular equipment stowage assembly (MESA) in the LM descent stage.

The LM Hasselblads have crew chest mounts that fit dovetail brackets on the crewman's remote control unit, thereby leaving both hands free. One of the LM motion picture cameras will be mounted in the right-hand window to record descent, landing, ascent and rendezvous. The 16mm camera stowed in the MESA will be carried aboard the lunar roving vehicle to record portions of the three EVAs.

Descriptions of the 24-inch panoramic camera and the 3inch mapping/stellar camera are in the orbital science section of this press kit.

TV and Photographic Equipment Table

Nomenclature	CSM at launch	LM at launch	CM to LM	LM to CM	CM at end
TV, color, zoom lens (monitor with CM system)	1	1			1
Camera, 35mm Nikon	1				1
Lens - 55mm	1				1
Cassette, 35mm	4				4
Camera, Data Acquisition, 16mm	1	1			1
Lens - 10mm	1	1			1
- 18mm	1				1
- 75mm	1				1
Film magazines	12				12
Camera, lunar surface, 16mm		1			
Battery operated lens - 10mm		1			
magazines	10		10	10	10
Camera, Hasselblad, 70mm	1				1
Electric lens - 80mm	1				1
- 250mm	1				1
- 105mm UV (4 bandpass filters)	1				1
film magazines	6				6
film magazine, 70mm UV	1				1
Camera, Hasselblad, 70mm lunar surface electric		3			
lens - 60mm		2			
- 500mm		1			
film magazines	13		13	13	13
Camera, 24-in. Panoramic (in Sim)	1				
film cassette (EVA transfer)	1				1
Camera, 3-in. mapping stellar (Sim)	1				
film magazine (EVA transfer)	1				1

ASTRONAUT EQUIPMENT

Space Suit

Apollo crewmen wear two versions of the Apollo space suit: the command module pilot version (CMP-A-7LB) for intravehicular operations in the command module and for extravehicular operations during SIM Bay film retrieval during transearth coast; and the extravehicular version (EV-A-7LB) worn by the commander and lunar module pilot for lunar surface EVAs. The CMP-A-7LB is the EV-A-7L suit used on Apollo 14, except as modified to eliminate lunar surface operations features not needed for Apollo 15 CMP functions and to alter suit fittings to interface with the Apollo 15 spacecraft.

The EV-A-7LB suit differs from earlier Apollo suits by having a waist joint that allows greater mobility while the suit is pressurized — stooping down for setting up lunar surface experiments, gathering samples and for sitting on the lunar roving vehicle.

From the inside out, the integrated thermal meteroid garment worn by the commander and lunar module pilot starts with rubber-coated nylon and progresses outward with layers of nonwoven Dacron, aluminized Mylar film and Beta marquisette for thermal radiation protection and thermal spacers, and finally with a layer of nonflammable Teflon-coated Beta cloth and an abrasion-resistant layer of Teflon fabric — a total of 18 layers.

Both types of the A-7LB suit have a central portion called a torso limb suit assembly consisting of a gas-retaining pressure bladder and an outer structural restraint layer.

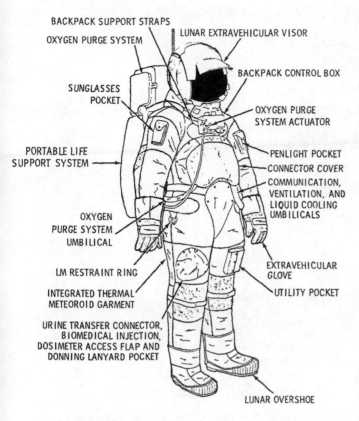

BACKPACK SUPPORT STRAPS
OXYGEN PURGE SYSTEM
LUNAR EXTRAVEHICULAR VISOR
BACKPACK CONTROL BOX
SUNGLASSES POCKET
OXYGEN PURGE SYSTEM ACTUATOR
PORTABLE LIFE SUPPORT SYSTEM
PENLIGHT POCKET
CONNECTOR COVER
COMMUNICATION, VENTILATION, AND LIQUID COOLING UMBILICALS
OXYGEN PURGE SYSTEM UMBILICAL
LM RESTRAINT RING
EXTRAVEHICULAR GLOVE
UTILITY POCKET
INTEGRATED THERMAL METEOROID GARMENT
URINE TRANSFER CONNECTOR, BIOMEDICAL INJECTION, DOSIMETER ACCESS FLAP AND DONNING LANYARD POCKET
LUNAR OVERSHOE

EXTRAVEHICULAR MOBILITY UNIT

The space suit, liquid cooling garment, portable life support system (PLSS), oxygen purge system, lunar extravehicular visor assembly, gloves and lunar boots make up the extravehicular mobility unit (EMU). The EMU provides an extravehicular crewman with life support for a seven-hour mission outside the lunar module without replenishing expendables.

Lunar extravehicular visor assembly — A polycarbonate shell and two visors with thermal control and optical coatings on them. The EVA visor is attached over the pressure helmet to provide impact, micrometeoroid, thermal and ultraviolet-infrared light protection to the EVA crewmen. After Apollo 12, a sunshade was added to the outer portion of the LEVA in the middle portion of the helmet rim.

Extravehicular gloves — Built of an outer shell of Chromel-R fabric and thermal insulation to provide protection when handling extremely hot and cold objects. The finger tips are made of silicone rubber to provide more sensitivity.

Constant-wear garment — A one-piece constant-wear garment, similar to "long johns", is worn as an undergarment for the space suit in intravehicular and on CSM EV operations, and with the inflight coveralls. The garment is porous-knit cotton with a waist-to neck zipper for donning. Biomedical harness attach points are provided.

Liquid Cooling garment — A knitted nylon-spandex garment with a network of plastic tubing through which cooling water from the PLSS is circulated. It is worn next to the skin and replaces the constant-wear garment during Lunar Surface EVA.

Portable life support system — A backpack supplying oxygen at 3.7 psi and cooling water to the liquid cooling garment. Return oxygen is cleansed of solid and gas contaminants by a lithium hydroxide and activated charcoal canister. The PLSS includes communications and telemetry equipment, displays and controls, and a power supply. The PLSS is covered by a thermal insulation jacket. (two stowed in LM.)

Oxygen purge system — Mounted atop the PLSS, the oxygen purge system provides a contingency 30-75 minute supply of gaseous oxygen in two bottles pressurized to 5,880 psia. The system may also be worn separately on the front of the pressure garment assembly torso for contingency EVA transfer from the LM to the CSM or behind the neck for CSM EVA. It serves as a mount for the VHF antenna for the PLSS. (Two stowed in LM).

During periods out of the space suits, crewmen wear two-piece Teflon fabric inflight coveralls for warmth and for pocket stowage of personal items.

Communications carriers — ("Snoopy Hats") with redundant microphones and earphones are worn with the pressure helmet; a light-weight headset is worn with the inflight coveralls.

SPACESUIT/PLSS APOLLO 15 - MAJOR DIFFERENCES

A7L-B SUIT
• ADDED MOBILITY – WAIST CONVOLUTE ADDED
                              NECK CONVOLUTE ADDED
                              ZIPPER RELOCATED FROM CROTCH - AIDS LEG ACTION
                              LESS SHOULDER FORCE

• DURABILITY/PERFORMANCE - FULL BLADDER/CONVOLUTE ABRASION PROTECTION INCREASED EVA CAPABILITY)
                              IMPROVED ZIPPER
                              IMPROVED THERMAL GARMENT
                              INCREASED DRINKING H_2O SUPPLY

-7 PLSS
• LONGER EVA CAPABILITY - INCREASED O_2 SUPPLY
                              INCREASED H_2O SUPPLY
                              LARGER BATTERY
                              ADD ITIONAL LiOH

OPS/BSLSS
• NO CHANGE

PLSS EXPENDABLES COMPARISON

APOLLO 14		APOLLO 15
1020 PSIA	OXYGEN	1430 PSIA
8.50 POUNDS	FEEDWATER	11.50 POUNDS
279 WATT-HOURS	BATTERY	390 WATT-HOURS
3.00 POUNDS	LiOH	3.12 POUNDS

Quart drinking water bags are attached to the inside neck rings of the EVA suits. The crewman can take a sip of water from the 6-by-8-inch bag through a 1/8-inch-diameter tube within reach of his mouth. The bags are filled from the lunar module potable water dispenser.

Buddy Secondary Life Support System — A connecting hose system which permits a crewman with a failed PLSS to share cooling water in the other crewman's PLSS. Flown for the first time on Apollo 14, the BSLSS lightens the load on the oxygen purge system in the event of a total PLSS failure in that the OPS would supply breathing and pressurizing oxygen while the metabolic heat would be removed by the shared cooling water from the good PLSS. The BSLSS will be stowed on the LRV.

Lunar Boots

The lunar boot is a thermal and abrasion protection device worn over the inner garment and boot assemblies. It is made up of layers of several different materials beginning with teflon coated Beta cloth for the boot liner to Chromel R metal fabric for the outer shell assembly. Aluminized Mylar, Nomex felt, Dacron, Beta cloth and Beta marquisette Kapton comprise the other layers. The lunar boot sole is made of high-strength silicone rubber.

Crew Food System

The Apollo 15 crew selected menus from a list of 100 food items qualified for flight. The balanced menus provide approximately 2,300 calories per man per day. Food packages are assembled into man-meal units for the first ten days of the mission. Items similar to those in the daily menu have been stowed in a pantry fashion which gives the crew some variety in making "real-time" food selection for later meals, snacks and beverages. Also, it allows the crew to supplement or substitute food items contained in the nominal man-meal package.

There are various types of food used in the menus. These include freeze-dried rehydratables in spoon-bowl packages; thermostabilized foods (wet packs) in flexible packages and metal easy-open cans, intermediate moisture and dry bite size cubes and beverages. New food items for this mission are thermostabilized beef steaks and hamburgers, an intermediate moisture apricot food bar and citrus flavored beverage.

Water for drinking and rehydrating food is obtained from two sources in the command module - a portable dispenser for drinking water and a water spigot at the food preparation station which supplies water at about 145 degrees and 55 degrees Fahrenheit. The portable water dispenser provides a continuous flow of water as long as the trigger is held down, and the food preparation spigot dispenses water in one-ounce increments.

A continuous flow water dispenser similar to the one in the command module is used aboard the lunar module for coldwater reconstitution of food stowed aboard the LM. Water is injected into a food package and the package is kneaded and allowed to sit for several minutes. The bag top is then cut open and the food eaten with a spoon. After a meal, germicide tablets are placed in each bag to prevent fermentation and gas formation. The bags are then rolled and stowed in waste disposal areas in the spacecraft.

Personal Hygiene

Crew personal hygiene equipment aboard Apollo 15 includes body cleanliness items, the waste management system and one medical kit.

Packaged with the food are a toothbrush and a two-ounce tube of toothpaste for each crewman. Each man-meal package contains a 3.5-by-4-inch wet-wipe cleansing towel. Additionally, three packages of 12-by-12-inch dry towels are stowed beneath the command module pilot's couch. Each package contains seven towels. Also stowed under the command module pilot's couch are seven tissue dispensers containing 53 three-ply tissues each.

Solid body wastes are collected in plastic defecation bags which contain a germicide to prevent bacteria and gas formation. The bags are sealed after use and stowed in empty food containers for post-flight analysis.

Urine collection devices are provided for use while wearing either the pressure suit or the inflight coveralls. The urine is dumped overboard through the spacecraft urine dump valve in the CM and stored in the LM.

Medical Kit

The 5-by-5-by-8-inch medical accessory kit is stowed in a compartment on the spacecraft right side wall beside the lunar module pilot couch. The medical kit contains three motion sickness injectors, three pain suppression injectors, one two-ounce bottle first aid ointment, two one-ounce bottles of eye drops, three bottles of nasal drops, two compress bandages, 12 adhesive bandages, one oral thermometer, and four spare crew biomedical harnesses. Pills in the medical kit are 60 antibiotic, 12 nausea, 12 stimulant, 18 pain killer, 60 decongestant, 24 diarrhea, 72 aspirin and 40 antacid. Additionally, a small medical kit containing four stimulant, eight diarrhea and four pain killer pills, 12 aspirin, one bottle eye drops, two compress bandages, eight decongestant pills, one automatic injector containing a pain killer, one bottle nasal drops is stowed in the lunar module flight data file compartment.

Survival Kit

The survival kit is stowed in two CM rucksacks in the right-hand forward equipment bay above the lunar module pilot.

Contents of rucksack No. 1 are: two combination survival lights, one desalter kit, three pairs of sunglasses, one radio beacon, one spare radio beacon battery and spacecraft connector cable, one knife in sheath, three water containers, two containers of Sun lotion, two utility knives, three survival blankets and one utility netting.

Rucksack No. 2: one three-man life raft with CO_2 inflater, one sea anchor, two sea dye markers, three sunbonnets, one mooring lanyard, three manlines and two attach brackets.

The survival kit is designed to provide a 48-hour postlanding (water or land) survival capability for three crewmen between 40 degrees North and South latitudes.

NATIONAL AERONAUTICS AND SPACE ADMINISTRATION
WASHINGTON, D. C. 10546

BIOGRAPHICAL DATA

NAME: David R. Scott (Colonel, USAF) Apollo 15 Commander NASA Astronaut

BIRTHPLACE AND DATE: Born June 6, 1932, in San Antonio, Texas. His parents, Brigadier General (USAF Retired) and Mrs. Tom W. Scott, reside in La Jolla, California.

PHYSICAL DESCRIPTION: Blond hair; blue eyes; height: 6 feet; weight: 175 pounds.

EDUCATION: Graduated from Western High School, Washington, D.C.; received a Bachelor of Science from the United States Military Academy and the degrees of Master of Science in Aeronautics and Astronautics and Engineer in Aeronautics and Astronautics from the Massachusetts Institute of Technology.

MARITAL STATUS: Married to the former Ann Lurton Ott of San Antonio, Texas. Her parents are Brigadier General (USAF Retired) and Mrs. Isaac W. Ott of San Antonio.

CHILDREN: Tracy L., March 25, 1961; Douglas W., October 8, 1963.

RECREATIONAL INTERESTS: His hobbies are swimming, handball, skiing, and photography.

ORGANIZATIONS: Associate Fellow of the American Institute of Aeronautics and Astronautics; and member of the Society of Experimental Test Pilots, and Tau Beta Pi, Sigma Xi and Sigma Gamma Tau.

SPECIAL HONORS: Awarded the NASA Distinguished Service Medal, the NASA Exceptional Service Medal, the Air Force Distinguished Service Medal, the Air Force Command Pilot Astronaut Wings and the Air Force Distinguished Flying Cross; and recipient of the AIAA Astronautics Award (1966) and the National Academy of Television Arts and Sciences Special Trustees Award (1969).

EXPERIENCE: Scott graduated fifth in a class of 633 at West Point and subsequently chose an Air Force career. He completed pilot training at Webb Air Force Base, Texas, in 1955 and then reported for gunnery training at Laughlin Air Force Base, Texas, and Luke Air Force Base, Arizona.

He was assigned to the 32nd Tactical Fighter Squadron at Soesterberg Air Base (RNAF), Netherlands, from April 1956 to July 1960. Upon completing this tour of duty, he returned to the United States for study at the Massachusetts Institute of Technology where he completed work on his Master's degree. His thesis at MIT concerned interplanetary navigation. After completing his studies at MIT in June 1962, he attended the Air Force Experimental Test Pilot School and then the Aerospace Research Pilot School.

He has logged more than 4,721 hours flying time — 4,011 hours in jet aircraft and 188 hours in helicopters.

CURRENT ASSIGNMENT: Colonel Scott was one of the third group of astronauts named by NASA in October 1963.

On March 16, 1966, he and command pilot Neil Armstrong were launched into space on the Gemini 8 mission — a flight originally scheduled to last three days but terminated early due to a malfunctioning OAMS thruster. The crew performed the first successful docking of two vehicles in space and demonstrated great piloting skill in overcoming the thruster problem and bringing the spacecraft to a safe landing.

He served as command module pilot for Apollo 9, March 3-13, 1969. This was the third manned flight in the Apollo series and the second to be launched by a Saturn V. The ten-day flight encompassed completion of the first comprehensive Earth-orbital qualification and verification tests of a "fully configured Apollo spacecraft" and provided vital information previously not available on the operational performance, stability and reliability of lunar module propulsion and life support systems.

Following a Saturn V launch into a near circular 102.3 x 103.9 nautical mile orbit, Apollo 9 successfully accomplished command/service module separation, transposition and docking maneuvers with the S-IVB-housed lunar module. The crew then separated their docked spacecraft from the S-IVB third stage and commenced an intensive five days of checkout operations with the lunar module, followed by five days of command/service module Earth orbital operations.

Highlight of this evaluation was completion of a critical lunar-orbit rendezvous simulation and subsequent docking, initiated by James McDivitt and Russell Schweickart from within the lunar module at a separation distance which exceeded 100 miles from the command/service module piloted by Scott.

The crew also demonstrated and confirmed the operational feasibility of crew transfer and extravehicular activity techniques and equipment, with Schweickart completing a 46-minute EVA outside the lunar module. During this period, Dave Scott completed a stand-up EVA in the open command module hatch photographing Schweickart's activities and also retrieving thermal samples from the command module exterior.

Apollo 9 splashed down less than four miles from the helicopter carrier USS GUADALCANAL. With the completion of this flight, Scott has logged 251 hours and 42 minutes in space.

He served as backup spacecraft commander for the Apollo 12 flight and is currently assigned as spacecraft commander for Apollo 15.

NAME: Alfred Merrill Worden (Major, USAF) Apollo 15 Command Module Pilot NASA Astronaut

BIRTHPLACE AND DATE: The son of Merrill and Helen Worden, he was born in Jackson, Michigan, on February 7, 1932. His parents reside in Jackson, Michigan.

PHYSICAL DESCRIPTION: Brown hair; blue eyes; height: 5 feet 10 1/2 inches; weight: 153 pounds.

EDUCATION: Attended Dibble, Griswold, Bloomfield and East Jackson grade schools and completed his secondary education at Jackson High School; received a Bachelor of Military Science Degree from the United States Military Academy in 1955 and Master of Science degrees in Astronautical/Aeronautical Engineering and Instrumentation Engineering from the University of Michigan in 1963.

CHILDREN: Merrill E., January 16, 1958; Alison P., April 6, 1960.

RECREATIONAL INTERESTS: He enjoys bowling, water skiing, swimming and handball.

EXPERIENCE: Worden, an Air Force Major, was graduated from the United States Military Academy in June 1955 and, after being commissioned in the Air Force, received flight training at Moore Air Base, Texas; Laredo Air Force Base, Texas; and Tyndall Air Force Base, Florida.

Prior to his arrival for duty at the Manned Spacecraft Center, he served as an instructor at the Aerospace Research Pilots School — from which he graduated in September 1965. He is also a graduate of the Empire Test Pilots School in Farnborough, England, and completed his training there in February 1965.

He attended Randolph Air Force Base Instrument Pilots Instructor School in 1963 and served as a pilot and armament officer from March 1957 to May 1961 with the 95th Fighter Interceptor Squadron at Andrews Air Force Base, Maryland.

He has logged more than 3,309 hours flying time-including 2,804 hours in jets and 107 in helicopters.

CURRENT ASSIGNMENT: Major Worden is one of the 19 astronauts selected by NASA in April 1966. He served as a member of the astronaut support crew for the Apollo 9 flight and as backup command module pilot for the Apollo 12 flight.

He is currently assigned as command module pilot for Apollo 15.

NAME: James Benson Irwin (Lieutenant Colonel, USAF? Apollo 15 Lunar Module Pilot NASA Astronaut

BIRTHPLACE AND DATE: Born March 17, 1930, in Pittsburgh, Pennsylvania, but he considers Colorado Springs, Colorado, as his home town. His parents, Mr. and Mrs. James Irwin, now reside in San Jose, California.

PHYSICAL DESCRIPTION: Brown hair; brown eyes; height: 5 feet 8 inches; weight: 160 pounds.

EDUCATION: Graduated from East High School, Salt Lake City, Utah; received a Bachelor of Science degree in Naval Sciences from the United States Naval Academy in 1951 and Master of Science degrees in Aeronautical Engineering and Instrumentation Engineering from the University of Michigan in 1957.

MARITAL STATUS: Married to the former Mary Ellen Monroe of Corvallis, Oregon; her parents, Mr. and Mrs. Leland F. Monroe, reside in Santa Clara, California.

CHILDREN: Joy C., November 26, 1959; Jill C., February 22, 1961; James B., January 4, 1963; Jan C., September 30 1964.

RECREATIONAL INTERESTS: Enjoys skiing and playing paddleball, handball, and squash; and his hobbies include fishing, diving, and camping.

ORGANIZATIONS: Member of the Air Force Association and the Society of Experimental Test Pilots.

SPECIAL HONORS: Winner of two Air Force Commendation Medals for service with the Air Force Systems Command and the Air Defense Command; and, as a member of the 4750th Training Wing, recipient of an outstanding Unit Citation.

EXPERIENCE: Irwin, an Air Force Lt. Colonel, was commissioned in the Air Force on graduation from the Naval Academy in 1951. He received his flight training at Hondo Air Base, Texas, and Reese Air Force Base, Texas.

Prior to reporting for duty at the Manned Spacecraft Center, he was assigned as Chief of the Advanced Requirements Branch at Headquarters Air Defense Command. He was graduated from the Air Force Aerospace Research Pilot School in 1963 and the Air Force Experimental Test Pilot School in 1961.

He also served with the F-12 Test Force at Edwards Air Force Base, California, and with the AIM 47 Project Office at Wright-Patterson Air Force Base, Ohio.

During his military career, he has accumulated more than 6,650 hours flying time—5,124 hours in jet aircraft and 387 in helicopters.

CURRENT ASSIGNMENT: Lt. Colonel Irwin is one of the 19 astronauts selected by NASA in April 1966. He was crew commander of lunar module (LTA-8) — this vehicle finished the first series of thermal vacuum tests on June 1, 1968. He also served as a member of the astronaut support crew for Apollo 10 and as backup lunar module pilot for the Apollo 12 flight.

Irwin is currently assigned as lunar module pilot for Apollo 15.

NAME: Richard F. Gordon, Jr. (Captain, USN), Backup Apollo 15 Commander NASA Astronaut

BIRTHPLACE AND DATE: Born October 5, 1929, in Seattle, Washington. His mother, Mrs. Angela Gordon, resides in Seattle.

PHYSICAL DESCRIPTION: Brown hair; hazel eyes; height: 5 feet 7 inches; weight: 150 pounds.

EDUCATION: Graduated from North Kitsap High School, Poulsbo, Washington; received a Bachelor of Science degree in Chemistry from the University of Washington in 1951.

MARITAL STATUS: Married to the former Barbara J. Field of Seattle, Washington. Her parents, Mr. and Mrs. Chester Field, reside in Freeland, Washington.

CHILDREN: Carleen, July 8, 1954; Richard, October 6, 1955; Lawrence, December 18, 1957; Thomas, March 25, 1959; James, April 26, 1960; Diane, April 23, 1961.

RECREATIONAL INTERESTS: He enjoys water skiing, sailing, and golf.

ORGANIZATIONS: Member of the Society of Experimental Test Pilots.

SPECIAL HONORS: Awarded two Navy Distinguished Flying Crosses, the NASA Exceptional Service Medal, the Navy Astronaut wings, the Navy Distinguished Service Medal, the Institute of Navigation Award for 1969, the Godfrey L. Cabot Award in 1970, and the Rear Admiral William S. Parsons Award for Scientific and Technical Progress in 1970.

EXPERIENCE: Gordon, a Navy Captain, received his wings as a naval aviator in 1953. He then attended All-weather Flight School and jet transitional training and was subsequently assigned to an all-weather fighter squadron at the Naval Air Station at Jacksonville, Florida.

In 1957, he attended the Navy's Test Pilot School at Patuxent River, Maryland, and served as a flight test pilot until 1960. During this tour of duty, he did flight test work on the F8U Crusader, F11F Tigercat, FJ Fury, and A4D Skyhawk, and was the first project test pilot for the F4H Phantom II.

He served with Fighter Squadron 121 at the Miramar, California, Naval Air Station as a flight instructor in the F4H and participated in the introduction of that aircraft to the Atlantic and Pacific fleets. He was also flight safety officer, assistant operations officer, and ground training officer for Fighter Squadron 96 at Miramar.

Winner of the Bendix Trophy Race from Los Angeles to New York in may 1961, he established a new speed record of 869.74 miles per hour and a transcontinental speed record of two hours and 47 minutes. He was also a student at the U.S. Naval Postgraduate School at Monterey, California.

He has logged more than 4,682 hours flying time — 3,775 hours in jet aircraft and 121 in helicopters.

CURRENT ASSIGNMENT: Captain Gordon was one of the third group of astronauts named by NASA in October 1963. He served as backup pilot for the Gemini 8 flight.

On September 12, 1966, he served as pilot for the 3-day Gemini 11 mission — on which rendezvous with an Agena was achieved in less than one orbit. He executed docking maneuvers with the previously launched Agena and performed two periods of extravehicular activity which included attaching a tether to the Agena and retrieving a nuclear emulsion experiment package. Other highlights accomplished by Gordon and command pilot Charles Conrad on this flight included the successful completion of the first tethered station-keeping exercise, establishment of a new altitude record of 850 miles, and completion of the first fully automatic controlled reentry. The flight was concluded on September 15, 1966, with the spacecraft landing in the Atlantic—2½ miles from the prime recovery ship USS GUAM.

Gordon was subsequently assigned as backup command module pilot for Apollo 9.

He occupied the command module pilot seat on Apollo 12, November 14-24, 1969. Other crewmen on man's second lunar landing mission were Charles Conrad (spacecraft commander) and Alan L. Bean (lunar module pilot). Throughout the 31-hour lunar surface stay by Conrad and Bean, Gordon remained in lunar orbit aboard the command module, "Yankee Clipper," obtaining desired mapping photographs of tentative landing sites for future missions. He also performed the final redocking maneuvers following the successful lunar orbit rendezvous which was initiated by Conrad and Bean from within "Intrepid"'after their ascent from the Moon's surface.

All of the mission's objectives were accomplished, and Apollo 12 achievements include: the first precision lunar landing with "Intrepid's" touchdown in the Moon's Ocean of Storms; the first lunar traverse by Conrad and Bean as they deployed the Apollo Lunar Surface Experiment Package (ALSEP), installed a nuclear power generator station to provide the power source for these long-term scientific experiments, gathered samples of the lunar surface for return to Earth, and completed a close up inspection of the Surveyor III spacecraft.

The Apollo 12 mission lasted 244 hours and 36 minutes and was concluded with a Pacific splashdown and subsequent recovery by the USS HORNET.

Captain Gordon has completed two space flights, logging a total of 315 hours and 53 minutes in space — 2 hours and 44 minutes of which were spent in EVA.

He is currently assigned as backup spacecraft commander for Apollo 15.

NAME: Vance DeVoe Brand (Mr.), Backup Apollo 15 Command Module Pilot NASA Astronaut

BIRTHPLACE AND DATE: Born in Longmont, Colorado, May 9, 1931. His parents, Dr. and Mrs. Rudolph W. Brand, reside in Longmont.

PHYSICAL DESCRIPTION: Blond hair; gray eyes; height: 5 feet 11 inches; weight: 175 pounds.

EDUCATION: Graduated from Longmont High School, Longmont, Colorado; received a Bachelor of Science degree in Business from the University of Colorado in 1953, a Bachelor of Science degree in Aeronautical Engineering from the University of Colorado in 1960, and a Master's degree in Business Administration from the University of California at Los Angeles in 1964.

MARITAL STATUS: Married to the former Joan Virginia Weninger of Chicago, Illinois. Her parents, Mr. and Mrs. Ralph D. Weninger, reside in Chicago.

CHILDREN: Susan N., April 30, 1954; Stephanie, August 6, 1955; Patrick R., March 22, 1958; Kevin S., December 1, 1963.

RECREATIONAL INTERESTS: Skin diving, skiing, handball, and jogging.

ORGANIZATIONS: Member of the Society of Experimental Test Pilots, the American Institute of Aeronautics and Astronautics, Sigma Nu, and Beta Gamma Sigma.

EXPERIENCE: Brand served as a commissioned officer and naval aviator with the U.S. Marine Corps from 1953 to 1957. His Marine Corps assignments included a 15-month tour in Japan as a jet fighter pilot. Following his release from active duty, he continued flying fighter aircraft in the Marine Corps Reserve and the Air National Guard until 1964, and he still retains a commission in the Air Force Reserve.

From 1960 to 1966, Brand was employed as a civilian by the Lockheed Aircraft Corporation. He first worked as a flight test engineer on the P3A "Orion" aircraft and later transferred to the experimental test pilot ranks.

In 1963, he graduated from the U.S. Naval Test Pilot School and was assigned to Palmdale, California, as an experimental test pilot on Canadian and German F-104 development programs. Immediately prior to his selection to the astronaut program, Brand was assigned to the West German F-104G Flight Test Center at Istres, France, as an experimental test pilot and leader of a Lockheed flight test advisory group.

He has logged 3,984 hours of flying time, which include 3,216 in jets and 326 hours in helicopters.

CURRENT ASSIGNMENT: Mr. Brand is one of the 19 astronauts selected by NASA in April 1966. He served as a crew member for the thermal vacuum test of 2TV-1, the prototype command module; and he was a member of the astronaut support crews for the Apollo 8 and 13 missions.

Currently he is backup command module pilot for Apollo 15.

NAME: Harrison H. Schmitt (PhD), Backup Apollo 15 Lunar Module Pilot NASA Astronaut

BIRTHPLACE AND DATE: Born July 3, 1935, in Santa Rita, New Mexico. His mother, Mrs. Harrison A. Schmitt, resides in Silver City, New Mexico.

PHYSICAL DESCRIPTION: Black hair; brown eyes; height: 5 feet 9 inches; weight: 165 pounds.

EDUCATION: Graduated from Western High School, Silver City, New Mexico; received a Bachelor of Science degree in Science from the California Institute of Technology in 1957; studied at the University of Oslo in Norway during 1957-58; received Doctorate in Geology from Harvard University in 1964.

MARITAL STATUS: Single.

RECREATIONAL INTERESTS: His hobbies include skiing, hunting, fishing, carpentry, and hiking.

ORGANIZATIONS: Member of the Geological Society of America, American Geophysical Union, and Sigma Xi.

SPECIAL HONORS: Winner of a Fulbright Fellowship (1957-58); a Kennecott Fellowship in Geology (1958-59); a Harvard Fellowship (1959-60); a Harvard Traveling Fellowship (1960); a Parker Traveling Fellowship (1961-62); a National Science Foundation Post-Doctoral Fellowship, Department of Geological Sciences, Harvard University (1963-64).

EXPERIENCE: Schmitt was a teaching fellow at Harvard in 1961; he assisted in the teaching of a course in ore deposits there. Prior to his teaching assignment, he did geological work for the Norwegian Geological Survey in Oslo, Norway, and for the U.S. Geological Survey in New Mexico and Montana. He also worked as a geologist for two summers in Southeastern Alaska.

Before coming to the Manned Spacecraft Center, he served with the U.S. Geological Survey's Astrogeology Branch at Flagstaff, Arizona. He was project chief for lunar field geological methods and participated in photo and telescopic mapping of the Moon; he was among the USGS astrogeologists instructing NASA astronauts during their geological field trips.

He has logged more than 1,329 hours flying time — 1,141 hours in jet aircraft and 177 in helicopters.

CURRENT ASSIGNMENT: Dr. Schmitt was selected as a scientist-astronaut by NASA in June 1965. He completed a 53-week course in flight training at Williams Air Force Base, Arizona and, in addition to training for future manned space flights, has been instrumental in providing Apollo flight crews with detailed instruction in lunar navigation, geology, and feature recognition.

Schmitt is currently assigned as backup lunar module pilot for Apollo 15.

APOLLO 15 FLAGS, LUNAR MODULE PLAQUE

The United States flag to be erected on the lunar surface measures 30 by 48 inches and will be deployed on a two-piece aluminum tube eight feet long. The flag, made of nylon, will be stowed in the lunar module descent stage modularized equipment stowage assembly.

Also carried on the mission and returned to Earth will be 25 United States flags, 50 individual state flags, flags of United States territories and flags of all United Nations member nations, each four by six inches.
A seven by nine-inch stainless steel plaque, similar to that flown on Apollo 14 will be fixed to the LM front leg. The plaque has on it the words "Apollo 15" with "Falcon" beneath, "July 1971," and the signatures of the three crewmen.

SATURN V LAUNCH VEHICLE

The Saturn V launch vehicle (SA-510) assigned to the Apollo 15 mission was developed under the direction of the Marshall Space Flight Center, Huntsville, Ala. The vehicle is similar to those vehicles used for the missions of Apollo 8 through Apollo 14.

First Stage

The first stage (S-1C) of the Saturn V was build by the Boeing Co. at NASA's Michoud Assembly Facility, New Orleans. The stage's five F-1 engines develop about 7.7 million pounds of thrust at launch. Major components of the stage are the forward skirt, oxidizer tank, inter-tank structure, fuel tank, and thrust structure. Propellant to the five engines normally flows at a rate of approximately 29,400 pounds (3,400 gallons) a second. One engine is rigidly mounted on the stage's centerline; the outer four engines are mounted on a ring at 90-degree angles around the center engine. These outer engines are gimbaled to control the vehicle's attitude during flight.

Second Stage

The second stage (S-II) was built by the Space Division of the North American Rockwell Corp. at Seal Beach, Calif. Five J-2 engines develop a total of about 1.15 million pounds of thrust during flight. Major structural components are the forward skirt, liquid hydrogen and liquid oxygen tanks (separated by an insulated common bulkhead), a thrust structure and an interstage section that connects the first and second stages. The engines are mounted and used in the same arrangement as the first stage's F-1 engines: four outer engines can be gimbaled; the center one is fixed.

Third Stage

The third stage (S-IVB) was built by the McDonnell Douglas Astronautics Co. at Huntington Beach, Calif. Major components are the aft interstage and skirt, thrust structure, two propellant tanks with a common bulkhead, a forward skirt, and a single J-2 engine. The gimbaled engine has a maximum thrust of 230,000 pounds, and can be restarted in Earth orbit.

Instrument Unit

The instrument unit (IU), built by the International Business Machines Corp. at Huntsville, Ala., contains navigation, guidance, and control equipment to steer the launch vehicle into Earth orbit and into translunar trajectory. The six major systems are structural, environmental control, guidance and control, measuring and telemetry, communications, and electrical.

The instrument unit's inertial guidance platform provides space-fixed reference coordinates and measures acceleration along three mutually perpendicular axes of a coordinate system. In the unlikely event of platform failure during boost, systems in the Apollo spacecraft are programmed to provide guidance for the launch

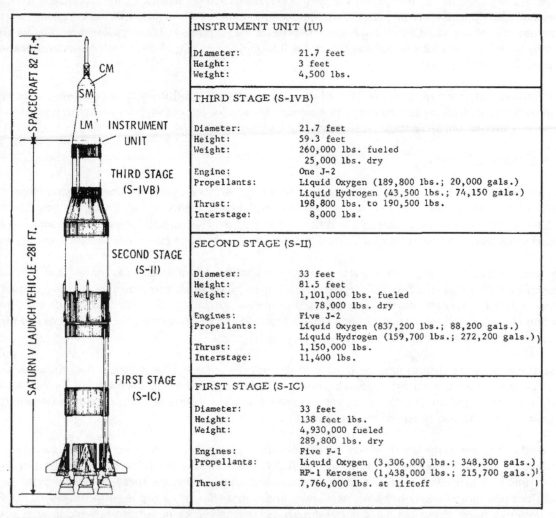

INSTRUMENT UNIT (IU)

Diameter:	21.7 feet
Height:	3 feet
Weight:	4,500 lbs.

THIRD STAGE (S-IVB)

Diameter:	21.7 feet
Height:	59.3 feet
Weight:	260,000 lbs. fueled
	25,000 lbs. dry
Engine:	One J-2
Propellants:	Liquid Oxygen (189,800 lbs.; 20,000 gals.)
	Liquid Hydrogen (43,500 lbs.; 74,150 gals.)
Thrust:	198,800 lbs. to 190,500 lbs.
Interstage:	8,000 lbs.

SECOND STAGE (S-II)

Diameter:	33 feet
Height:	81.5 feet
Weight:	1,101,000 lbs. fueled
	78,000 lbs. dry
Engines:	Five J-2
Propellants:	Liquid Oxygen (837,200 lbs.; 88,200 gals.)
	Liquid Hydrogen (159,700 lbs.; 272,200 gals.)
Thrust:	1,150,000 lbs.
Interstage:	11,400 lbs.

FIRST STAGE (S-IC)

Diameter:	33 feet
Height:	138 feet lbs.
Weight:	4,930,000 fueled
	289,800 lbs. dry
Engines:	Five F-1
Propellants:	Liquid Oxygen (3,306,000 lbs.; 348,300 gals.)
	RP-1 Kerosene (1,438,000 lbs.; 215,700 gals.)
Thrust:	7,766,000 lbs. at liftoff

NOTE: Weights and measures given above are for the nominal vehicle configuration for Apollo.   The figures may vary slightly due to changes before launch
to meet changing conditions. Weights of dry stages and propellants do not equal
total weight because frost and miscellaneous smaller items are not included in
chart.

SATURN V LAUNCH VEHICLE

vehicle. After second stage ignition, the spacecraft commander can manually steer the vehicle if the launch vehicle's stable platform were lost.

Propulsion

The Saturn V has 27 propulsive units, with thrust ratings ranging from 70 pounds to more than 1.5 million pounds. The large main engines burn liquid propellants; the smaller units use solid or hypergolic (self-igniting) propellants.

The five F-1 engines give the first stage a thrust range of from 7,765,852 pounds at liftoff to 9,155,147 pounds at center engine cutoff. Each F-1 engine weighs almost 10 tons, is more than 18 feet long, and has a nozzle exit diameter of nearly 14 feet. Each engine consumes almost three tons of propellant a second.

The first stage has four solid-fuel retro-rockets that fire to separate the first and second stages. Each retrorocket produces a thrust of 75,800 pounds for 0.54 seconds.

Thrust of the J-2 engines on the second and third stages is 205,000 pounds during flight, operating through a range of 180,000 to 230,000 pounds. The 3,500-pound J-2 engine provides higher thrust for each pound of propellant burned per second than the F-1 engine because the J-2 burns high-energy, low molecular weight, liquid hydrogen. F-1 and J-2 engines are built by the Rocketdyne Division of the North American Rockwell Corp. at Canoga Park, Calif.

Four retro-rockets, located in the S-IVB's aft interstage, separate the S-II from the S-IVB. Two jettisonable ullage rockets settle propellants before engine ignition. Six smaller engines in the two auxiliary propulsion system modules on the S-IVB stage provide three-axis attitude control.

Significant Vehicle Changes

Saturn V vehicle SA-510 will be able to deliver a payload that is more than 4,000 pounds heavier than the Apollo 14 payload. The increase provides for the first Lunar Roving Vehicle and for an exploration time on the lunar surface almost twice that of any other Apollo mission. The payload increases were achieved by revising some operational aspects of the Saturn V and through minor changes to vehicle hardware.

The major operational changes are an Earth parking orbit altitude of 90 nautical miles (rather than 100), and a launch azimuth range of 80 to 100 degrees (rather than 72 to 96). Other operational changes include slightly reduced propellant reserves and increased propellant loading for the first opportunity translunar injection (TLI). A significant portion of the payload increase is due to more favorable temperature and wind effects for a July launch versus one in January.

Most of the hardware changes have been made to the first (S-IC) stage. They include reducing the number of retro-rocket motors (from eight to four), reorificing the F-1 engines, burning the outboard engines nearer to LOX depletion, and burning the center engine longer than before. Another change has been made in the propellant pressurization system of the second (S-II) stage.

Three other changes to the launch vehicle were first made to the Apollo 14 vehicle: a helium gas accumulator is installed in the S-II's center engine liquid oxygen (LOX) line, a backup cutoff device is in the same engine, and a simplified propellant utilization valve is installed on all J-2 engines. These changes prevent high oscillations (the "pogo" effect) in the S-II stage and provide more efficient J-2 engine performance. For Apollo 15 a defective cutoff device can be remotely deactivated on the pad or in flight to prevent an erroneous "vote" for cutoff.

APOLLO SPACECRAFT

The Apollo spacecraft for the Apollo 15 mission consists of the command module, service module, lunar module, a spacecraft lunar module adapter (SLA), and a launch escape system. The SLA houses the lunar module and serves as a mating structure between the Saturn V instrument unit and the SM.

Launch Escape System (LES) — The function of the LES is to propel the command module to safety in an aborted launch. It has three solid-propellant rocket motors: a 147,000-pound-thrust launch escape system motor, a 2,400-pound-thrust pitch control motor, and a 31,500-pound-thrust tower jettison motor. Two canard vanes deploy to turn the command module aerodynamically to an attitude with the heatshield forward. The system is 33 feet tall and four feet in diameter at the base, and weighs 9,108 pounds.

Command Module (CM) — The command module is a pressure vessel encased in heat shields, cone-shaped, weighing 12,831 pounds at launch.

The command module consists of a forward compartment which contains two reaction control engines and components of the Earth landing system; the crew compartment or inner pressure vessel containing crew accommodations, controls and displays, and many of the spacecraft systems; and the aft compartment housing ten reaction control engines, propellant tankage, helium tanks, water tanks, and the CSM umbilical cable. The crew compartment contains 210 cubic feet of habitable volume.

Heat-shields around the three compartments are made of brazed stainless steel honeycomb with an outer layer of phenolic epoxy resin as an ablative material.

The CSM and LM are equipped with the probe-and-drogue docking hardware. The probe assembly is a powered folding coupling and impact attentuating device mounted in the CM tunnel that mates with a conical drogue mounted in the LM docking tunnel. After the 12 automatic docking latches are checked following a docking maneuver, both the probe and drogue are removed to allow crew transfer between the CSM and LM.

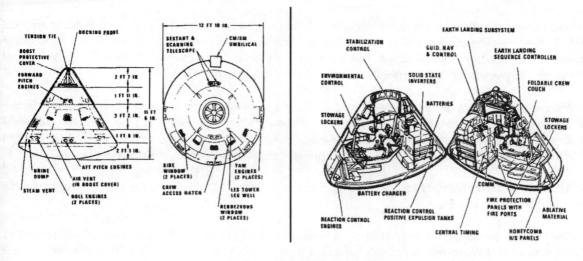

COMMAND MODULE

SERVICE MODULE

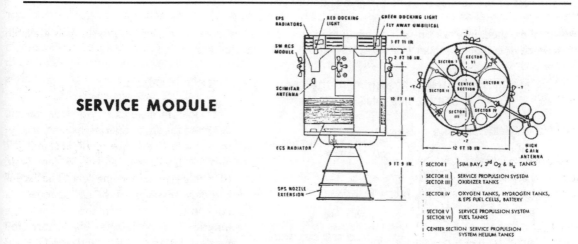

Service Module (SM) — The Apollo 15 service module will weigh 54,063 pounds at launch, of which 40,593 pounds is propellant for the 20,500-pound thrust service propulsion engine: (fuel: 50/50 hydrazine and unsymmetrical dimethyl-hydrazine; oxidizer: nitrogen tetroxide). Aluminum honeycomb panels one-inch thick form the outer skin, and milled aluminum radial beams separate the interior into six sections around a central cylinder containing service propulsion system (SPS) helium pressurant tanks. The six sectors of the service module house the following components: Sector I — oxygen tank 3 and hydrogen tank 3, J-mission SIM bay; Sector II — space radiator, +Y RCS package, SPS oxidizer storage tank; Sector III — space radiator, +Z RCS package, SPS oxidizer storage tank; Sector IV — three fuel cells, two oxygen tanks, two hydrogen tanks, auxiliary battery; Sector V — space radiator, SPS fuel sump tank, -Y RCS package; Sector VI — space radiator, SPS fuel storage tank, -Z RCS package.

Spacecraft-LM adapter (SLA) Structure — The spacecraft-LM adapter is a truncated cone 28 feet long tapering from 260 inches in diameter at the base to 154 inches at the forward end at the service module mating line. The SLA weighs 4,061 pounds and houses the LM during launch and the translunar injection maneuver until CSM separation, transposition, and LM extraction. The SLA quarter panels are jettisoned at CSM separation.

Command-Service Module Modifications

Following the Apollo 13 abort in April 1970, several changes were made to enhance the capability of the CSM to return a flight crew safely to Earth should a similar incident occur again.

These changes included the addition of an auxiliary storage battery in the SM, the removal of destratification fans in the cryogenic oxygen tanks, and the removal of thermostat switches from the oxygen tank heater circuits. The auxiliary battery added was a 415-ampere hour, silver oxide/zinc, non-rechargeable type, weighing 135 pounds which is identical to the five lunar module descent batteries.

Additional changes incorporated were a third 320-pound capacity cryogenic oxygen tank in the SM (Sector I), a valve which allows the third oxygen tank to be isolated from the fuel cells and from the other two tanks in an emergency so as to feed only the command module environmental control system. These latter changes had already been planned for the Apollo 15 J-series spacecraft to extend the mission duration.

Additionally, a third 26-pound capacity hydrogen tank was added to complement the third oxygen tank to provide for additional power.

Other changes to the CSM include addition of handrails and foot restraints for the command module pilot's transearth coast EVA to retrieve film cassettes from the SIM bay cameras. The SIM bay and its experiment packages, described in more detail in the orbital science section of the press kit, have been thermally isolated from the rest of the service module by the addition of insulation material.

Additional detailed information on command module and lunar module systems and subsystems is available in reference documents at query desks at KSC and MSC News Centers.

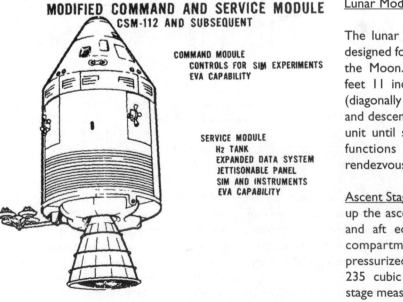

MODIFIED COMMAND AND SERVICE MODULE
CSM-112 AND SUBSEQUENT

COMMAND MODULE
CONTROLS FOR SIM EXPERIMENTS
EVA CAPABILITY

SERVICE MODULE
H₂ TANK
EXPANDED DATA SYSTEM
JETTISONABLE PANEL
SIM AND INSTRUMENTS
EVA CAPABILITY

Lunar Module (LM)

The lunar module is a two-stage vehicle designed for space operations near and on the Moon. The lunar module stands 22 feet 11 inches high and is 31 feet wide (diagonally across landing gear). The ascent and descent stages of the LM operate as a unit until staging, when the ascent stage functions as a single spacecraft for rendezvous and docking with the CM.

Ascent Stage — Three main sections make up the ascent stage: the crew midsection, and aft equipment bay. Only the crew compartment and midsection are pressurized (4.8 psig). The cabin volume is 235 cubic feet (6.7 cubic meters). The stage measures 12 feet 4 inches high by 14 feet 1 inch in diameter. The ascent stage has six substructural areas: crew compartment, midsection, aft equipment bay, thrust chamber assembly cluster supports, antenna supports, and thermal and micrometeoroid shield.

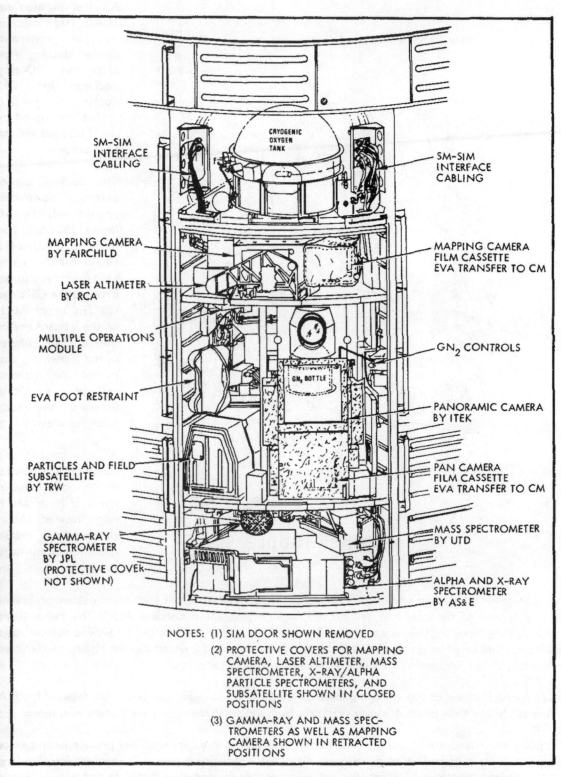

CRYOGENIC OXYGEN TANK

SM-SIM INTERFACE CABLING

SM-SIM INTERFACE CABLING

MAPPING CAMERA BY FAIRCHILD

MAPPING CAMERA FILM CASSETTE EVA TRANSFER TO CM

LASER ALTIMETER BY RCA

MULTIPLE OPERATIONS MODULE

GN₂ CONTROLS

GN₂ BOTTLE

EVA FOOT RESTRAINT

PANORAMIC CAMERA BY ITEK

PARTICLES AND FIELD SUBSATELLITE BY TRW

PAN CAMERA FILM CASSETTE EVA TRANSFER TO CM

GAMMA-RAY SPECTROMETER BY JPL (PROTECTIVE COVER NOT SHOWN)

MASS SPECTROMETER BY UTD

ALPHA AND X-RAY SPECTROMETER BY AS& E

NOTES: (1) SIM DOOR SHOWN REMOVED

(2) PROTECTIVE COVERS FOR MAPPING CAMERA, LASER ALTIMETER, MASS SPECTROMETER, X-RAY/ALPHA PARTICLE SPECTROMETERS, AND SUBSATELLITE SHOWN IN CLOSED POSITIONS

(3) GAMMA-RAY AND MASS SPEC-TROMETERS AS WELL AS MAPPING CAMERA SHOWN IN RETRACTED POSITIONS

Mission SIM Bay Science Equipment Installation

The cylindrical crew compartment is 92 inches (2.35 meters) in diameter and 42 inches (1.07 m) deep. Two flight stations are equipped with control and display panels, armrests, body restraints, landing aids, two front windows, an overhead docking window, and an alignment optical telescope in the center between the two flight stations. The habitable volume is 160 cubic feet.

A tunnel ring atop the ascent stage meshes with the command module docking latch assemblies. During docking, the CM docking ring and latches are aligned by the LM drogue and the CSM probe.

The docking tunnel extends downward into the midsection 16 inches (40 cm). The tunnel is 32 inches (81 cm) in diameter and is used for crew transfer between the CSM and LM. The upper hatch on the inboard end of the docking tunnel opens inward and cannot be opened without equalizing pressure on both hatch surfaces.

A thermal and micrometeoroid shield of multiple layers of Mylar and a single thickness of thin aluminum skin encase the entire ascent stage structure.

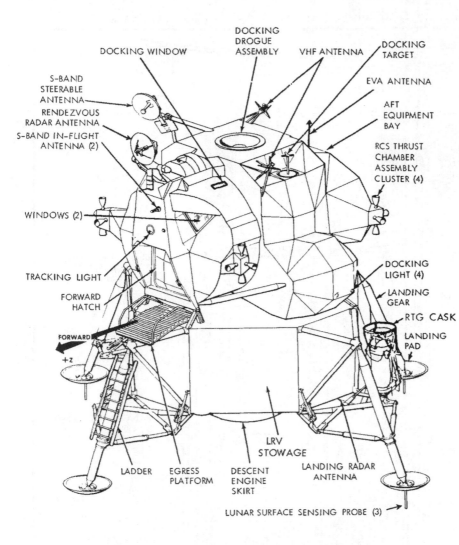

DOCKING WINDOW
DOCKING DROGUE ASSEMBLY
VHF ANTENNA
DOCKING TARGET
S-BAND STEERABLE ANTENNA
RENDEZVOUS RADAR ANTENNA
S-BAND IN-FLIGHT ANTENNA (2)
EVA ANTENNA
AFT EQUIPMENT BAY
RCS THRUST CHAMBER ASSEMBLY CLUSTER (4)
WINDOWS (2)
TRACKING LIGHT
FORWARD HATCH
FORWARD
+Z
DOCKING LIGHT (4)
LANDING GEAR
RTG CASK
LANDING PAD
LADDER   EGRESS PLATFORM   DESCENT ENGINE SKIRT   LANDING RADAR ANTENNA
LRV STOWAGE
LUNAR SURFACE SENSING PROBE (3)

LUNAR MODULE

Descent Stage — The descent stage center compartment houses the descent engine, and descent propellant tanks are housed in the four bays around the engine. Quadrant II contains ALSEP. The radioisotope thermoelectric generator (RTG) is externally mounted. Quadrant IV contains the MESA. The descent stage measures ten feet seven inches high by 14 feet 1 inch in diameter and is encased in the Mylar and aluminum alloy thermal and micrometeoroid shield. The LRV is stowed in Quadrant I.

The LM egress platform or "porch" is mounted on the forward outrigger just below the forward hatch. A ladder extends down the forward landing gear strut from the porch for crew lunar surface operations.

The landing gear struts are released explosively and are extended by springs. They provide lunar surface landing impact attenuation. The main struts are filled with crushable aluminum honeycomb for absorbing compression loads. Footpads 37 inches (0.95 m) in diameter at the end of each landing gear provide vehicle support on the lunar surface.

Each pad (except forward pad) is fitted with a 68-inch long lunar surface sensing probe which upon contact with the lunar surface signals the crew to shut down the descent engine.

The Apollo LM has a launch weight of 36,230 pounds. The weight breakdown is as follows:

Ascent stage, dry	4,690 lbs.	Includes water and oxygen; no crew
Descent stage, dry	6,179 lbs.	
RCS propellants (loaded)	633 lbs.	
DPS propellants (loaded)	19,508 lbs.	
APS propellants (loaded)	5,220 lbs.	
Total	36,230 lbs.	

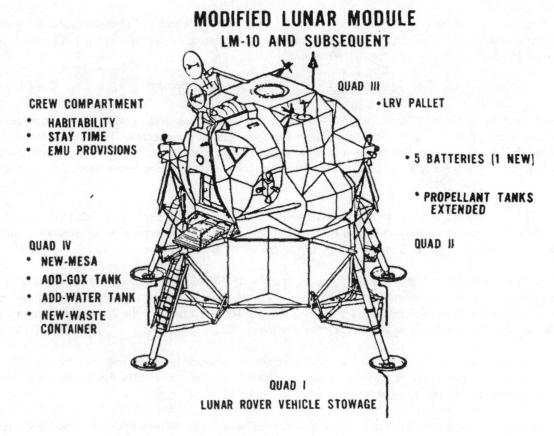

MODIFIED LUNAR MODULE
LM-10 AND SUBSEQUENT

QUAD III
• LRV PALLET

CREW COMPARTMENT
• HABITABILITY
• STAY TIME
• EMU PROVISIONS

• 5 BATTERIES (1 NEW)

• PROPELLANT TANKS EXTENDED

QUAD IV
• NEW-MESA
• ADD-GOX TANK
• ADD-WATER TANK
• NEW-WASTE CONTAINER

QUAD II

QUAD I
LUNAR ROVER VEHICLE STOWAGE

Lunar Module Changes

Although the lunar module exhibits no outward significant change in appearance since Apollo 14, there have been numerous modifications and changes to the spacecraft in its evolution from the H mission to the longer-duration J mission model. Most of the changes involve additional consumables required for the longer stay on the lunar surface and the additional propellant required to land the increased payload on the Moon

Significant differences between LM-8 (Apollo 14) and LM-10 (Apollo 15) are as follows:

*Fifth battery added to descent stage for total 2075 amp hours. Batteries upgraded from 400 AH each to 415 AH.

*Second descent stage water tank added for total 377 pounds capacity.

*Second descent stage gaseous oxygen (GOX) tank added for total 85 pounds capacity. Permits six 1410 psi PLSS recharges. (LM-8: six 900 psi recharges.)

*Addition of system capable of storing 1200 cc/man/day urine and 100 cc/man/hour PLSS condensate.

*Additional thermal insulation for longer stay time (67 hours instead of 35 hours on LM-8).

*Additional descent stage payload: Lunar Roving Vehicle in Quad I previously occupied by erectable S-band antenna and laser reflector; two pallets in Quad III — one 64.6 pounds for LRV (holds hand tool carrier) and one 100-pound payload pallet for laser ranging retro reflector; 600-pound gross weight capability of enlarged modular equipment stowage assembly (MESA) in Quad IV, compared to 200-pound capacity in LM-8; Quad II houses ALSEP.

*Changes to descent engine include quartz-lined engine chamber instead of silica-lined, and a 10-inch nozzle extension; a 3.36-inch extension to propellant tanks increase total capacity by 1150 pounds to yield 157 seconds hover time (LM-8=140 seconds).

MANNED SPACE FLIGHT NETWORK SUPPORT

NASA's worldwide Manned Space Flight Network (MSFN) will track and provide nearly continuous communications with the Apollo astronauts, their launch vehicle and spacecraft. This network also will continue the communications link between Earth and the Apollo experiments left on the lunar surface, and track the Particles and Fields Subsatellite to be ejected into lunar orbit from the Apollo service module SIM bay.

The MSFN is maintained and operated by the NASA Goddard Space Flight Center, Greenbelt, Md., under the direction of NASA's Office of Tracking and Data Acquisition. Goddard will become the emergency control center if the Houston Mission Control Center is impaired for an extended time.

The MSFN employs 11 ground tracking stations equipped with 30- and 85-foot antennas, an instrumented tracking ship, and four instrumented aircraft. For Apollo 15, the network will be augmented by the 210-foot antenna system at Goldstone, Calif. (a unit of NASA's Deep Space Network), and the 210-foot radio antenna of the National Radio Astronomy observatory at Parkes, Australia.

NASA Communications Network (NASCOM). The tracking network is linked together by the NASA Communications Network. All information flows to and from MCC Houston and the Apollo spacecraft over this communications system.

The NASCOM consists of more than two million circuit miles, using satellites, submarine cables, land lines, microwave systems, and high frequency radio facilities. NASCOM control center is located at Goddard. Regional communication switching centers are in Madrid; Canberra, Australia; Honolulu; and Guam.

Three Intelsat communications satellites will be used for Apollo 15. One satellite over the Atlantic will link Goddard with Ascension Island and the Vanguard tracking ship. Another Atlantic satellite will provide a direct link between Madrid and Goddard for TV signals received from the spacecraft. The third satellite over the mid-Pacific will link Carnarvon, Canberra, Guam and Hawaii with Goddard through a ground station at Jamesburg, Calif.

Mission Operations: Prelaunch tests, liftoff, and Earth orbital flight of the Apollo 15 are supported by the MSFN station at Merritt Island, Fla., four miles from the launch pad.

During the critical period of launch and insertion of the Apollo 15 into Earth orbit, the USNS Vanguard

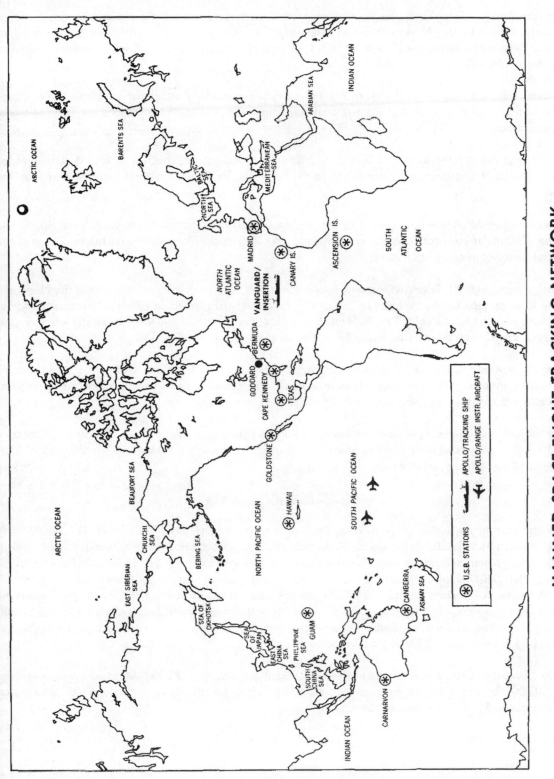

MANNED SPACE FLIGHT TRACKING NETWORK

provides tracking, telemetry, and communications functions. This single sea-going station of the MSFN will be stationed about 1,000 miles southeast of Bermuda.

When the Apollo 15 conducts the TLI maneuver to leave Earth orbit for the Moon, two Apollo Range Instrumentation Aircraft (ARIA) will record telemetry data from Apollo and relay voice communications between the astronauts and the Mission Control Center at Houston. These aircraft will be airborne between Australia and Hawaii.

Approximately one hour after the spacecraft has been injected into a translunar trajectory, three prime MSFN stations will take over tracking and communicating with Apollo. These stations are equipped with 85-foot antennas.

Each of the prime stations, located at Goldstone, Madrid and Honeysuckle is equipped with dual systems for tracking the command module in lunar orbit and the lunar module in separate flight paths or at rest on the Moon.

For reentry, two ARIA will be deployed to the landing area to relay communications between Apollo and Mission Control at Houston. These aircraft also will provide position information on the Apollo after the blackout phase or reentry has passed.

Television Transmissions: Television from the Apollo spacecraft during the journey to and from the Moon and on the lunar surface will be received by the three prime stations, augmented by the 210-foot antennas at Goldstone and Parkes. The color TV signal must be converted at the MSC Houston. A black and white version of the color signal can be released locally from the stations in Spain and Australia.

TV signals originating from the TV camera stationary on the Moon will be transmitted to the MSFN stations via the lunar module. While the camera is mounted on the LRV, the TV signals will be transmitted directly to the tracking stations as the astronauts tour the Moon.

Once the LRV has been parked near the lunar module, its batteries will have about 80 hours of operating life. This will allow ground controllers to position the camera for viewing the lunar module liftoff, post lift-off geology, and any other desired scenes.

APOLLO PROGRAM COSTS

Apollo manned lunar landing program costs through the first landing, July 1969, totaled $21,349,000,000. These included $6,939,000,000 for spacecraft development and production; $7,940,000,000 for Saturn launch vehicle development and production; $854,000,000 for engine development; $1,137,000,000 for operations support; $541,000,000 for development and operation of the Manned Space Flight Network; $1,810,000,000 for construction of facilities; and $2,128,000,000 for operation of the three Manned Space Flight Centers. At its peak, the program employed about 300,000 people, more than 90 per cent of them in some 20,000 industrial firms and academic organizations. Similarly, more than 90 per cent of the dollars went to industrial contractors, universities and commercial vendors.

Apollo 15 mission costs are estimated at $445,000,000: these include $185,000,000 for the launch vehicle; $65,000,000 for the command/service module; $50,000,000 for the lunar module; $105,000,000 for operations; and $40,000,000 for the science payload.

A list of major prime contractors and subcontractors for Apollo 15 is available in the News Centers at KSC and MSC.

Distribution of Apollo Estimated Program by Geographic Location Fiscal Year 1971

Based on Prime Contractor Locations Amounts in Millions of Dollars

State	Value	State	Value
Alabama	97	Nebraska	*
Alaska	*	Nevada	*
Arizona	2	New Hampshire	*
Arkansas	*	New Jersey	11
California	202	New Mexico	5
Colorado	3	New York	85
Connecticut	11	North Carolina	*
Delaware	5	North Dakota	*
Florida	161	Ohio	4
Georgia	1	Oklahoma	*
Hawaii	1	Oregon	*
Idaho	*	Pennsylvania	5
Illinois	2	Rhode Island	*
Indiana	*	South Carolina	*
Iowa	*	South Dakota	*
Kansas	*	Tennessee	2
Kentucky	*	Texas	140
Louisiana	31	Utah	*
Maine	*	Vermont	*
Maryland	5	Virginia	4
Massachusetts	37	Washington	6
Michigan	35	West Virginia	*
Minnesota	2	Wisconsin	11
Mississippi	20	Wyoming	*
Missouri	2	District of Columbia	21
Montana	*	*Less than one million dollars	

ENVIRONMENTAL IMPACT OF APOLLO/SATURN V MISSION

Studies of NASA space mission operations have concluded that Apollo does not significantly effect the human environment in the areas of air, water, noise or nuclear radiation.

During the launch of the Apollo/Saturn V space vehicle, products exhausted from Saturn first stage engines in all cases are within an ample margin of safety. At lower altitudes, where toxicity is of concern, the carbon monoxide is oxidized to carbon dioxide upon exposure at its high temperature to the surrounding air. The quantities released are two or more orders of magnitude below the recognized levels for concern in regard to significant modification of the environment. The second and third stage main propulsion systems generate only water and a small amount of hydrogen. Solid propellant ullage and retro rocket products are released and rapidly dispersed in the upper atmosphere at altitudes above 43.5 miles (70 kilometers). This material will effectively never reach sea level and, consequently, poses no toxicity hazard.

Should an abort after launch be necessary, some RP-1 fuel (kerosene) could reach the ocean. However, toxicity of RP-1 is slight and impact on marine life and waterfowl are considered negligible due to its dispersive characteristics. Calculations of dumping an aborted SIC stage into the ocean showed that spreading and evaporating of the fuel occurred in one to four hours.

There are only two times during a nominal Apollo mission when above normal overall sound pressure levels are encountered. These two times are during vehicle boost from the launch pad and the sonic boom experienced when the spacecraft enters the Earth's atmosphere. Sonic boom is not a significant nuisance since it occurs over the mid-Pacific Ocean.

NASA and the DOD have made a comprehensive study of noise levels and other hazards to be encountered for launching vehicles of the Saturn V magnitude. For uncontrolled areas the overall sound pressure levels are well below those which cause damage or discomfort. Saturn launches have had no deleterious effects on wildlife which has actually increased in the NASA-protected areas of Merritt Island.

A source of potential radiation hazard is the fuel capsule of the radioisotope thermoelectric generator (supplied by the AEC) which provides electric power for Apollo lunar surface experiments. The fuel cask is designed so that no contamination can be released during normal operations or as a result of the maximum credible accident.

PROGRAM MANAGEMENT

The Apollo Program is the responsibility of the office of Manned Space Flight (OMSF), National Aeronautics and Space Administration, Washington, D.C. Dale D. Myers is Associate Administrator for Manned Space Flight.

NASA Manned Spacecraft Center (MSC), Houston, is responsible for development of the Apollo spacecraft, flight crew training, and flight control. Dr. Robert R. Gilruth is Center Director.

NASA Marshall Space Flight Center (MSFC), Huntsville, Ala., is responsible for development of the Saturn launch vehicles. Dr: Eberhard F. M. Rees is Center Director.

NASA John F. Kennedy Space Center (KSC), Fla., is responsible for Apollo/Saturn launch operations. Dr. Kurt H. Debus is Center Director.

The NASA office of Tracking and Data Acquisition (OTDA) directs the program of tracking and data flow on Apollo. Gerald M. Truszynski is Associate Administrator for Tracking and Data Acquisition.

NASA Goddard Space Flight Center (GSFC), Greenbelt, Md., manages the Manned Space Flight Network and Communications Network. Dr. John F. Clark is Center Director.

The Department of Defense is supporting NASA during launch, tracking, and recovery operations. The Air Force Eastern Test Range is responsible for range activities during launch and down-range tracking. Recovery operations include the use of recovery ships and Navy and Air Force aircraft.

Apollo/Saturn Officials

NASA Headquarters
Dr. Rocco A. Petrone Apollo Program Director, OMSF
Chester M. Lee (Capt., USN, Ret.) Apollo Mission Director, OMSF
John K. Holcomb (Capt., USN, Ret.) Director of Apollo Operations, OMSF
Lee R. Scherer (Capt., USN, Ret.) Director of Apollo Lunar Exploration, OMSF Kennedy Space Center
Miles J. Ross Deputy Center Director
Walter J. Kapryan Director of Launch Operations
Raymond L. Clark Director of Technical Support
Robert C. Hock Apollo/Skylab Program Manager
Dr. Robert H. Gray Deputy Director, Launch Operations
Dr. Hans F. Gruene Director, Launch Vehicle Operations
John J. Williams Director, Spacecraft Operations
Paul C. Donnelly Launch Operations Manager
Isom A. Rigell Deputy Director for Engineering

Manned Spacecraft Center
Dr. Christopher C. Kraft, Jr. Deputy Center Director
Col. James A. McDivitt (USAF) Manager, Apollo Spacecraft Program
Donald K. Slayton Director, Flight Crew operations
Sigurd A. Sjoberg Director, Flight Operations
Milton L. Windler Flight Director
Gerald D. Griffin Flight Director
Eugene F. Kranz Flight Director
Glynn S. Lunney Flight Director
Dr. Charles A. Berry Director, Medical Research and Operations
Marshall Space Flight Center
Dr. Eberhard Rees Director
Dr. William R. Lucas Deputy Center Director, Technical
R. W. Cook Deputy Center Director, Management

James T. Shepherd Director (acting), Program Management
Herman F. Kurtz manager (acting), Mission operations office
Richard G. Smith Manager, Saturn Program office
Matthew W. Urlaub Manager, S-IC Stage, Saturn Program office
William F. LaHatte Manager, S-II Stage, Saturn Program Office
Charles H. Meyers Manager, S-IVB Stage, Saturn Program Office
Frederich Duerr Manager, Instrument Unit, Saturn Program Office
William D. Brown Manager, Engine Program Office
S. F. Morea Manager, LRV Project, Saturn Program Office

Goddard Space Flight Center
Ozro M. Covington Director, Networks
William P. Varson Chief, Network Computing & Analysis Division
H. William Wood Chief, Network Operations Division
Robert Owen Chief, Network Engineering Division
L. R. Stelter Chief, NASA Communications Division

Department of Defense
Maj. Gen. David M. Jones (USAF) DOD Manager for Manned Space Flight Support Operations
Col. Kenneth J. Mask (USAF) Deputy DOD Manager for Manned Space Flight Support Operations, and Director, DOD Manned Space Flight Support Office
Rear Adm. Thomas B. Hayward (USN) Commander, Task Force 130, Pacific Recovery Area
Rear Adm. Roy G. Anderson (USN) Commander Task Force 140, Atlantic Recovery Area
Capt. Andrew F. Huff Commanding Officer, USS Okinawa, LPH-3 Primary Recovery Ship
Brig. Gen. Frank K. Everest, Jr. Commander Aerospace Rescue (USAF) and Recovery Service

CONVERSION TABLE

Multiply	By	To Obtain
Distance:		
feet	0.3048	meters
meters	3.281	feet
kilometers	3281	feet
kilometers	0.6214	statute miles
statute miles	1.609	kilometers
nautical miles	1.852	kilometers
nautical miles	1.1508	statute miles
statute miles	0.86898	nautical miles
statute miles	1760	yards
Velocity:		
feet/sec	0.3048	meters/sec
meters/sec	3.281	feet/sec
meters/sec	2.237	statute mph
feet/sec	0.6818	statute miles/hr
feet/sec	0.5925	nautical miles/hr
statute miles/hr	1.609	km/hr
nautical miles/hr (knots)	1.852	km/hr
km/hr	0.6214	statute miles/hr
Liquid measure, weight:		
gallons	3.785	liters
liters	0.2642	gallons
pounds	0.4536	kilograms
kilograms	2.205	pounds
Volume:		
cubic feet	0.02832	cubic meters
Pressure:		
pounds/sq. inch	70.31	grams/sq. cm

ON THE MOON WITH APOLLO 15
A Guidebook to Hadley Rille and the Apennine Mountains

NATIONAL AERONAUTICS AND SPACE ADMINISTRATION

June 1971

PREFACE

Never before in man's history has it been possible for more than a few people to witness major scientific discoveries. Yet with each Apollo mission to the Moon's surface, millions of people throughout the world can watch through television the activities of the astronauts. The understanding by the viewer of those activities and his sense of sharing in the scientific excitement of the mission are greatly increased when there is a general understanding of the scientific and engineering aspects. Yet for most of us, the usual discussions are clouded with jargon.

My purpose in writing this guidebook is to give in simple terms information about the Apollo 15 mission to the Moon's surface so you can share with me the excitement of the scientific exploration of the Hadley-Apennine region of the Moon.

Many people helped me prepare this guidebook. Richard Baldwin and Gordon Tevedahl collected background material. George Gaffney coordinated all art work. Jerry Elmore, Norman Tiller, Ray Bruneau and Boyd Mounce drew most of the original sketches. Andrew Patnesky provided several new photographs. The manuscript was improved greatly as a result of comments by Jack Schmitt, George Abbey, Verl Wilmarth, James Head, Donald Beattie, Rosemary Wang, Herbert Wang, Ruth Zaplin, Mary Jane Tipton, and my seventeen-year-old daughter Debra. My secretary Jean Ellis helped with many revisions. To all of these people, I express my thanks.

GENE SIMMONS May 1971 Nassau Bay

HOW TO USE THIS GUIDEBOOK

Excellent commentaries have been available over television for each previous Apollo mission. However, because of the increased complexity of the surface operations on Apollo 15 and because of the greater amount of time devoted to science-activities, I believe that a written guide would be welcomed by the interested viewer of Apollo 15. The material in this guidebook is intended to be used in conjunction with the other material shown over commercial TV.

The science-activities of the astronauts on the surface are divided between "experiments" and "traverses". For the experiments, the astronauts set up equipment on the Moon that collects data and (generally) transmits the data back to Earth. These experiments are described briefly in the section "Lunar Surface Scientific Experiments and Hardware". The reader need not read about all the details of each experiment on first reading. Quite frankly, even I find that section somewhat tedious to read, probably because it is rather complete, but I have chosen to keep it in the present form so that you may refer to the individual experiments as you wish. I do recommend scanning this section before the first EVA in order to understand something about each of the experiments.

Most of the astronaut's time on the lunar surface will be spent on the traverses. The section "Traverse Descriptions" is a guide to those activities. It tells in general terms the things the astronauts will do on each traverse and indicates what they will do next. It should be used in the same way, that a flexible itinerary for a vacation trip through New England would be used. Refer to it during the traverse. But do not try to read it in great detail before the traverse.

The section "Lunar Geology Experiment" should be read before the traverses begin. There you will find descriptions of the tools that are used, the various kinds of photographs taken, and so on.

Finally, you should know that a glossary and list of acronyms are included in the rear of the guidebook. I expect the definitions and short discussions to be found there will help in understanding some of the terms and concepts now in common use in the scientific exploration of the Moon.

Introduction

The Apollo 15 mission to the Moon's surface is expected to be launched from Cape Kennedy on 26 July 1971 and to land a few days later near a very large and majestic mountain range, the Apennine Mountains. A sketch of the front side of the Moon is shown in figure 1 and the location of the landing site is shown in relation to other sites. This landing site is extremely attractive from the viewpoint of lunar science. It will give the astronauts their first chance to collect rocks from lunar mountains and to study at first-hand a feature, termed rille, which resembles in many ways the channels cut on Earth by meandering streams. The origin of rilles is probably not the same as that of the familiar terrestrial stream-cut channels because no water is present now on the Moon's surface and probably never existed there. The origin of rilles is a puzzle.

Near the landing site are Hadley Mountain, which rises about 14,000 feet above the surrounding lowlands and Mount Hadley Delta, which rises about 11,000 feet. The actual surface on which the Lunar Module or LM* will land is everywhere pock-marked by craters of various sizes. The smallest craters known are less than 1/1000 inch across; the largest exceed 50 miles. The craters were produced during the past few million years when objects from space struck the Moon. The craters are still being produced but there is no danger to the astronaut because collisions with the Moon are very infrequent. For example, an object larger than birdseed would strike the landing site only once every few years. But because erosion is so slow on the Moon, the craters produced millions of years ago are still preserved and appear as seen in photographs throughout this guidebook. The mechanisms of erosion, the process by which rocks and soil are removed from a particular spot, are very different on the Earth and the Moon. Most terrestrial erosion is accomplished by running water and is relatively rapid. Most lunar erosion is the result of impacting objects and the resulting craters destroy previously existing ones.

Since the first manned lunar landing, Apollo 11, in July 1969, significant improvements in both equipment and procedures have increased dramatically the capabilities of Apollo 15 over those of previous missions. Total duration of the mission has increased from 9 days to a planned time of about 12½ days and a maximum of 16 days. Actual time for the LM to remain on the lunar surface has doubled, from 33.5 hours previously to a planned 67.3 hours. The amount of time spent by the astronauts on the lunar surface outside the LM, which has become known as Extravehicular Activity or EVA, has more than doubled from a maximum of 9.3 hours previously to a planned 20 hours. The EVA time will be spent in three periods of 7, 7, and 6 hours duration. The weight of the scientific equipment that will be used in lunar orbit has increased from 250 pounds to 1,050 pounds. The weight of the scientific equipment to be landed on the lunar surface has increased from 510 pounds to about 1200 pounds. And finally, the astronauts will have with them for the first time a small, four-wheeled vehicle for travel over the Moon's surface. It is termed Rover and can carry two astronauts, equipment, and rocks. Unlike the Russian vehicle Lunokhod that was recently landed and is still operating, it cannot be operated remotely from Earth.

A summary of major events for the entire Apollo 15 mission is shown in Table 1. Scientific activities while the spacecraft is in orbit around the Earth, consist mainly in photographing the Earth with film that is sensitive to ultraviolet (uv) radiation for the purpose of examining various terrestrial, cloud and water features. By using uv, we hope to "see" these features more clearly than we could see them with visible light. From space, the atmosphere gets in the way of seeing. The situation is somewhat akin to that of using sunglasses to reduce glare, so the wearer can see better. The uv photography will be continued during the journey to the Moon and pictures will be obtained at various distances from the Earth. During this journey and before the landing on the Moon, one of the spent stages of the rockets that were used to lift the spacecraft from the Earth, and designated S-IVB, will be crashed into the Moon. The sound waves generated by the S-IVB impact travel through the Moon and will be detected by sensitive receivers (seismometers) now operating at the Apollo 12 and 14 sites. (This experiment is discussed more fully later in this guidebook.)

*Abbreviations and acronyms are very useful in situations where time is limited, such as a mission to the Moon's surface. Common ones are noted in this book where first used. An extensive list is given at the end of the text.

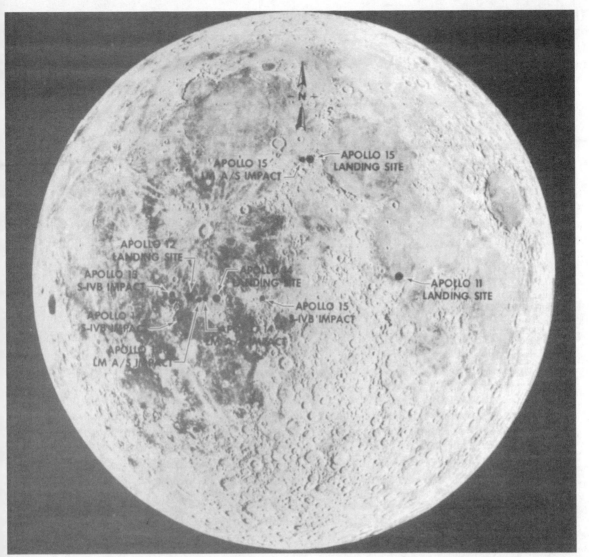

Figure I. - Front side of the Moon. This side always faces the Earth. Shown here are locations of the previous Apollo landings and of the impacts on the Moon of spent S-IVB stages and LM ascent stages. The impacts create sound waves in the Moon that are used to study the interior of the Moon.

Shortly after placing their spacecraft in orbit about the Moon, the astronauts separate it into two parts. One part, the combined Command and Service Modules (CSM), remains in lunar orbit while the other part, the Lunar Module (LM), descends to the surface.

One astronaut remains in the CSM and performs many scientific experiments. These orbital experiments will obtain data over a large part of both front and back sides of the Moon because the path of the point directly beneath the spacecraft, termed ground track, is different for each revolution of the spacecraft. See figure 2. Notice that the orbit of the CSM is not parallel to the equator. If the Moon did not rotate about its axis, the ground track would change very little on each successive revolution of the CSM. However, the Moon does rotate slowly about its axis. It completes one full revolution every 28 earth-days and therefore the ground track is different for each CSM revolution.

Several of these orbital experiments will measure the approximate chemical composition of the Moon's surface materials. Others are intended to measure the variations of gravity and of the magnetic field around the Moon. A laser altimeter will be used to obtain precise elevations of features that lie on the Moon's surface beneath the orbiting CSM. An extensive set of photographs will be obtained. The pilot will observe and photograph many features on the Moon never before available to astronauts.

The other two astronauts descend to the surface of the Moon in the LM. The rest of this guidebook is a discussion of their equipment and of their activities.

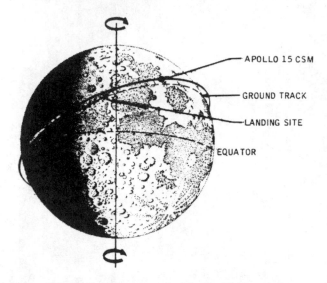

Figure 2. - Trajectory and ground track of Apollo 15. Because the Moon rotates, the ground track is different for each revolution of the CSM.

The LM, illustrated in figure 3, lands two astronauts on the Moon's surface. It has two parts, a descent stage and an ascent stage. The descent stage contains a rocket engine, fuel necessary to land both stages, a four-wheeled battery-powered vehicle to be used on the Moon, water and oxygen, and scientific equipment to be left on the Moon when the astronauts return to Earth. The other part, the ascent stage, contains the following items: (1) equipment for communications with the Earth and with the CSM, (2) navigational equipment, (3) a computer, (4) food, oxygen, and other life-support supplies, and (5) another rocket engine and fuel needed to leave the Moon and rendezvous with the CSM. All three astronauts return to Earth in the Command Module.

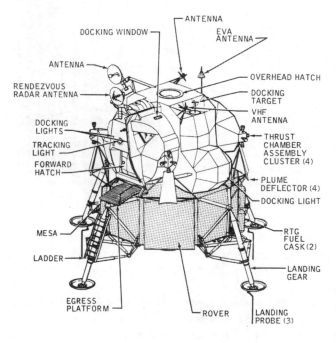

Figure 3. - The Lunar Module (LM). The shaded portion, the descent stage, remains on the Moon when the astronauts leave in the ascent stage to rendezvous with the CM and return to Earth.

Soon after the LM lands on the Moon, about 1½ hours, the astronauts will spend a half hour describing and photographing the surrounding area. The commander will open the upper hatch and stand with his head and shoulders outside the LM. During this Standup Extravehicular Activity (SEVA) the LM cabin will be open to the lunar atmosphere and will therefore be under vacuum conditions. Both astronauts must wear their space suits. Because the commander's head will be above the LM, he will have excellent visibility of the landing site. If the LM lands within 100 yards, the length of a football field, of the planned spot, then the commander will see the panoramic view sketched in figure 4. He will shoot photographs, which will include panoramas, with both 500 mm and 60 mm lens. His verbal descriptions during the SEVA will help Mission Control to accurately pinpoint, the actual landing site. Of equal importance is the fact that the descriptions will assist in the continuing evaluation of the surface science plans. It is likely that the astronauts will draw attention during the SEVA to some surface features, previously overlooked, that we will wish to examine sometime during the three EVA's.

NORTH   EAST   SOUTH   WEST   NORTH

Figure 4. - SEVA Panorama. Artist Jerry Elmore has depicted here the panorama that the Commander will see from his vantage point above the LM during the SEVA. Mount Hadley Delta, due south stands about 11,000 feet above the landing site. Hadley Mountain, the large dark mountain situated northeast of the site is about 3,000 feet higher than Mount Hadley Delta.

When the astronauts leave the LM, a process appropriately termed egress and shown in figure 5, they must wear a suit that protects them from the Moon's high vacuum. This suit is illustrated in figure 6. Although it was designed to allow freedom of movement, it still restricts considerably the motion of the astronauts. An example may be useful. Think how difficult it is to run, chop wood, or work outdoors on an extremely cold day in winter when you wear many layers of clothes. The astronauts suits are even more restrictive. The Portable Life Support System (PLSS) contains the oxygen needed by the astronaut and radios for communication. It also maintains the temperature inside the suit at a comfortable level for the astronaut.

Figure 5. Egress. Apollo 11 astronaut Aldrin is shown egressing from the LM. Note the ladder that leads down one leg from the platform.

Landing Site Description

The Apollo 15 landing area, termed Hadley-Apennine, is situated in the north central part of the Moon (latitude 26° 04' 54" N. longitude 03° 39' 30" E) at the western foot of the majestic Apennine Mountains, and by the side of Hadley Rille. See figure 1. The Apennines rise 12,000 to 15,000 feet above the lunar surface and ring the southeastern edge of Mare Imbrium (Sea of Rains). For comparison with Earth features, the steep western edge of the Apennine Mountains is higher than either the eastern face of the Sierra Nevadas in the western U.S. or the edge of the Himalayan Mountains that rises several thousand feet above the plains of India. The actual landing point was selected so the astronauts could study the sinuous Hadley Rille, the Apennine Mountains and several other geological features. A beautiful perspective view of the local landing site, as seen from an angle of about 30 degrees, is shown in figure 1. In drawing this figure, we have combined the precision that is available from modern-day digital computers and the insights that can come only from an artist. Thus the features are very accurately drawn but they are displayed in a way that the human eye will see them.

In the rest of this section, I will discuss the several geologic features present at the landing site: The Apennine Mountains, Hadley Rille, the cluster of craters at the foot of Hadley Mountain, and the North Complex. All of them are clearly visible in figure 7.

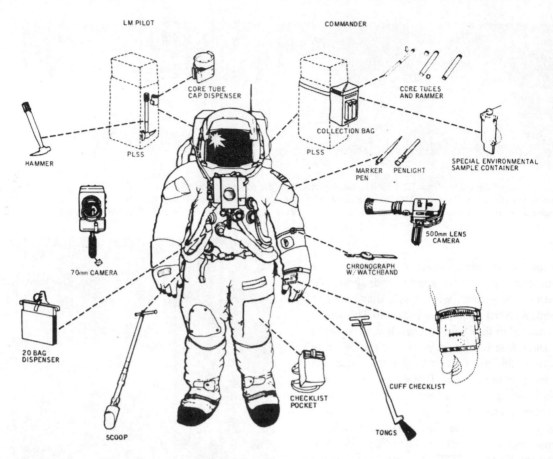

Figure 6. - Astronaut suit. The suit prevents exposure of the astronaut to the Moon's vacuum. It incorporates mans improvements over the suits used on previous Apollo flights. Sketched also are several items of equipment.

THE APENNINE MOUNTAINS

These mountains form part of the southeastern boundary of Mare Imbrium and are believed to have been formed at the same time as the Imbrium basin.*

The general relations of the Apennine Mountains Hadley Rille, and a branch of the Apennine Mountain chain, termed Apennine Ridge, are seen in figure 8. Most lunar scientists agree that the Imbrium basin was formed by impact of a large object but there is no general agreement on the details of the processes involved in the origin of the rille or the mountains. One possible process of basin formation is shown schematically in figure 9. The impact of the object causes material to be thrown out in much the same way that material is splashed when a large rock is dropped in soft mud. From a study of the samples of material that is ejected from the crater, we can measure the age of the material and obtain the date at which the impact occurred. We can also determine the nature of the material at depth in the Moon. I think it is very likely that most of the material available for sampling at the Apollo 15 landing site consists of rocks and soil ejected from Imbrium basin. Some material older than the Imbrium impact may be found at the base of Mount Hadley Delta. Thus one of the main geological goals of this mission is to sample those rocks.

Our understanding of the details of crater formation has been improved by the study of impact craters on Earth. One such crater that is generally well-known is Meteor Crater, near Flagstaff, Arizona. Other impact craters, less well-known but intensely studied by geologists, exist in Tennessee, Canada, Australia, Germany, and elsewhere. An oblique photograph of Meteor Crater is shown in figure 10.

*To the scientist, the distinction between a mare, which is the surface material, and the associated basin, which includes the shape and distribution of materials at depth beneath the mare, is very important.

HADLEY RILLE

Hadley Rille is a V-shaped sinuous rille that roughly parallels the Apennine Mountains along the eastern boundary of Mare Imbrium (figures 7 and 8). It originates in an elongated depression in an area of low domes that are probably volcanic in nature. It has an average width of about 1 mile, a depth that varies generally from 600 to 900 feet but at the landing site is 1,300 feet, and is about 80 miles long. Such sinuous rilles are very common on the surface of the Moon. Their origin in general, and of Hadley Rille in particular, is very puzzling to lunar scientists and has been debated for many years. It has been attributed by various scientists to flowing water (although as a result of studying rocks returned on previous missions we no longer believe this hypothesis), the flow of hot gases associated with volcanism, the flow of lava (in much the same way that lava flows down the sides of the Hawaiian volcanoes and to collapsed lava tubes. Today, most scientists agree that the origin of such rilles is associated with fluid flow or with faulting yet we now know that water is generally absent from the Moon and probably never existed there in large quantities. The visit by the Apollo 15 crew to Hadley Rille will undoubtedly shed some light on the origin of rilles.

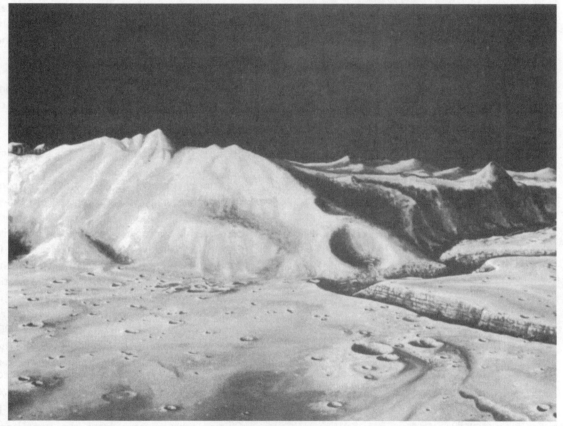

Figure 7. - Landing site for Apollo 15 as seen from the north. The large mountain, Hadley Delta, rises about 11,000 feet above the nearby plain. The valley, Hadley Rille, is about one mile wide and 1,200 feet deep. Origin of the rille is a puzzle and its study is one of the objectives of this mission. (Artwork by Jerry Elmore.)

The approximate slope of the sides of Hadley Rille, near the landing site, is about 33 degrees. The depth is about 1,300 feet. Many fresh outcrops of rock that are apparently layered are seen along and just below the rille rim. The layers probably represent lava flows. Many large blocks have rolled downslope to settle on the floor of the rille. An analogous terrestrial feature that clearly has its origin in the flow of water is seen in figure 11, the Rio Grande Gorge near Taos, New Mexico. An important part of the Apollo 15 astronauts training was a study for two days of this feature.

Examination of the rille floor and sampling of the rocks located there would be extremely valuable evidence on the origin of the rille would almost certainly be found. But perhaps more importantly, rocks from a depth of about 1,200 feet would be collected. The study of the vertical changes in rocks, termed stratigraphy, provides the basic data necessary to construct the history of the Moon. (For example, many facts about the geological history of the Earth have been read from the rocks exposed in the walls and bottom of the Grand Canyon.) The scientific need to examine rocks from the bottom of the rille is so great that many people have tried to solve the problem of how to get them. One prominent scientist suggested that the astronauts use a crossbow with string attached to the arrow for retrieval. Of course the arrow would have been modified so that it adhered to rocks in some way rather than pierce them. This idea, as well as others, was abandoned because it was not practical. Even though access to the rille floor is not possible, sampling of the rocks that occur along the rim and photographing the walls are planned and may aid lunar scientists in determining the origin of rilles.

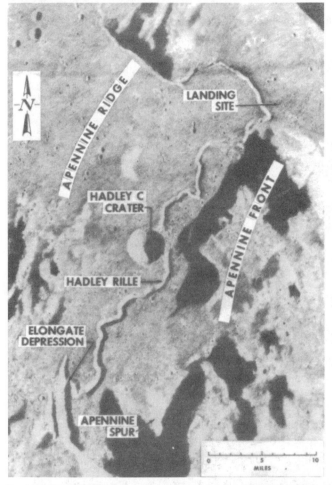

Figure 8. - General geography of the Apennine-Hadley landing site. Note that Hadley Rille is about 80 miles long.

SECONDARY CRATER CLUSTER

A group of craters, labeled "South Cluster" on figure 16, will be observed and photographed during the second EVA. This group or cluster of craters is thought to have formed from impact of a group of objects that struck the Moon at the same time. Those objects were, in turn, thrown out from some other spot on the Moon by the impact of a single object from space. Hence the term secondary impact crater is applied to such craters.

Because the objects that created secondary impact craters came from some other spot on the Moon, the rock samples collected from such features may include samples from other parts of the Moon, and perhaps from considerable distance. Most of the material present in the vicinity of the craters is undoubtedly the material that was present before the craters were formed. The exotic material, that which came from elsewhere, is probably quite rare and the amount present at any crater may be less than 1 part per 1,000. Only after extensive investigation of the samples back in the laboratory on Earth can we be reasonably sure about the origin of a particular sample. Some lunar scientists believe that the objects that produced the South Cluster craters came from the very large crater Autolycus, situated about 100 miles to the northwest.

An additional reason for collecting rocks in, or near, crater's is that the impact, in forming the crater, always exhumes rocks from the bottom of the crater. Therefore the material that surrounds a crater includes material that originally was located at the bottom of the crater, at the top, and at all intermediate depths. If the material changes within the depth of the crater, then a study of the rock samples will very likely indicate that change.

Thus, for these two reasons, that we may sample distant localities and that we may see changes in the rocks with depth, the collection of samples at South Cluster is an important objective of the Apollo 15 mission.

NORTH COMPLEX

Not all features on the Moon's surface were formed by impacting objects. Some were formed by internal processes. It is never easy on the basis of photographs or telescopic observations to distinguish between an internal and an external origin for a particular feature. In fact Galileo, the first man to look at the Moon through a telescope, about 300 years ago, suggested that all the craters on the Moon were due to volcanoes. His hypothesis stood unchallenged for two centuries until someone suggested the impact hypothesis. As so often happens in science, long, and sometimes bitter, arguments over which hypothesis was correct raged for about 100 years. Today, we believe that most lunar features have resulted from impacts but some have been caused by internal processes.

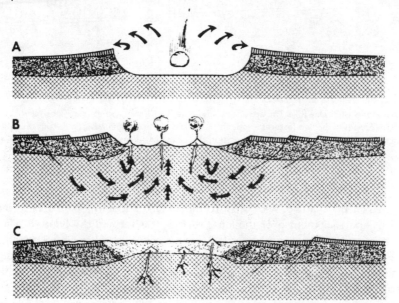

Figure 9. - Possible explanation of the origin of large basins on the Moon. In A, impact of an object from space, a large meteorite, created a hole and splashed material great distances. Some rocks were thrown hundreds of miles. The presence of the hole and the high temperatures generated by the heat from the impact create volcanoes. Material is transported to the surface of the basin. In the final stage, Part C, the basin has been largely filled again but with less dense rock at the surface. This explanation was suggested originally by Professor Donald Wise.

Figure 10. - Meteor crater. This crater, about a half mile across, 600 feet deep, and located near Flagstaff, Arizona was caused by the impact of a large meteorite with the Earth in prehistoric time. Thousands of pieces of the meteorite have been found in the surrounding area. This feature has been studied extensively by members of the U.S. Geological Survey and has shed light on the detail of crater formation. Note the raised rim, a characteristic of many lunar craters. The crater, readily accessible by automobile, is well worth the small time required to visit if one is nearby. Photo courtesy of U.S. Geological Survey,

The features in the North Complex appear to have resulted from internal processes. Apollo 15 will give us the first opportunity to study at first hand the form of such features and the nature of their rocks and soils.

Figure 11. - A terrestrial model of Hadley Rille. The astronauts study the Rio Grande Gorge, near Taos New Mexico in preparation for their study of Hadley Rille. Shapes of the two features are similar and the rocks may possibly be similar but the origins of the two features are almost certainly different. During field training exercises, both Scott and Irwin shown here, carried mocked-up PLSS's.

Lunar Roving Vehicle

Inside the LM the astronauts will take with them to the surface a four-wheeled vehicle that can be used to transport themselves and equipment over the lunar surface. It is termed the Lunar Roving Vehicle (LRV) or Rover (figure 12). It is powered by two silver-zinc, 36-volt batteries and has an individual electric motor for each of the four wheels. An early version of the Rover, used for astronaut training, is shown in figure 13. The Rover deployment scheme is shown in figure 14. There is a navigation system that contains a directional gyroscope and provides information as to distance traversed as well as heading. In addition to the astronaut's oral descriptions, television pictures are telemetered back to Mission Control in Houston from the Rover. These pictures will be shown over the commercial TV networks.

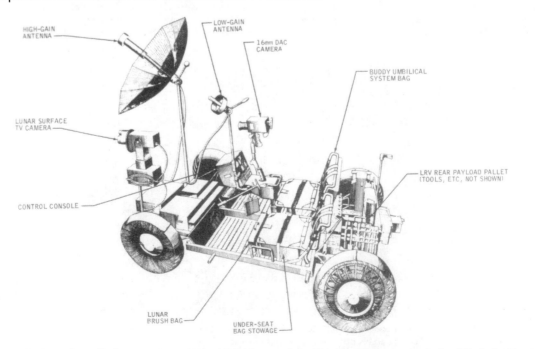

Figure 12. - The Lunar Rover. Both astronauts sit in seats with safety seat belts. About 7 minutes are required to fully deploy Rover. The capacity of the Rover is 370 pounds. The vehicle will travel about 10 miles per hour on level ground. The steps necessary to remove it from the LM and to ready it for use are shown in Figure 14.

Figure 13. - This model of the lunar Rover, nicknamed "Grover", was used in astronaut training exercises. The canyon in the background is the Rio Grande Gorge, a natural terrestrial model of Hadley Rille, near Taos, New Mexico. Shown on Grover are Irwin and Scott, the "surface" astronauts of Apollo 15.

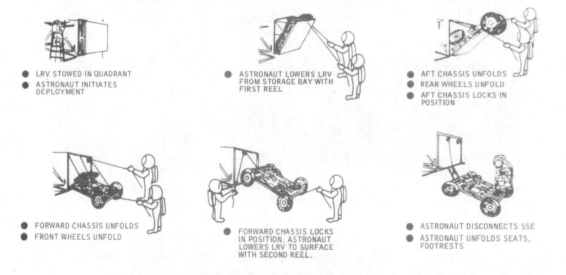

Figure 14. - Deployment sequence for the Lunar Roving Vehicle.

Surface Science Activities

Each of the two astronauts that descend to the lunar surface in the LM will spend about 20 hours in three periods of 7, 7, and 6 hours outside the LM working on the lunar surface. Most of that time will be used to study geological features, collect and document samples of rocks and soil, and set up several experiments that will be left behind on the lunar surface when the astronauts return to Earth.

The surface traverses described in this guidebook, which was written about 3 months before launch, should be considered as general guides for the astronauts to follow. From previous Apollo missions, we have learned that although some minor changes in plans are likely to occur, major changes are unlikely. On each mission a few changes were made by the crew because of unforeseen conditions. Instructions to the astronauts have always been "to use their heads" in following the detailed plans and this mission is no exception. In addition, the astronauts may consult over the radio with a group of scientists located in Mission Control at Houston and decide during the mission to make some changes. Undoubtedly, some details of the traverses will change. Equipment changes, on the other hand, are very unlikely to occur because all of the equipment has been built and is now being stowed in the spacecraft.

TRAVERSE DESCRIPTIONS

The planned Rover traverses are shown in figures 15 and 16. The activities at each of the stops on all three traverses and along each traverse between stops are shown in Table 2. In order to use Table 2 effectively, the reader must have scanned most of the next section, "Surface Scientific Experiments and Hardware", and to have read the section "Lunar Geology Experiment".

Figure 15. - The traverses planned for use with the Lunar Roving Vehicle. The Roman numerals indicate the three EVA's. The numbers are station stops. The station stops are keyed to the information given in Table 2. These same traverses are shown in figure 16, an overhead view of the landing site.

The numbers assigned to each of the traverse stations shown in the figures and tables of this guidebook will not change. However, extra stations may be added before, as well as during the mission. These extra Stations will be termed "supplementary sample stations" to avoid confusing them with the existing stations.

In the event that the Rover becomes inoperative sometime during the mission, a series of walking traverses has been planned. (Figure 17.)

LUNAR SURFACE SCIENTIFIC EXPERIMENTS AND HARDWARE

In addition to the observations made by the astronauts and the collection of samples of lunar material to be returned to Earth, several scientific experiments will be set out by the astronauts on the lunar surface. The equipment for these experiments will remain behind on the Moon after the astronauts return to Earth. Data from these experiments will be sent to Earth over microwave radio links, similar to the ones used extensively for communications on Earth.

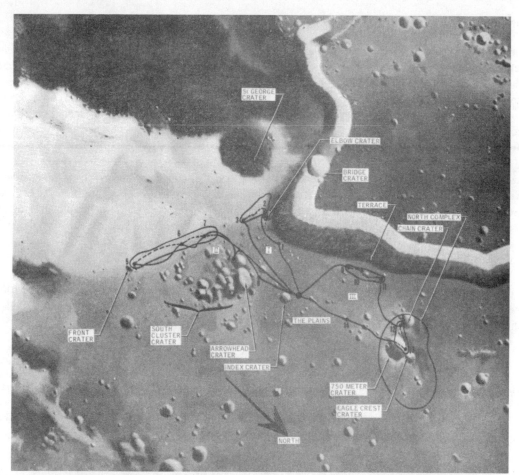

Figure 16. - Rover traverses. See explanation of Figure 15.

Figure 17. - Walking traverses. These alternate traverses can be done on foot. They will be used if the Rover becomes inoperative.

Apollo Lunar Surface Experiments Package (ALSEP)

Several of these experiments are a part of the Apollo Lunar Surface Experiments Package (ALSEP). General layout of the equipment on the lunar surface is shown in figure 18. A photograph of the Apollo 14 ALSEP is shown in figure 19. The ALSEP central station, figure 20, although obviously not an experiment, provides radio communications with the Earth and a means for control of the various experiments. The experiments connected electrically to the central station are the Passive Seismic Experiment, the Lunar Surface Magnetometer, the Solar Wind Spectrometer, Suprathermal Ion Detector, Heat Flow Experiment, Cold Cathode Ion Gauge, and Lunar Dust Detector. I discuss briefly each of these experiments.

Electrical power for the experiments on the lunar surface is provided by the decay of radio isotopes in a device termed Radioisotope Thermoelectric Generator (RTG), shown in figure 21. A total of roughly 70 watts is delivered. Let me draw special attention to this power of 70 watts. It is truly incredible that all of the experiments together use approximately the amount of power that is consumed by an ordinary 75 watt light bulb. The electrical wires are flat, ribbon-like cables that may be seen in figure 19. The RTG is filled with fuel after the astronauts place it on the lunar surface.

During EVA 1, the astronauts remove the ALSEP equipment from the LM, carry it to a site at least 300 feet from the LM, and place it on the lunar surface. A summary of these ALSEP operations is given in Table 3. A list of the principal investigators and their institutions is included in Table 4.

Heat Flow Experiment (HFE)

Heat flows from hot regions to cold regions. There is no known exception to this most general law of nature. We are certain that the interior of the Moon is warm. It may be hot. Therefore heat flows from the interior of the Moon to the surface where it is then lost into cold space by radiation. It is the function of the Heat Flow Experiment (HFE) to measure the amount of heat, flowing to the surface at the Hadley-Apennine site.

Figure 18. - General layout of the ALSEP. The sizes of the astronaut, equipment, and lunar features are drawn to different scales. Locations are shown in true relation to the surface features of the Moon.

At the present time, the heat flowing to the surface of the Moon from the interior has been produced mostly by decay of the natural radioactive elements thorium, uranium, and potassium. Measurements made directly on the lunar samples returned to Earth by Apollo 11, 12, and 14 have revealed the presence of significant amounts of these elements. The normal spontaneous decay of these elements into other elements slowly releases energy. The decay process is similar to that used in a nuclear reactor on Earth to generate electrical power from uranium. In the Moon, most of the energy appears in the form of heat which raises the temperature of the interior of the Moon.

In addition to the amount of radioactive material present, the internal temperature of the Moon depends on other parameters. The thermal properties of lunar rocks are equally important. The thermal conductivity of a material is a measure of the relative ease with which thermal energy flows through it. Rather well-known is the fact that metals are good conductors and that, fiberglass, asbestos, and bricks are poor conductors. Most of us would never build a refrigerator with copper as the insulation. Values of the thermal properties of rocks are closer to the values of fiberglass than to those of copper and other metals. Rocks are fairly good insulators.

The Heat Flow Experiment has been designed to measure the rate of heat loss from the interior of the Moon. To obtain this measurement, two holes are to be drilled into the surface of the Moon by one of the astronauts to a depth of about 10 feet by means of the drill sketched in figure 22. After each hole is drilled, temperature sensors (platinum resistance thermometers) are placed at several points in the lower parts of the holes and several thermocouples (for measuring temperatures with lower precision) are placed in the upper portions of the holes. See figure 23. The thermal properties of the rocks will be measured by the equipment that is placed in the hole, they will also be measured on samples that are returned to the Earth.

Because the temperature of the rock is disturbed by the drilling process, the various measurements for heat flow will be taken at regular intervals over several months. As the residual heat left around the hole from drilling dissipates with time, the temperatures measured in the experiment will approach equilibrium.

The great importance of the HFE derives from the fact that knowledge of the amount of heat flowing from the interior of the Moon will be used to set limits on the amount of radioactivity now present in the Moon and to set limits on models of the thermal history of the Moon.

Figure 19. - Apollo 14 ALSEP. Individual items are readily identified by comparison with the sketches included in this guidebook.

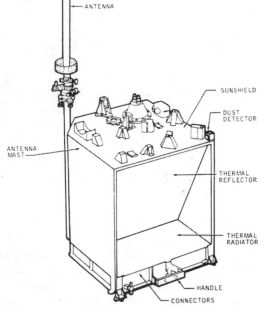

Figure 20. - The ALSEP central station. This equipment's connected electrically to each of the other ALSEP experiments. It is a maze of electronics that accepts the electrical signals from various experiments and converts them into a form suitable for transmission by radio back to the Earth. The pole-like feature on top of the central station is a high-gain antenna. It is pointed towards the Earth. Commands may be sent from the Earth to the central station to accomplish various electronic tasks.

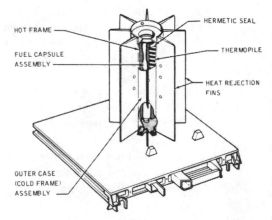
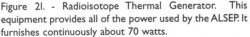

Figure 21. - Radioisotope Thermal Generator. This equipment provides all of the power used by the ALSEP. It furnishes continuously about 70 watts.

Passive Seismic Experiment (PSE)

The Passive Seismic Experiment (PSE) is used to measure extremely small vibrations of the Moon's surface. It is similar to seismometers used on the Earth to study the vibrations caused by earthquakes and by man-made explosions. The PSE equipment is seen in figure 24. The principle of operation is indicated in figure 25. As the instrument is shaken, the inertia of the mass causes the boom to move relative to the case. This relative motion is detected electrically by the capacitor and the electrical signal is then transmitted by radio to the Earth.

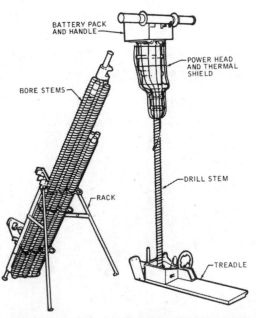

Figure 22. - Lunar Surface Drill. This drill will be used to drill holes on the Moon to a depth of about 10 feet. It is electrically powered and operates from batteries. The treadle is used to steady the drill stem and to deflect cuttings from striking the astronaut. Two holes are used for the heat flow experiment and a third one is used to obtain samples for study back on Earth.

The data from the PSE, in conjunction with similar data from Apollo 12 and 14 sites, are especially valuable. They will be used to study the nature of the interior of the Moon, to determine the location of moonquakes and to detect the number and size of meteoroids that strike the lunar surface.

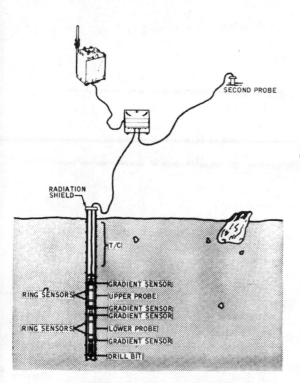

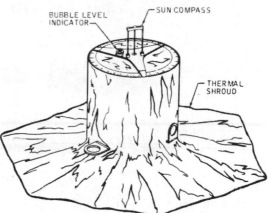

Figure 23. - Heat Flow Experiment. Probes are placed in two holes drilled in the lunar surface with the drill shown in figure 22. One hole is shown in the figure as a section to show the various parts. The gradient is the difference of temperature at two points divided by the distance between the points. Heat flow is measured by measuring the gradient and independently measuring the thermal conductivity; heat flow is the product of gradient and thermal conductivity. The symbol T/C indicates thermocouples that are present in the upper part of the holes.

Figure 24. - Passive Seismometer. The instrument is covered with a blanket of superinsulation to protect it from the extreme variations of temperature on the Moon (-400° to +200°F) . The principle of operation is shown in figure 25. The level, used on the Moon in exactly the same way as on the Earth, indicates whether the instrument is level. The Sun compass is used to indicate direction. A typical seismic signal for the Moon is seen in figure 26. Such signals are detected at the Apollo 12 and 14 sites at the rate of about one per day. There is usually increased activity when the Moon is farthest from the Earth and also when it is nearest the Earth.

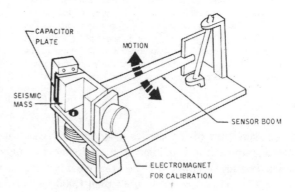

Figure 25. - Principle of operation of passive seismometer. See text for details.

The Moon is still being bombarded by small objects; most of them are microscopic in size. The Earth is also being bombarded but most small objects completely disintegrate in the Earth's atmosphere; they are the familiar shooting stars.

Lunar Surface Magnetometer (LSM)

The Lunar Surface Magnetometer (LSM) is used to measure the variations with time of the magnetic field at the surface of the Moon. A similar instrument was left at the Apollo 12 site. It is still sending data to Earth. None was left at the Apollo 14 site although two measurements of the magnetic field were made there with a smaller, portable magnetometer. The LSM equipment is shown in figure 27. Because the magnetic field at the surface of the Moon can change in amplitude, frequency, and direction, the LSM is used to measure the magnetic field in three directions. The sensors are located at the ends of three booms.

The magnetic field of the Moon (and also the Earth) has two parts, one that changes with time and one that is steady and does not change rapidly with time. The part that changes with time is caused by travelling electromagnetic waves.

PASSIVE SEISMIC EXPERIMENT

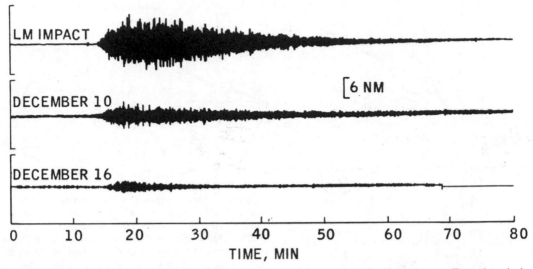

Figure 26. - Typical seismic signals for the Moon. These events were sensed at the Apollo 12 seismometer. To produce the largest signals shown here, the Moon's surface moved about 2 ten-thousandths of an inch.

Figure 27. - Lunar Surface Magnetometer. Measurements are obtained as a function of time of the magnetic field at the surface of the Moon by the lunar surface magnetometer. The actual sensors are located in the enlarged parts at the end of the three booms. The plate located in the center of the instrument is a sun shade to protect the electronics in the box at the junction of the three booms from direct sunlight.

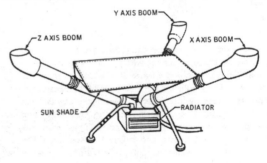

The steady part of the Earth's magnetic field, that part which does not, change rapidly with time, is about 50,000 gamma (the usual unit of magnetic field employed by Earth scientists). It causes compasses to point approximately north-south. The steady part of the lunar magnetic field measured at the Apollo 12 site was about 35 gamma somewhat more than 1,000 times smaller than the Earth's field. Yet the 35 gamma field was several times larger than we had expected. Similar measurements obtained at the Apollo 14 site with the smaller portable magnetometer revealed a magnetic field in two different spots of about 65 gamma and 100 gamma. The steady part of the lunar magnetic field is undoubtedly due to the presence of natural magnetism in lunar rocks. The natural magnetism was probably inherited early in the Moon's history (perhaps several billion years ago) when the Moon's magnetic field was many times larger than today.

The LSM is also used to measure the variation with time of the magnetic field at the surface of the Moon. The variations are caused by electromagnetic waves that emanate from the Sun and propagate through space. The largest change in the magnetic field ever measured in space, about 100 gammas, was detected by the Apollo 12 LSM.

Variations with time in the magnetic field at the surface of the Moon are influenced greatly by the electrical properties of the interior of the Moon. Therefore, a study of the variations with time of the magnetic field will reveal the electrical properties of the Moon as a function of depth. Because the electrical properties of rocks are influenced by the temperature, we hope to use the data from the LSM to measure indirectly temperatures in the interior of the Moon.

Lunar Atmosphere and Solar Wind Experiments

(1) Solar Wind Spectrometer (SWS)

(2) Solar Wind Composition (SWC)

(3) Suprathermal Ion Detector (SIDE) /Cold Cathode Ion Gauge (CCIG)

Matter is ejected, more or less continuously, by the Sun and spreads throughout the solar system. It is called the solar wind. It is very tenuous. It moves with a speed of a few hundred miles per second. The energy, density, direction of travel, and variations with time of the electrons and protons in the solar wind that strike the surface of the Moon will be measured by the Solar Wind Spectrometer. This equipment is shown in figure 28. The seven sensors are located beneath the seven dust shields seen in the picture. The data allow us to study the existence of the solar wind at the lunar surface, the general properties of the solar wind and its interaction with the Moon. The solar wind "blows" the Earth's magnetic field into the form of a long tail that extends past the Moon. Thus the SWS is also used to study the Earth's magnetic tail.

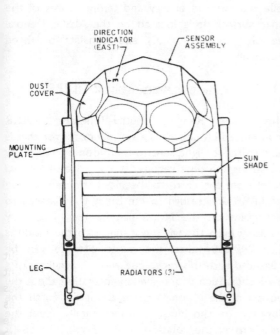

Some equipment carried to the Moon to determine the composition of the solar wind is extremely simple. The Solar Wind Composition (SWC) experiment is essentially a sheet of aluminum foil like the familiar household item used to wrap food. It is seen in figure 29. Exposed on the lunar surface to the solar wind, it traps in the foil the individual particles of the solar wind. The foil is returned to Earth and the individual elements are examined in the laboratory. Sponsored by the Swiss government, this experiment is international in scope.

Two experiments, the Suprathermal Ion Detector Experiment (SIDE) and the Cold Cathode Ion Gauge (CCIG) are used to measure the number and types of ions on the Moon. An ion is an electrically charged molecule. It may be either positive or negative which depends on whether one or more electrons are lost or gained, respectively. Those ions on the Moon are chiefly hydrogen and helium and are largely derived from the solar wind but several others are present also. The hardware is illustrated in figure 30.

Figure 28. - Solar Wind Spectrometer. With this instrument, the solar wind will be studied. It measures energy, density, directions of travel, and the variations with time of the solar wind that strikes the surface at the Moon. During the journey to the Moon, this equipment is carried with the legs and sun shade folded so that it will occupy less space. The astronaut unfolds the legs and sun shade before setting it out on the Moon.

The SIDE is used to measure the flux, number, density, velocity, and the relative energy of the positive ions near the lunar surface. The CCIG, although a separate experiment, is electronically integrated with the SIDE. It, is used to measure the pressure of the lunar atmosphere. It operates over the pressure range of 10^{-6} to 10^{-12} torr. (For comparison of these units, the Earth's atmosphere at sea level produces a pressure of about 760 torr and the pressure in the familiar Thermos vacuum bottles is about 10^{-3} torr). The lowest pressures obtainable on Earth in vacuum chambers is about 10^{-13} torr. The pressure measured at the Apollo 14 site by CCIG was about 10^{-12} torr. At that pressure only 500,000 molecules of atmosphere would be present in a volume of 1 cubic inch. Although that number may seem like many molecules, remember that 10^{15} times as many exist in each cubic inch near the surface of the Earth! The lunar pressure varies slightly with time.

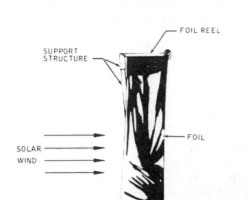

Figure 29. - Solar Wind Composition Experiment. Particles in the solar wind strike the aluminum foil, are trapped in it, and finally brought back to Earth by the astronauts for examination. This experiment is sponsored by the Swiss government.

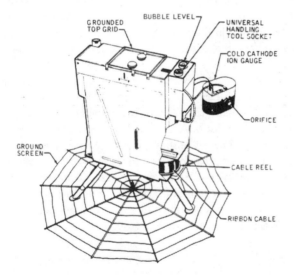

Figure 30. - Suprathermal Ion Detector (SIDE) and Cold Cathode Ion Gauge (CCIG). These two experiments are used to study the atmosphere of the Moon.

Astronauts continually release gas molecules from their suit and PLSS. The molecules are chiefly water and carbon dioxide. These additional molecules increase locally the atmospheric pressure and the CCIG readily shows the presence of an astronaut in the immediate vicinity. It is expected that the Apollo 14 CCIG will "see" the arrival of the Apollo 15 LM on the Moon from the exhaust gases.

Lunar Dust Detector (LDD)

The main purpose of the Lunar Dust Detector (LDD) is to measure the amount of dust accumulation on the surface of the Moon. It also measures incidentally the damage to solar cells caused by high energy radiation and it measures the reflected infrared energy and temperatures of the lunar surface. It is located on the ALSEP Central Station (see figure 20) and consists of three photocells.

Laser Ranging Retro-Reflector (LRRR)

The Laser Ranging Retro-Reflector (LRRR pronounced LR-cubed) is a very fancy mirror that is used to reflect light, sent to the Moon by a laser. By measuring the time required for a pulse of light to travel from the Earth to the Moon, be reflected by the LRRR, and return to the Earth, the distance to that point on the Moon can be measured very precisely. Even though the distance to the Moon is abut 240,000 miles, the exact distance can be measured with this technique with an accuracy of a few inches. Such data provide information about the motion of the Moon in space about the Earth, the vibrations of the Moon, and incidentally about the variations in the rotation of the Earth. The LRRR equipment is shown in figure 31. It consists of 300 individual fused silica optical corner reflectors. Obviously, scientists in any country on Earth can use this equipment to return their own laser beams. Similar ones, though smaller, were left at the Apollo 11, 12, and 14 sites.

Lunar Geology Experiment (LGE)

Most of the time spent by the astronauts during the three EVA's will be devoted to investigation of various geologic features at the landing site and to collecting samples of rocks. Many detailed photographs will be obtained to supplement the verbal descriptions by the astronauts. Samples of the rocks present at the site will be bagged and brought back to Earth. The astronauts will use several individual pieces of equipment to help them with their tasks. In this section I describe briefly the individual items used in studying the geology of the Hadley-Apennine region and in collecting samples for return to Earth.

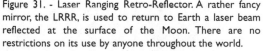

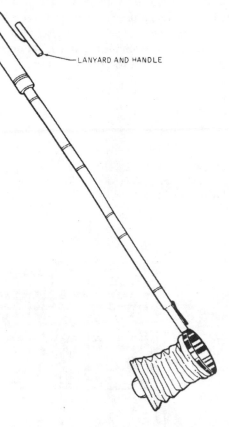

Figure 31. - Laser Ranging Retro-Reflector. A rather fancy mirror, the LRRR, is used to return to Earth a laser beam reflected at the surface of the Moon. There are no restrictions on its use by anyone throughout the world.

Figure 32. - Contingency Sampler. During the flight to the Moon, the handle is folded. Tension on the rope stiffens it so that the astronauts can scoop quickly a few rocks and some soil. The bag is made of Teflon, will hold about 2 pounds of rocks and soil, and is detached from the handle before stowage in the LM. This tool is used to collect material from the immediate area of the LM very soon after the astronauts first egress from the LM so that some samples will have been obtained if the surface activities must be curtailed and the mission aborted.

Soon after the astronauts first set foot on the surface of the Moon, one will use the tool shown in figure 32 to collect a small (about 1-2 lbs.) sample of rock and soil. That sample is termed the contingency sample. It is stowed immediately on board the LM to insure that at least some material would be obtained in the unlikely event that the surface activities had to be terminated abruptly and prematurely for any reason.

Observations made on the lunar surface of the various geological features are very important. The television camera allows us on Earth to follow the astronauts and to "see" some of the same features, though not nearly so well, as the astronauts see. The TV camera used on Apollo 14, similar to the one on this mission, is shown in figure 33. The Apollo 15 TV camera will be mounted on the Rover during the traverses.

Figure 33. - Apollo 14 television camera. The astronaut is adjusting the TV camera to obtain the best possible viewing of activities around the LM during the Apollo 14 mission. A similar television camera will be carried aboard Apollo 15; it will be mounted sometimes on the Rover. Note the many craters in the foreground and the boulders in the distance.

Other tools used by the astronauts are shown in figure 34 together with an aluminum frame for carrying them. The hammer is used to drive core tubes into the soil, to break small pieces of rocks from larger ones, and in general for the same things that any hammer might be used on Earth. Because the astronaut cannot conveniently bend over and reach the lunar surface in his space suit, an extension handle is used with most tools. The scoop (figures 35 and 36) is used to collect lunar soil and occasionally small rocks. The tongs, sketched in figure 34 and shown in figure 37, an Apollo 12 photograph, are used to collect small rocks while standing erect.

Figure 34. - Lunar Geological Hand Tools. This equipment is used to collect samples of rock and soil on the Moon. See text and subsequent figure for details.

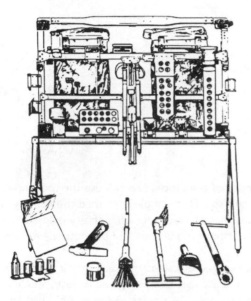

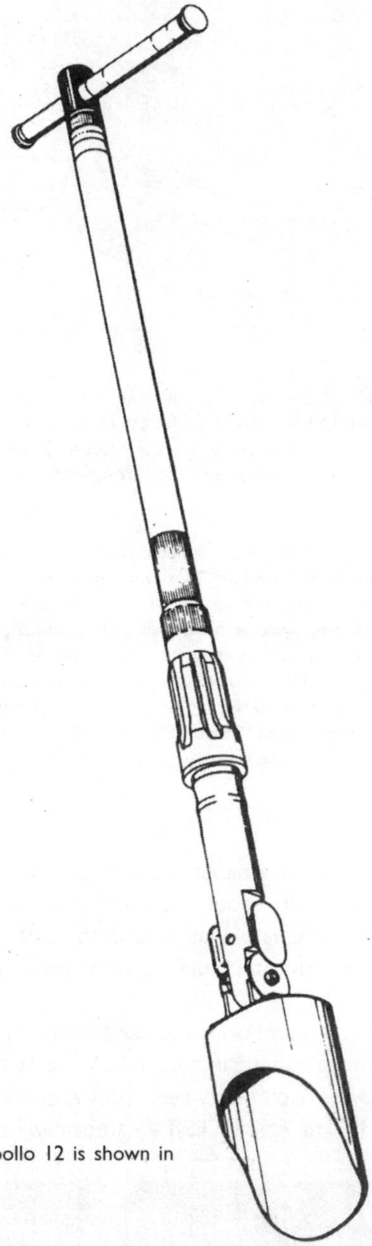

Figure 35. - Scoop with extension handle. Its use in Apollo 12 is shown in Figure 36.

j

The drive tubes (figure 38) are used to collect core material from the surface to depths of 1 to 4½ feet. The core remains in the tubes for return to Earth. Preservation of the relative depths of the core material is especially important. The drive tubes were originally suggested about 6 years ago by the late Dr. Hoover Mackin, a geologist. Shown in figure 39 is a drive tube that was driven into the Moon's surface on Apollo 14. The individual tubes are about 18 inches long. As many as three tubes can be used together for a total length of about 4½ feet.

After the surface samples are collected, they are placed in numbered sample bags made of Teflon (figure 40). These bags are about the size of the familiar kitchen storage bags. After a sample is bagged, the thin aluminum strip is folded to close the bag and prevent the samples from becoming mixed with others. The bags are finally placed in the sample return containers, sketched in figure 41, for return to Earth. The Apollo Lunar Sample Return Container (ALSRC) is about the size of a small suitcase. It is made of aluminum and holds 30 to 50 lbs of samples.

Figure 36. - Use of scoop in Apollo 12. Note the small rock in the scoop.

Figure 37. - Tongs shown in use on Apollo 12 to collect a small rock.

A special container, termed Special Environmental Sample Container (SESC), is used to collect material on the surface of the Moon for specific purposes. (See figure 42.) This container has pressure seals to retain the extremely low pressures of the Moon. It is made of stainless steel. The sample to be collected on Apollo 15 and returned in this container will be collected in such a manner that it will have very little contamination with materials, either organic or inorganic, from Earth. The largest sources of biological contamination are the astronauts themselves: the suits leak many micro-organisms per minute and the lunar rocks collected on previous missions have all contained some organic material (a few parts per billion).

Whether any of the organic material was present on the Moon before the astronauts' landing is uncertain; this question is currently being intensely investigated

The Hasselblad cameras used by the astronauts (figure 43), were made especially for this use. The film is 70 mm wide, exactly twice as wide as the familiar 35 mm film. The color film is similar in characteristics to Ektachrome-EF daylight-type. The black and white film has characteristics like Plus X. The primary purpose of the cameras is that of documenting observations made by the astronauts. Especially important is the careful documentation of rocks that are collected for study back on Earth. Ideally, several photographs are taken of the rocks: (1) before collection with the Sun towards the astronaut's back, (2) before collection with the Sun to the side of the astronaut, (3) before collection a third photo to provide a stereo pair, and (4) after collection a single photo to permit us to see clearly which sample was collected. A device, termed gnomon and illustrated in figure 44, is included with these pictures to provide a scale with which to measure size and a calibration of the photometric properties of the Moon's surface. In addition to these photographs, a fifth one is desirable to show the general location of the sample with respect to recognizable features of the lunar surface. An example from Apollo 14 is seen in figure 45. The photos taken before collection and after collection show clearly which rock was removed.

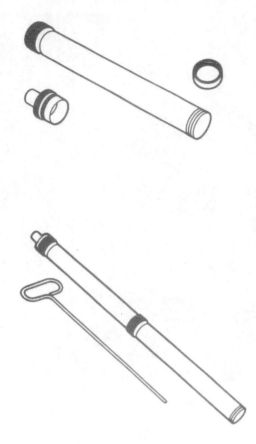

Figure 39. - Drive tube in lunar surface at Apollo 14 site. Note in addition the footprints, rocks, and small craters.

Figure 38. - Drive Tubes. These tubes, about 18 inches long, are pushed or driven into the lunar surface to collect samples as a function of depth. Two, three, or even four of them may be joined together to obtain a longer core. Their use in Apollo 14 may be seen in Figure 39.

Figure 40. - Lunar sample bag. The bag resembles the familiar kitchen item "Baggies". It is made of Teflon. A strip of aluminum is used to close the bag.

At some stations, still more documentation is desirable. Panoramic views, also called pans, are obtained by shooting many photographs of the horizon while turning a few degrees between snapping each successive photo. The photos have considerable overlap. After return to Earth the overlap is eliminated and the photos pieced together to yield a composite view of the Moon's surface as seen from a particular spot. One example from Apollo 14 is shown in figure 46. Others may be seen in the July issue of National Geographic Magazine. In addition, the overlapped regions are used for stereoscopic viewing of the surface. Truly three-dimensional views are obtained in this way.

Marble-sized rocks from the Moon have proven to be especially valuable in lunar science. They are large enough to allow an extensive set of measurements to be made, yet small enough that many of them can be collected. Accordingly, we have designed and built a tool for Apollo 15 to collect many such samples. It is termed a rake, although the resemblance to the familiar garden tool is now slight. It is illustrated in figure 47.

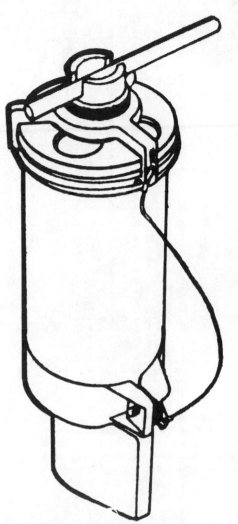

Figure 41. - Apollo Lunar Sample Return Container. Made of aluminum, this bag is used to return lunar samples to Earth. It is about the size of a small suitcase but is many times stronger.

Figure 42. - Special Environmental Sample Container. This container has special vacuum seals to prevent gases and other materials from entering the container and being adsorbed on the surfaces during the journey to the Moon. They also prevent contamination of the samples by rocket exhaust gases and the Earth's atmosphere during the return journey.

The Apollo Lunar Surface Drill (ALSD), used to drill the two holes for the Heat Flow Experiment and illustrated in figure 32, is used also to drill a third hole from which the samples are saved. The drill bit is hollow and allows rock to pass into the hollow drill stem. These samples, referred to as core, are about 0.8 inch in diameter. Individual pieces of rock are likely to be buttonshaped and ¼ inch thick. A few pieces may be harder. Most of the material will probably consist of lunar soil. These samples should not be confused with the samples obtained with the drive tubes which are also termed core. This equipment can drill and collect solid rock, if any is encountered, whereas the drive tubes can collect only material that is small enough to enter the tube.

Soil Mechanics Experiment

The mechanical properties of the lunar soil are important for both engineering and scientific reasons. Future design of spacecraft, surface vehicles, and shelters for use on the Moon will be based, in part at least on the data collected in the soil mechanics experiment of this mission. To obtain data, many observations will be made during the performance of the other experiments. Such items as the quantity of dust ejected by the exhaust, from the descending LM, the amount of dust thrown up by the wheels on the Rover and the depth to which the astronauts sink while walking, are all important factors in estimating the properties of the lunar soil. In addition to these qualitative observations, the astronauts will carry equipment with them with which to measure quantitatively the bearing strength of the soil, a recording penetrometer. It is illustrated in figure 48.

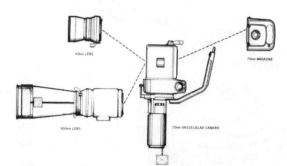

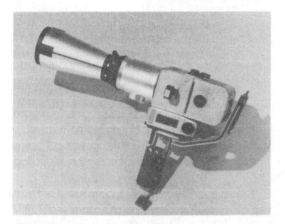

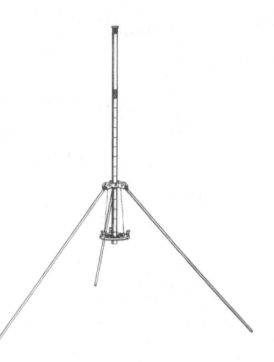

Figure 44. - Gnomon. This device is used to provide a physical scale and to calibrate the photometric properties of the samples on the Moon. It can also be seen in figure 45, an Apollo 14 photograph.

Figure 43. - Hasselblad camera. The film, which may be black and white or color, is 70 mm wide. Two separate lenses are used with this camera on the surface of the Moon. The 500 mm lens, a telephoto lens, shown attached to the camera in the photograph will be used to photograph the walls of Hadley Rille.

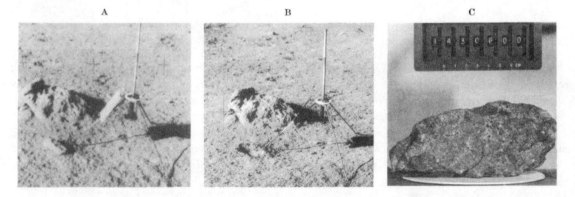

Figure 45. - Photographic documentation of lunar samples. These three Apollo 14 photographs indicate clearly the method used to identify the rocks that were collected. The shadows in A, together with knowledge of the time that the photo was taken have been used to orient the specimen. A location photograph (not shown) allows us to determine the relative location of this sample with respect to others collected during the mission. Photo A was taken before the rock was collected. Photo B was taken after collection. Photo C was taken in the laboratory after the Apollo 14 mission had returned to Earth. The Field Geology Team led by Dr. Gordon Swann, identified the rock in photos A and B as sample 14306 and deduced from photo A the orientation on the lunar surface.

Figure 46. - Panoramic view obtained on Apollo 14. The method of piecing together several photos is clearly shown. Other panoramas may be seen in the July 1971 issue of National Geographic magazine. The tracks toward the upper left lead to ALSEP. The Mobile Equipment Transport is seen in the foreground. The elliptic shadow near the MET was cast by the S-band radio antenna used for communication with Earth. The TV camera is seen on the right-hand side of the pan. The inverted cone seen just below the TV camera is part of a small rocket engine used to turn the spacecraft in flight.

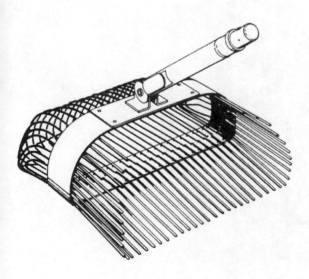

Figure 47. - Rake. This tool will be used on Apollo 15 to collect marble-size rocks.

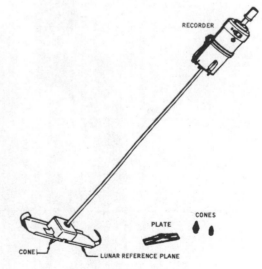

Figure 48. - Self-recording penetrometer.

The Crew

The prime crew consists of Dave Scott, Commander, Jim Irwin, LM pilot, and Al Worden, CM pilot. Scott and Neil Armstrong during the Gemini 8 mission performed the first successful docking of two vehicles in space. Scott was the CM pilot on Apollo 9 in 1969, the third-manned flight in the Apollo series and backup commander for Apollo 12. Jim Irwin served as backup crew member for the Apollo 12 flight. Al Worden served as the backup Command Module pilot for Apollo 12.

The Apollo 15 backup crew consists of Dick Gordon, Commander, Vance Brand, CM pilot, and H. H. (Jack) Schmitt, LM pilot. The prime surface crew is shown in figures 49, 50 and 51. The backup crew is shown in figures 52 and 53.

This crew, like previous ones, has undergone intensive training during the past few months and somewhat more casual training during the last few years. In addition to the many exercises needed to learn to fly proficiently their spacecraft, the astronauts have learned much about science, and in particular, about lunar science. After all, they will each spend many hours on the Moon or in orbit around the Moon performing scientific research. The surface astronauts have had tutorial sessions with many of the nation's best scientists. They are able to set up experiments, such as those of ALSEP, but more importantly, they understand the scientific purposes behind the various experiments.

Most of the time on the lunar surface during Apollo 15 will be spent observing geologic features and collecting samples. Obviously anyone can pick up rocks with which to fill boxes and bags. Only a person highly trained in the geosciences, however, can properly select those few rocks from many that are likely to yield the greatest amount of scientific return when examined in minute detail in the laboratory back on Earth. The Apollo 15 crew has spent many hours in the field studying rocks under the guidance of geologists from the U.S. Geological Survey, several universities, and NASA's Manned Spacecraft Center. The prime crew has been especially fortunate in having the constant geologic tutelage of Astronaut Jack Schmitt, a geologist himself.

Figure 49. - Apollo 15 astronauts Jim Irwin and Dave Scott studies geology on a field trip near Taos, New Mexico in March 1971. In the left background, shown in profile with Texas-style hat, is Professor Lee Silver, a field geologist from the California Institute of Technology who has contributed significantly toward geological training of the crew.

Figure 50. - Apollo 15 astronauts Jim Irwin (left) and Dave Scott during the field trip to study geology near Taos, New Mexico. During such training exercises the astronauts typically carry backpacks that simulate the PLSS. Note the Hasselblad cameras, the scoop, and a gnomon.

Figure: 52. - Apollo 15 backup surface crew, Dick Gordon (left) and Jack Schmitt. They are using the self-recording penetrometer to measure soil properties near Taos New Mexico.

Figure 51 - CM Pilot Al Worden. A major part of the crew's training in science is the study of geology. Worden is recording his observations of the rocks at this training site.

Figure 53.-Astronaut Vance Brand collects rocks on a training trip to Iceland. During the training exercises, the astronauts record their observations on the rocks geological features.

Bibliography

This bibliography is not intended to be extensive. It is a guide to simply-written, and mostly inexpensive, books that I believe useful for additional reading.

Alter, Dinsmore, editor, Lunar Atlas, Dover Publications, Inc., New York, 1968. Excellent and very inexpensive. Contains many photographs of various features on the Moon. Strongly recommended for the interested layman. Paper-bound $3.00.

American Association for the Advancement of Science, Washington, D.C., Apollo 11 Lunar Science Conference, McCall Printing Company, 1970. Historic milepost in lunar science. Contains the first public release of information obtained on the Apollo 11 samples by several hundred scientists. Written for fellow scientists. Obtain from AAAS, 1515 Massachusetts Avenue, N.W., Washington, D.C. 20003, Hardback $14.00, Paperback $3.00.

Baldwin, Ralph B., The Measure of the Moon, The University of Chicago Press, 1963. Exhaustive study of the Moon. Important summary of knowledge of the Moon that existed before the lunar flights began. Although in places, the reading may be a little difficult, it is generally accessible to the layman, $13.50.

Cortright, Edgar M., ed., Exploring Space with a Camera, NASA SP-168, NASA, Washington. D.C. Inexpensive. Contains many beautiful photographs obtained from space. Well worth the small investment for the layman with even mild interests in space. Government Printing Office, Washington, D.C. $4.25.

Hess, Wilmot, Robert Kovach, Paul W. Gast and Gene Simmons, The Exploration of the Moon, Scientific American, Vol. 221, No. 4, October 1969, pp. 54-72. General statement of plans for lunar exploration. Written before first lunar landing. Suitable for layman. The authors were all instrumental in planning the lunar surface scientific operations of the Apollo program. Reprint available from W. H. Freeman and Company, 600 Market St., San Francisco, Calif. 9410-1, 25¢ postage paid.

Kopal, Zdenek, An Introduction to the Study of the Moon, Cordon and Breach, New York, N.Y., 1966. For the mathematically inclined person, this book is an excellent introduction.

Kosofsky, L. J. and Farouk El-Baz, The Moon as Viewed by Lunar Orbiter, NASA SP-200, NASA. Washington, D.C.. 1970. Excellent reproductions of beautiful photographs of the Moon obtained from the Lunar Orbiter spacecraft. U.S. Government Printing Office, Washington D.C., $7.75.

Mason, Brian, Meteorites, John Wiley and Sons, Inc., 1962. Excellent introduction to a subject related closely to lunar science.

Mutch, Thomas A., Geology of the Moon, A Stratigraphic View, Princeton University Press, Princeton, New Jersey, 1970, $17.00. Excellent introduction to lunar geology. Written before Apollo 11 landing but still quite current. Previous geological training not necessary.

NASA. Ranger IX Photographs of the Moon, NASA SP-112, NASA, Washington, D.C., 1966. Beautiful close-up photographs of the Moon, obtained on the final mission of the Ranger series, U.S. Government Printing Office, Washington, D.C., $6.50.

NASA, Earth Photographs from Gemini VI through XII, NASA SP-171. NASA, Washington, D.C., 1968. Contains many beautiful photographs of the Earth from space. In color. U.S. Government Printing Office, Washington, D.C. $8.00.

NASA, Surveyor Program Results, NASA SP-184, NASA, Washington, D.C., 1969. Final report of the results obtained in the Surveyor Program. Surveyor was the first soft-landed spacecraft on the Moon and provided many important data. Because only one of the Surveyor sites has been revisited, the data given in this book are very important to our current understanding of the Moon. Part is easily readable by the layman: some is more difficult. U.S. Government Printing Office, Washington, D.C., $4.75.

Wood, John A., Meteorites and the Origin of Planets, McGraw-Hill Book Company, 1968. Inexpensive. Suitable for layman. Good introduction to meteorites.

APOLLO 15
SCOTT WORDEN IRWIN

Apollo 15 Prime crew (l to r)
Commander David R. Scott,
Command Module Pilot Alfred M. Worden,
Lunar Module Pilot James B. Irwin.

(Below) Apollo 15 crew discuss egress
contingencies with UDT Leader Schmidt.

(Above) Irwin and Scott hold a press conference to discuss
the intricacies of the first lunar rover (LRV). The lightweight
car used electric batteries to traverse the harsh lunar
terrain. A redesigned television system was devised to
transmit pictures from various points on the EVA. The
camera can be seen in the foreground wrapped in gold foil
to protect it from the solar heat.

Irwin and Scott load the LRV with equipment during preflight training May 1971(above). Scott demonstrates the lunar drill during a time line simulation (left). The crew stand alongside their experiments and equipment at the Manned Spacecraft Center in Texas, February 1971(below).

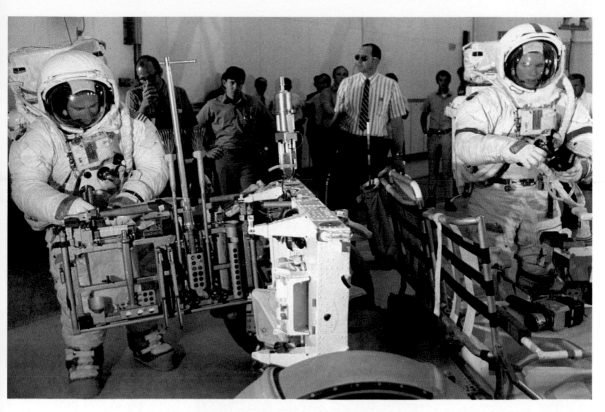

Irwin and Scott load the Lunar Roving Vehicle (LRV) (above) with equipment at the Kennedy Space Center May 1971.
The tool rack is retracted showing the various geology tools used at Hadley-Apennine.
Irwin and Scott ride the Lunar Rover trainer before departure to the moon.
The chevron wheels are visible as is the television camera and battery packs (below).

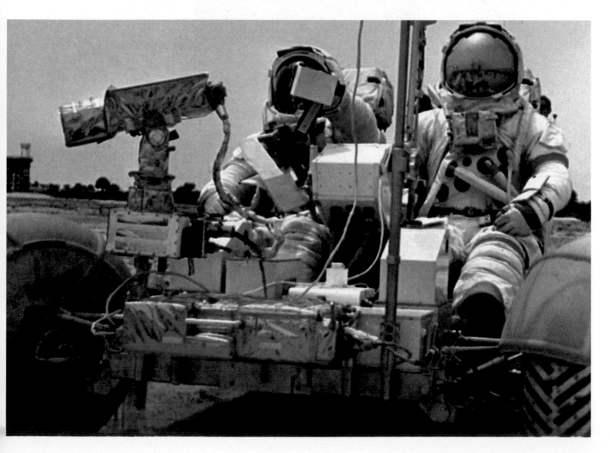

Apollo 15 launch 9:34 EDT
July 26 1971. (left)
Apollo 15 CSM in lunar orbit
(above) showing an excellent view of
the SIM bay. The complex new
experiment package can be seen
before loading onto the Saturn
launch vehicle (below).

Irwin works at the LRV at Hadley-Apennine. St George Crater can be seen three miles in the background. (above)
Scott holds a gnomon leveling device in this picture taken from the LRV camera. Mt Hadley Delta is in the background (below)

Scott (above) and Irwin (below) take time out to salute the flag in these spectacular pictures. Mount Hadley Delta rises thousands of feet behind them, the LRV can be seen bottom right.

Jim Irwin works at the Lunar Rover. The time spent at Hadley-Apennine was an unparalleled success due to the time spent by the crew in geology training . Many new challenges were faced not least the daring landing amongst the lunar mountains. Due to careful planning and redesign Apollo 15 was able to carry over 4000 lbs more payload than previous flights.

The moment of high tension is captured for the first time on TV as Falcon lifts off. (below)

The lunar module "Falcon" returns to dock with the orbiting CSM "Endeavour".

Hadley Rille weaves its way around the Apennine mountains. St George Crater is left center.

CM Pilot Worden flew a sophisticated experiment package as well as making the first EVA between the Earth and the moon.

Apollo 15 returns to the Pacific Ocean with one main parachute deflated. The impact was heavier than normal and can be seen inset at left. The crew look tired but elated on board the recovery raft as they await helicopter pick-up. Splashdown was at 4:46 EDT August 7th 1971.

ALBEDO *al-bee-doh*

Relative brightness. It is the ratio of the amount of electromagnetic radiation reflected by a body to the amount of incident radiation.

ANGSTROM UNIT *ang-strom*

A unit of length equal to 10^{-10} meters or 10^{-4} microns. It is approximately four billionths of an inch. In solids, such as salt, iron, aluminum, the distance between atoms is usually a few Angstroms.

APERTURE *a-per-ture*

A small opening such as a camera shutter through which light rays pass to expose film when the shutter is open.

ATTENUATION *a-ten-u-eh-shun*

Decrease in intensity usually of such wave phenomena as light or sound.

BASALT *ba-salt*

A type of dark gray rock formed by solidification of molten material. The rocks of Hawaii are basalts.

BISTATIC RADAR *bi-sta-tic ray-dar*

The electrical properties of the Moon's surface can be measured by studying the characteristics of radio waves reflected from the Moon. If the radio transmitter and receiver are located at the same place, the term monostatic radar is used. If they are located at different places, then bistatic is used. In the study of the Moon with bistatic radar, the transmitter is aboard the CSM and the receiver is on the Earth.

BRECCIA *brech-ya*

A coarse-grained rock composed of angular fragments of pre-existing rocks.

CASSETTE *kuh-set*

Photographic film container.

CISLUNAR *sis-lune-ar*

Pertaining to the space between the Earth and Moon or the Moon's orbit.

COLORIMETRIC

Pertaining to the measurement of the intensities of different colors as of lunar surface materials.

COSMIC RAYS *kos-mik*

Streams of very high energy nuclear particles, commonly protons, that bombard the Earth and Moon from all directions.

COSMOLOGY *kos-mol-uh-gee*

Study of the character and origin of the universe.

CRATER *cray-ter*

A naturally occurring hole. On Earth, a very few craters are formed by meteorites striking the Earth; most are caused by volcanoes. On the Moon, most craters were caused by meteorites. Some lunar craters were apparently formed by volcanic processes. In the formation of lunar craters, large blocks of rock (perhaps as large as several hundred meters across) are thrown great distances from the crater. These large blocks in turn form craters also — such craters are termed secondary craters.

CROSS-SUN

A direction approximately 90 degrees to the direction to the Sun and related to lunar surface photography.

CROSSTRACK

Perpendicular to the instantaneous direction of a spacecraft's ground track.

CRYSTALLINE ROCKS

Rocks consisting wholly or chiefly of mineral crystals. Such rocks on the Moon usually formed by cooling from a liquid melt.

DIELECTRIC *dye-ee-lek-trik*

A material that is an electrical insulator. Most rocks are dielectrics.

DIURNAL *dye-err-nal*

Recurring daily. Diurnal processes on Earth repeat themselves every 24 hours but on the Moon repeat every 28 earth days. The length of a lunar day is 28 earth days.

DOPPLER TRACKING *dopp-lur*

A system for measuring the trajectory of spacecraft from Earth using continuous radio waves and the Doppler effect. An example of the Doppler effect is the change in pitch of a train's whistle and a car's horn on passing an observer. Because of this effect, the frequency of the radio waves received on Earth is changed slightly by the velocity of the spacecraft in exactly the same way that the pitch of a train's whistle is changed by the velocity of the train.

DOWN-SUN

In the direction of the solar vector and related to lunar surface photography.

EARTHSHINE

Illumination of the Moon's surface by sunlight reflected from the Earth. The intensity is many times smaller than that of the direct sunlight.

ECLIPTIC PLANE *e-klip-tik*

The plane defined by the Earth's orbit about the Sun.

EFFLUENT *eff-flu-ent*

Any liquid or gas discharged from a spacecraft such as waste water, urine, fuel cell purge products, etc. also any material discharged from volcanoes.

EGRESS *e-gress*

A verb meaning to exit or to leave. The popularization of this word has been attributed to the great showman, P. T. Barnum, who reportedly discovered that a sign marked exit had almost no effect on the large crowds that accumulated in his exhibit area but a sign marked "to egress" led the crowds outdoors. In space terminology it means simply to leave the spacecraft.

EJECTA *e-jek-tuh*

Lunar material thrown out (as resulting from meteoroid impact or volcanic action).

ELECTRON *e-lek-tron*

A small fundamental particle with a unit of negative electrical charge, a very small mass, and a very small diameter. Every atom contains one or more electrons. The proton is the corresponding elementary particle with a unit of positive charge and a mass of 1837 times as great as the mass of the electron.

FIELD

A region in which each point has a definite value such as a magnetic field.

FIELD OF VIEW

The region "seen" by the camera lens and recorded on the film.
The same phrase is applied to such other equipment as radar and radio antennas.

FILLET *fill-it*

Debris (soil) piled against a rock; several scientists have suggested that the volume of the fillet may be directly proportional to the time the rock has been in its present position and to the rock size.

FLUORESCENCE *flur-es-ence*

Emission of radiation at one wavelength in response to the absorption of energy at a different wavelength. Some lunar materials fluoresce. Most do not. The process is identical to that of the familiar fluorescent lamps.

FLUX

The rate of flow per unit area of some quantity such as the flux of cosmic rays or the flux of particles in the solar wind.

FRONT

The more or less linear outer slope of a mountain range that rises above a plain or plateau. In the U.S., the Colorado Front Range is a good example.

GALACTIC *ga-lah-tik*

Pertaining to a galaxy in the universe such as the Milky Way.

GAMMA

A measure of magnetic field strength; the Earth's magnetic field is about 50,000 gamma. The Moon's magnetic field is only a few gamma.

GAMMA RAY

One of the rays emitted by radioactive substances. Gamma rays are highly penetrating and can traverse several centimeters of lead.

GEGENSCHEIN *geg-en-schine*

A faint light covering a 20-degree field-of-view projected on the celestial sphere about the Sun-Earth vector (as viewed from the dark side of the Earth).

GEOCHEMICAL GROUP

A group of three experiments especially designed to study the chemical composition of the lunar surface remotely from lunar orbit.

GEODESY *ge-odd-eh-see*

Originally, the science of the exact size and shape of the Earth; recently broadened in meaning to include the Moon and other planets.

GEOPHYSICS *g-e-oh-phys-ics*   Physics of planetary bodies, such as the Earth and Moon, and the surrounding environment; the many branches include gravity, magnetism, heat flow, seismology, space physics, geodesy,meteorology, and sometimes geology.

GNOMON *know-mon*   A rod mounted on a tripod in such a way that it is free to swing in any direction and indicates the local vertical; it gives sun position and serves as a size scale. Color and reflectance scales are provided on the rod and a colorimetric reference is mounted on one leg.

GRADIENT *gray-dee-unt*   The rate of change of something with distance. Mathematically, it is the space rate of change of a function. For example, the slope of a mountain is the gradient of the elevation.

INGRESS *in-gress*   A verb meaning to enter. It is used in connection with entering the LM. See also "egress."

IN SITU *in-sit-u*   Literally, "in place", "in its original position". For example, taking photographs of a lunar surface rock sample "in situ" (as it lays on the surface).

LIMB   The outer edge of the apparent disk of a celestial body, as the Moon or Earth, or a portion of the edge.

MANTLE   An intermediate layer of the Moon between the outer layer and the central core.

MARE *maar*   A large dark flat area on the lunar surface (Lunar Sea). May be seen with the unaided eye.

MARIA *maar-ya*   Plural of mare.

MASCONS *mass-conz*   Large mass concentrations beneath the surface of the Moon. They were discovered only three years ago by changes induced by them in the precise orbits of spacecraft about the Moon.

MASS SPECTROMETER *mass spek-trom-a-tur*   An instrument which distinguishes chemical species in terms of their different isotopic masses.

METEORITE *me-te-oh-rite*   A solid body that has arrived on the Earth or Moon from outer space. It can range in size from microscopic to many tons. Its composition ranges from that of silicate rocks to metallic iron-nickel. For a thorough discussion see Meteorites by Brian Mason, John Wiley and Sons, 1962.

MICROSCOPIC   Of such a size as to be invisible to the unaided eye but readily visible through a microscope.

MINERALOGY   The science of minerals; deals with the study of their atomic structure and their general physical and chemical properties.

MONOPOLE *Mon-oh-pole*   All known magnets have two poles, one south pole and one north pole. The existence of a single such pole, termed a monopole, has not yet been established but is believed by many physicists to exist on the basis of theoretical studies. Lunar samples have been carefully searched on Earth for the presence of monopoles.

MORPHOLOGY *mor-fol-uh-ge*   The external shape of rocks in relation to the development of erosional forms or topographic features.

NADIR   That point on the Earth (or Moon) vertically below the observer.

OCCULTATION *ah-cull-tay-shun*   The disappearance of a body behind another body of larger apparent size. For example the occultation of the Sun by the Moon as viewed by an earth observer to create a solar eclipse.

OZONE *oh-zone*   Triatomic oxygen (O_3); found in significant quantities in the Earth's atmosphere.

PANORAMA *pan-uh-ram-a*   A series of photographs taken from a point to cover 360 degrees around that point.

PENUMBRAL *pe-num-bral*   Referring to the part of a shadow in which the light (or other rays such as the solar wind) is only partially masked, in contrast to the umbra in which light is completely masked, by the intervening object.

PETROGRAPHY *pe-trog-rah -fy*   Systematic description of rocks based on observations in the field, on hand specimens, and on microscopic examination.

PLASMA *plaz-muh*   An electrically conductive gas comprised of neutral particles, ionized particles and free electrons but which, when taken as a whole, is electrically neutral.

PRIMORDIAL *pry-mor-dee-uhl*   Pertaining to the earliest, or original, lunar rocks that were created during the time between the initial and final formation stages of the Moon.

PROTON *prow-ton*   The positively charged constituent of atomic nuclei. For example, the entire nucleus of a hydrogen atom having a mass of 1.67252×10^{-27} kilograms.

RAY   Bright material that extends radially from many craters on the Moon; believed to have been formed at the same time as the associated craters were formed by impacting objects from space; usually, but not always, arcs of great circles. They may be several hundred kilometers long.

REGOLITH *reg-oh-lith*   The unconsolidated residual material that resides on the solid surface of the Moon (or Earth).

RETROGRADE   Lunar orbital motion opposite the direction of lunar rotation.

RILLE/RILL   A long, narrow valley on the Moon's surface.

RIM   Elevated region around craters and rilles.

SAMPLE   Small quantities of lunar soil or rocks that are sufficiently small to return them to Earth. On each mission several different kinds of samples are collected. Contingency sample consists of 1 to 2 pounds of rocks and soil collected very early in the surface operations so that at least some material will have been returned to Earth in the event that the surface activities are halted abruptly and the mission aborted. Documented sample is one that is collected with a full set of photographs to allow positive identification of the sample when returned to Earth with the sample in situ together with a complete verbal description by the astronaut. Comprehensive sample is a documented sample collected over an area of a few yards square.

S-BAND   A range of frequencies used in radar and communications that extends from 1.55 to 0.2 kilomegahertz.

SCARP   A line of cliffs produced by faulting or erosion.

SEISMIC *size-mik*   Related to mechanical vibration within the Earth or Moon resulting from, for example, impact of meteoroids on the surface.

SOLAR WIND   Streams of particles (mostly hydrogen and helium) emanating from and flowing approximately radially outward from the Sun.

SPATIAL   Pertaining to the location of points in three-dimensional space; contrasted with temporal (pertaining to time) locations.

SPECTROMETER   An instrument which separates radiation into energy bands (or, in a mass spectrometer, particles into mass groups) and indicates the relative intensities in each band or group.

SPUR   A ridge of lesser elevation that extends laterally from a mountain or mountain range.

STELLAR	Of or pertaining to stars.
STEREO	A type of photography in which photographs taken of the same area from different angles are combined to produce visible features in three-dimensional relief.
SUPPLEMENTARY SAMPLE STOP	A stop added to a traverse after the stations are numbered. Mission planning continues through launch and the supplementary sample stops are inserted between normal traverse stations.
SUPRATHERMAL soup-rah-therm-al	Having energies greater than thermal energy.
SUBSATELLITE	A small unmanned satellite, deployed from the spacecraft while it is in orbit, designed to obtain various types of solar wind, lunar magnetic, and S-band tracking data over an extended period of time.
TALUS tail-us	Rock debris accumulated at the base of a cliff by erosion of material from higher elevation.
TEMPORAL	Referring to the passage or measurement of time.
TERMINATOR term-ugh-nay-tor	The lIne separating the illuminated and the darkened areas of a body such as the Earth or Moon which is not self-luminous.
TERRA terr - ugh	Those portions of the lunar surface other than the maria; the lighter areas of the Moon. They are visible to the unaided eye.
TIDAL	Referring to the very small movement of the surface of the Moon or the Earth due to the gravitational attraction of other planetary bodies. Similar to the oceanic tides, the solid parts of the Earth's crust rise and fall twice daily about three feet. Lunar tides are somewhat larger. The tides of solid bodies are not felt by people but are easily observed with instruments.
TIMELINE	A detailed schedule of astronaut or mission activities indicating the activity and time at which it occurs within the mission.
TOPOGRAPHIC top-oh-gra-fick	Pertaining to the accurate graphical description, usually on maps or charts, of the physical features of an area on the Earth or Moon.
TRANSEARTH	During transit from the Moon to the Earth.
TRANSIENT tran-she-unt	A short-lived, random event; often occurring in a system when first turned-on and before reaching operating equilibrium. For example, the initial current surge that occurs when an electrical system is energized.
TRANSLUNAR	During transit from the Earth to the Moon.
TRANSPONDER trans-pon-der	A combined receiver and transmitter whose function is to transmit signals automatically when triggered by a suitable radio signal.
UMBRA um-brah	The dark central portion of the shadow of a large body such as the Earth or Moon ; compare penumbra.
UP-SUN	Into the direction of the Sun and related to lunar surface photography.
URANIUM your-rain-nee-um	One of the heavy-metallic elements that are radioactive.
VECTOR	A quantity that requires both magnitude and direction for its specification, as velocity, magnetic force field and gravitational acceleration vectors.
WAVELENGTH	The distance between peaks (or minima) of waves such as ocean waves or electromagnetic waves.
X-RAY	Electromagnetic radiation of non-nuclear origin within the wavelength interval of 0.1 to 100 Angstroms (between gamma-ray- and ultra-violet radiation). X-rays are used in medicine to examine teeth, lungs, bones, and other parts of the human body; they also occur naturally.
ZODIACAL LIGHT zow-dye-uh-cal	A faint glow extending around the entire zodiac but showing most prominently in the neighborhood of the Sun. (It may be seen in the west after twilight and in the east before dawn as a diffuse glow. The glow may be sunlight reflected from a great number of particles of meteoritic size in or near the ecliptic in the planetoid belt).

Acronyms

ALSEP	Apollo lunar surface experiments package	LRV	lunar roving vehicle
ALSRC	Apollo lunar sample return container	LSM	lunar surface magnetometer
CCIG	cold cathode ion gauge	MIT	Massachusetts Institute of Technology
CM	command module	MSC	Manned Spacecraft Center
CSM	command and service module	NASA	National Aeronautics and Space Administration
DPS	descent propulsion system	PLSS	portable life support system
e.s.t.	eastern standard time	PSE	passive seismic experiment
EVA	extravehicular activity	RCS	reaction control system
HFE	heat flow experiment	RTG	radioisotope thermoelectric generator
g.e.t.	ground elapsed time	SESC	surface environment sample container
G.m.t.	Greenwich mean time	SIDE	suprathermal ion detector experiment
IR	infrared	S-IVB	Saturn IVB (rocket stage)
JPL	Jet Propulsion Laboratory	SME	soil mechanics experiment
LDD	lunar dust detector	SWC	solar wind composition experiment
LGE	lunar geology experiment	SWS	solar wind spectrometer experiment
LM	lunar module	TV	television
LRL	Lunar Receiving Laboratory, NASA Manned Spacecraft Center	USGS	U.S. Geological Survey
LRRR	laser ranging retro-reflector		

TABLE 1.-Timeline of Apollo 15 Mission Events*

Event	Time from liftoff (hr/min)	CDT/date
Launch		8:34 am July 26
Earth Orbit Insertion	00:12	8:46 am
Trans Lunar Injection	2:50	11:24 am
Lunar Orbit Insertion	78:31	3:05 pm July 29
Descent Orbit Insertion	82:40	7:14 pm
Spacecraft Separation	100:14	12:48 pm July 30
Lunar Landing	104:42	5:15 pm
Stand Up EVA	106:10	6:43 pm
EVA 1	119:50	8:24 am July 31
EVA 2	141:10	5:44 am August 1
EVA 3	161:50	2:24 am August 2
Lunar Liftoff	171:38	12:12 pm
Spacecraft Docking	173:30	2:04 pm
Trans Earth Injection	223:44	4:18 pm August 4
Trans Earth EVA	242:00	10:34 am August 5
Pacific Ocean Splashdown.	295:12	3:46 pm August 7
(26° N. Lat./ 158° W Long.)		

*These times are exact for launch on 26 July 71. They change somewhat for other launch dates.

TABLE 2.-LRV Exploration Traverse

[The entries in this table are brief. They are explained in the text and in the glossary. The table should be considered a general guide only; not every item is mandatory at each stop. The times are especially likely to change during the mission. The reader may wish to mark the actual times for himself on the table]

Station/activity	Elapsed time at start (hr:min)	Segment time (hr:min)	Geological features	Observations and activities
				EVA I
LM		1:25	Smooth mare	Observe LEI, prepare for departure from moon, contingency sample, deploy LRV
Travel	1:25	0:17	Across typical smooth mare material toward rim of Hadley Rille	Observe and describe traverse over smooth mare material Describe surface features and distribution of large boulders. Note any difference between mare and rille rim material
Check Point	1:42	0:02		
Travel	1:44	0:07	Around Elbow Crater	Observe low ridge around Elbow Crater Observe any differences between rille rim material and mare material
I	1:51	0:15	Southern part of Elbow Crater ejecta blanket	Observe distribution of ejecta around Elbow Crater Radial sampling of rocks at Elbow Crater Panoramic photography
Travel	2:06	0:08	To Apennine Front slope north of St. George Crater	Look for changes in rocks or ground that indicate the base of the mountain Compare the material of the Front with mare and rille rim material Observe character and distribution of St. George ejecta blanket
2	2:14	0:45	Near base of Apennine Front north of St. George Crater	Radial sampling of rocks at St. George Crater Comprehensive sample in area at Front Double drive core tube 500mm lens camera photography of blocks on rim of St. George and of rille Stereo pan from high point Fill SESC at Apennine Front Penetrometer
Travel	2:59	0:09	Across base of Apennine Front to edge of possible debris flow	Observe Apennine material and its relation to mare surface
Area Stop 3	3:08	0:14	At base of Apennine Front adjacent to possible debris flow	Examine flow and compare with mare and Front Documented samples of Apennine Front and flow material Observe and describe vertical and lateral changes in Apennine Front: compare with previous stop Panoramic photography Observe characteristics of EVA II route
Travel	3:22	0:28	From base of Apennine Front across mare to LM	Observe characteristics and extent of possible debris flow Observe area to be traversed on EVA II Compare mare material with Apennine Front and rille rim Observe possible ray material
LM	3:50	3:10	Smooth mare	ALSEP deployment-see Table 3 for details Store samples and records Ingress LM
				EVA II
LM		0:49	Smooth mare	Egress LM, prepare for traverse
Travel	0:49	0:11	South along smooth mare SW of secondary crater cluster to base of Apennine Front	Observe smooth mare characteristics Observe secondary- crater cluster characteristics Traverse along Apennine Front; determine position of base of Front and search for optimum sampling areas for stops on return leg of traverse Photography as appropriate
Check Point	1:00	0:02		
Travel	1:02	0:15		
4	1:17	0:20	East along Apennine Front Secondary crater cluster south of 400m crater	Same as above Soil/rake sample Documented samples Panoramic photography 500mm photography of Apennine Front Exploratory trench Possibly drive core tube through secondary ejecta Observe crater interior and ejecta Sample both typical and exotic rock types Compare secondary crater material with other geologic units at the site
Travel	1:37	0:10	South along smooth mare SW at secondary crater cluster to base of Apennine Front	Observe smooth mare characteristics Observe secondary crater cluster characteristics and crater forms Photography as appropriate

Check Point	1:47	0:04		
Travel	1:51	0:10	East along Apennine Front	Traverse along Apennine Front; determine position of base of Front and search for optimum sampling areas for stops on return leg of traverse Photography as appropriate Observe possible debris flows, downslope movement; look for source
Check Point	2:01	0:04	Apennine Front	Same as above
Check Point	2:10	0:04		
Travel	2:14	0:12	Along Apennine Front to area stop 5	Same as above
Area Stop 5	2:26	0:51	At base of Apennine Front near rim of Front Crater	Documented samples from upslope side of Front Crater in Apennine Front Documented samples from northern rim of Front Crater; particularly at sharp 80-m crater on rim Stereo pan Exploratory trench upslope of Front Crater 500mm photography of any interesting targets Stereo pairs upslope of any interesting targets
Travel	3:19	0:15	Along base of Apennine Front	Observe lateral variations in material and surface textures Search for blocky areas along Apennine Front which are suitable for sampling (craters, etc.) Photography as appropriate
6	3:32	0:44	Along base of Apennine Front on slope in intercrater areas or on crater rims; chosen at crew's discretion, based on previous observations	Include the following activities which should be modified according to the local geology: Description of Apennine Front in sampling area Comparison of Apennine Front and of the material there with other surface units Documented samples of Apennine Front material Panoramic photography Exploratory trench Possible drive core tube 500mm photography Stereo pairs of interesting features upslope
Travel	4:160	0:08	Along base of Apennine Front	Observe lateral variations in material and surface textures Search for blocky areas along Apennine Front which are suitable for sampling (craters, etc.) Photography- as appropriate
7	4:24	0:44	Along base of Apennine Front on slope in intercrater areas or on crater rims; chosen at crew's discretion, based on previous observations	Observe lateral variations in material and surface textures Search for blocky areas along Apennine Front which are suitable for sampling (craters, etc.) Photography as appropriate At the last Apennine Front stop, based on previous observations along Front, crew uses discretion to complete sampling
Travel	5:08	0:22	From base of Apennine Front along southwestern edge of secondary crater cluster	Observe secondary crater deposits and relation to other terrain Observe eastern edge of possible debris flow from Apennine Front Photography as appropriate
8	5:30	0:37	Mare material near crater	Comprehensive sample area Documented sampling of large mare crater Possible fillet/rock sample Possible large and small equidimensional rock samples Panoramic photography Trench Possible buried rock sample Fill SESC Penetrometer
Travel	6:07	0:13	Across smooth mare	Compare mare material with other lunar material Observe possible ray material
LM	6:20	0:40	Smooth mare	Store samples and records Ingress LM
			EVA III	
LM		0:42	Smooth mare	Egress LM, prepare for traverse
Travel	0:42	0:07	Across smooth mare between LM and rim of Hadley Rille	Compare smooth mare material with rille rim material
Supplementary Sample Stop	0:49	0:05	Smooth mare between LM and rim of Hadley Rille	Soil/rock sample Panoramic photography
Travel	0:54	0:12	Across smooth mare to rille rim turning N W at rille rim to the Terrace	Compare smooth mare material to rille rim material
9	1:06	0:50	At rim of Hadley Rille at southern end of the Terrace	Observe and describe rille and far wall 500 mm lens camera photography Comprehensive sample Single or double drive core tube Panoramic photography Documented sampling of crater at edge of rille Possible pan on edge of crater Penetrometer
Travel	1:56	0:03	Along rille rim at the Terrace	Continued description of rille and rim material Photography- as appropriate
10	1:59	0:10	Along rille rim at the Terrace	500mm lens camera panoramic photography provides stereo base for station 9; same targets should be photographed Documented sample from crater on rille rim Panoramic photography
Travel	2:09	0:06	Along rille rim to north end of the Terrace	Continued description of rille and rille rim material Photography as appropriate
1	2:15	0:10	At rim of Hadley Rille at N W end of the Terrace	Observe and describe rille and far rille wall; compare with previous observations 500mm lens camera photography Documented samples of rille rim and crater at edge of rille Panoramic photography Compare rille rim material with other terrain
Travel	2:34	0:07	From rille rim and traverse across mare toward North Complex	Observe changes in material from rille rim to mare to North Complex
Supplementary Sample Stop	2:41	0:05	Between rille rim and North Complex	Soil/rock sample Panoramic photography
Travel	2:46	0:12	Toward Chain Crater in the North Complex	Observe changes in material from rille rim to mare to North Complex Observe characteristics of crater chain originating in Chain Crater Observe possible secondary craters
1	22:58	0:23	Southeastern rim of Chain Crater in North Complex at junction with elongate depression	Documented sample of crater ejecta Documented sample of North Complex material Panoramic photography Possible drive core tube Describe wall of crater and its relation to elongate depression Attempt to determine whether crater was caused by impact
Travel	3:21	0:08	Between large craters in North Complex	Observe area between craters in North Complex and compare ejecta with other materials at the site Continue to compare North Complex with other terrain types
13	3:29	0:53	Multiple objective stop at end of North Complex between Chain Crater and 700-m crater	The more interesting features in the North Complex are the following: 160-m crater on western rim of the 700-m crater 700-m crater Eaglecrest Crater Scarps Based on the characteristics and accessibility of each of these features, the following tasks should be completed at the discretion of the crew: Documented sampling Panoramic or stereo panoramic photography Possible drive core tube Exploratory trench Soil sample Targets for 500mm photography Penetrometer
Travel	4:22	0:19	From North Complex into the mare region with possible secondaries from ray	Observe and describe differences in material and surface textures between North Complex and mare Note amount of secondary cratering Photography as appropriate
14	4:41	0:20	180-m crater in mare south of North Complex	Compare blocks and mare material with North Complex Documented sample of mare material Possible fillet/rock sample Possible large and small equidimensional rock samples Possible radial sampling of fresh 5- 10m. crater Panoramic photography Exploratory trench in ray material
Travel	5:01	0:15	Across mare between North Complex and LM	Describe differences between this area and other mare areas Note distribution of possible secondaries
LM	5:16	0:44	Smooth mare fill	Store samples and records Ingress LM

TABLE 3.-Summary Timeline for ALSEP Deployment

Approximate Time at start of activity (minutes)	Commander's activity	LM pilot's activity
0		Both remove ALSEP from LM and stow it on ROVER
10		Both remove ALSEP from LM and stow it on ROVER
20	Drive Rover to ALSEP site	Walk to ALSEP site
30	Heat Flow Experiment Removes equipment from Rover & sets up on moon.	Make electrical connections to ALSEP
40	HFE-Continues to deploy equipment	Deploy Passive Seismic Experiment (PSE)
50	HFE-Assemble drill	Solar Wind Experiment Lunar Surface Magnetometer (LSM)
60	HFE-Drill first hole, place probes in first hole	LSM
70	HFE (drill second hole)	Install sunshield Install ALSEP antenna
80	HFE-Place probes in second hole	ALSEP antenna SIDE/CCIG
90	Drill core sampling	Activate ALSEP Central Station
100	Drill core sampling	LRRR Photos of ALSEP
110	Drill core sampling	Photos of ALSEP

TABLE 4.-Principal Investigators for the Apollo 15 Lunar Surface Scientific Experiments

Experiment	Principal investigator	Institution
Passive Seismic	Dr. Gary V. Latham	Lamont-Doherty Geological Observatory Columbia University Palisades, New York 10964
Lunar Surface Magnetometer	Dr. Palmer Dyal	Space Science Division NASA Ames Research Center Moffett Field, California 94034
Solar Wind Spectrometer	Dr. Conway W. Snyder	Jet Propulsion Laboratory Pasadena, California 91103
Suprathermal Ion Detector	Dr. John W. Freeman	Department of Space Science Rice University Houston, Texas 77001
Heat Flow	Dr. Marcus E. Langseth	Lamont-Doherty Geological Observatory Columbia University Palisades, New York 10964
Cold Cathode Ion Gauge	Dr. Francis S. Johnson	University of Texas at Dallas Dallas, Texas 75230
Lunar Geology Experiment	Dr. Gordon A. Swann	Center of Astrogeology U.S. Geological Survey Flagstaff, Arizona 86001
Laser Ranging Retro-Reflector	Dr. James E. Faller	Wesleyan University, Middletown, Connecticut 06457
Solar Wind Composition	Dr. Johannes Geiss	University of Berne, Berne, Switzerland
Soil Mechanics	Dr. James K. Mitchell	Department of Civil Engineering, University of California, Berkeley, California 94726
Lunar Dust Detector	Mr. James R. Bates	Science Missions Support Division, Manned Spacecraft Center, Houston, Texas 77058

Report No. M-933-71-15

MISSION OPERATION REPORT

APOLLO 15 MISSION

OFFICE OF MANNED SPACE FLIGHT

Prelaunch Mission Operation Report

No. M-933-71-15

MEMORANDUM   17 July 1971

TO: A/Administrator

FROM: MA/Apollo Program Director

SUBJECT: Apollo 15 Mission (AS-510)

We plan to launch Apollo 15 from Pad A of Launch Complex 39 at the Kennedy Space Center no earlier than July 26, 1971. This will be the fourth manned lunar landing and the first of the Apollo "J" series missions which carry the Lunar Roving Vehicle for surface mobility, added Lunar Module consumables for a longer surface stay time, and the Scientific Instrument Module for extensive lunar orbital science investigations.

Primary objectives of this mission are selenological inspection, survey, and sampling of materials and surface features in a pre-selected area of the Hadley-Apennine region of the moon; emplacement and activation of surface experiments; evaluation of the capability of Apollo equipment to provide extended lunar surface stay time, increased EVA operations, and surface mobility; and the conduct of in-flight experiments and photographic tasks. In addition to the standard photographic documentation of operational and scientific activities, television coverage is planned for selected periods in the spacecraft and on the lunar surface. The lunar surface TV coverage will include remote controlled viewing of astronaut activities at each major science station on the three EVA traverses and the eclipse of the sun by the earth on August 6, 1971.

The 12-day mission will be terminated with the Command Module landing in the Pacific Ocean near Hawaii. Recovery and transportation of the crew and lunar samples to the Manned Spacecraft Center will be without the quarantine procedures previously employed.

Rocco A. Petrone

APPROVAL:

Dale D. Myers
Associate Administrator for Manned Space Flight

FOREWORD

MISSION OPERATION REPORTS are published expressly for the use of NASA Senior Management, as required by the Administrator in NASA Management Instruction HQMI 8610. I, effective 30 April 1971. The purpose of these reports is to provide NASA Senior Management with timely, complete, and definitive information on flight-mission plans, and to establish official Mission Objectives which provide the basis for assessment of mission accomplishment.

Prelaunch reports are prepared and issued for each flight project just prior to launch. Following launch, updating (Post Launch) reports for each mission are issued to keep General Management currently informed of definitive mission results as provided in NASA Management Instruction HQMI 8610.I.

Primary distribution of these reports is intended for personnel having program/project management responsibilities which sometimes results in a highly technical orientation. The Office of Public Affairs publishes a comprehensive series of reports on NASA flight missions which are available for dissemination to the Press.

APOLLO MISSION OPERATION REPORTS are published in two volumes: the MISSION OPERATION REPORT (MOR); and the MISSION OPERATION REPORT, APOLLO SUPPLEMENT. This format was designed to provide a mission-oriented document in the MOR, with supporting equipment and facility description in the MOR, APOLLO SUPPLEMENT. The MOR, APOLLO SUPPLEMENT is a program-oriented reference document with a broad technical description of the space vehicle and associated equipment, the launch complex, and mission control and support facilities.

Published and Distributed by PROGRAM and SPECIAL REPORTS DIVISION (XP) EXECUTIVE SECRETARIAT - NASA HEADQUARTERS

SUMMARY APOLLO/SATURN FLIGHTS

Mission	Launch Date	Launch Vehicle	Payload	Description
AS-201	2/26/66	SA-201	CSM-009	Launch vehicle and CSM development. Test of CSM subsystems and of the space vehicle. Demonstration of reentry adequacy of the CM at earth orbital conditions.
AS-203	7/5/66	SA-203	LH$_2$ in S-IVB	Launch vehicle development. Demonstration of control of LH$_2$ by continuous venting in orbit.
AS-202	8/25/66	SA-202	CSM-011	Launch vehicle and CSM development. Test of CSM subsystems and of the structural integrity and compatibility of the space vehicle. Demonstration of propulsion and entry control by G&N system. Demonstration of entry at 28,500 fps.
APOLLO 4	11/9/67	SA-501	CSM-017 LTA-10R	Launch vehicle and spacecraft development. Demonstration of Saturn V Launch Vehicle performance and of CM entry at lunar return velocity.
APOLLO 5	1/22/68	SA-204	LM-1 SLA-7	LM development. Verified operation of LM subsystems: ascent and descent propulsion systems (including restart) and structures. Evaluation of LM staging. Evaluation of S-IVB/IU orbital performance.
APOLLO 6	4/4/68	SA-502	CM-020 SM-014 LTA-2R SLA-9	Launch vehicle and spacecraft development. Demonstration of Saturn V Launch Vehicle performance.
APOLLO 7	10/11/68	SA-205	CM-101 SM-101 SLA-5	Manned CSM operations. Duration 10 days 20 hours.

APOLLO 8	12/21/68	SA-503	CM-103 SM-103 LTA-B SLA-11	Lunar orbital mission. Ten lunar orbits. Mission duration 6 days 3 hours. Manned CSM operations.
APOLLO 9	3/3/69	SA-504	CM-104 SM-104 LM-3 SLA-12	Earth orbital mission. Manned CSM/LM operations. Duration 10 days 1 hour.
APOLLO 10	5/18/69	SA-505	CM-106 SM-106 LM-4 SLA-13	Lunar orbital mission. Manned CSM/LM operations. Evaluation of LM performance in cislunar and lunar environment, following lunar landing profile. Mission duration 8 days.
APOLLO 11	7/16/69	SA-506	CM-107 SM-107 LM-5 SLA-14	First manned lunar landing mission. Lunar surface stay time 21.6 hours. One dual EVA (5 man hours). Mission duration 8 days 3 hours.
APOLLO 12	11/14/69	SA-507	CM-108 SM-108 LM-6 SLA-15	Second manned lunar landing mission. Demonstration of point landing capability. Deployment of ALSEP I. Surveyor III investigation. Lunar surface stay time 31.5 hours. Two dual EVA's (15.5 manhours). Mission duration 10 days 4.6 hours.
APOLLO 13	4/11/70	SA-508	CM-109 SM-109 LM-7 SLA-16	Planned third lunar landing. Mission aborted at approximately 56 hours due to loss of SM cryogenic oxygen and consequent loss of capability to generate electrical power and water.
APOLLO 14	1/13/71	SA-509	CM-110 SM-110 LM-8 SLA-17	Third manned lunar landing mission. Selenological inspection, survey and sampling of materials of Fra Maura Formation. Deployment of ALSEP. Lunar Surface Staytime 33.5 hours. Two dual EVA's (18.8 man hours). Mission duration 9 days.

NASA OMSF MISSION OBJECTIVES FOR APOLLO 15
PRIMARY OBJECTIVES

Perform selenological inspection, survey, and sampling of materials and surface features in a preselected area of the Hadley-Apennine region.

Emplace and activate surface experiments.

Evaluate the capability of the Apollo equipment to provide extended lunar surface stay time, increased EVA operations, and surface mobility.

Conduct in-flight experiments and photographic tasks from lunar orbit.

Rocco A. Petrone
Apollo Program Director

Dale D. Myers
Associate Administrator for Manned Space Flight

Date: 16 July 1971

MISSION OPERATIONS

GENERAL

The following paragraphs contain a brief description of the nominal launch, flight, recovery, and post-recovery operations. For the third month launch opportunity, which may involve a T-24 hour launch, there will be a second flight plan. Overall mission profile is shown in Figure 1.

LAUNCH WINDOWS

The mission planning considerations for the launch phase of a lunar mission are, to a major extent, related to launch windows. Launch windows are defined for two different time periods: a "daily window" has a duration of a few hours during a given 24-hour period; a "monthly window" consists of a day or days which meet the mission operational constraints during a given month or lunar cycle.

Launch windows will be based on flight azimuth limits of 80° to 100° (earth-fixed heading of the launch vehicle at end of the roll program), on booster and spacecraft performance, on insertion tracking, and on lighting constraints for the lunar landing sites.

The Apollo 15 launch windows and associated lunar landing sun elevation angles are presented in Table 1.

TABLE 1
LAUNCH WINDOWS

LAUNCH DATE	WINDOWS (EST) OPEN	CLOSE	SUN ELEVATION ANGLE
July 26, 1971	0934	1211	12.0°
July 27, 1971	0937	1214	23.2°
August 24, 1971	0759	1038	11.3°
August 25, 1971	0817	1055	22.5°
September 22, 1971	0637	0917	12.0°
September 23, 1971	0720	1000	12.0°
September 24, 1971	0833	1112	23.0°

LAUNCH THROUGH TRANSLUNAR INJECTION

The space vehicle will be launched from Pad A of launch complex 39 at the Kennedy Space Center. The boost into a 90-NM earth parking orbit (EPO) will be accomplished by sequential burns and staging of the S-IC and S-II launch vehicle stages and a partial burn of the S-IVB stage. The S-IVB/IU and spacecraft will coast in a circular EPO for approximately 1.5 revolutions while preparing for the first opportunity S-IVB translunar injection (TLI) burn, or 2.5 revolutions if the second opportunity TLI burn is required. Both injection opportunities are to occur over the Pacific Ocean. The S-IVB TLI burn will place the S-IVB /IU and spacecraft on a translunar trajectory targeted such that transearth return to an acceptable entry corridor can be achieved with the use of the Reaction Control System (RCS) during at least five hours (7 hrs: 57 min. Ground Elapsed Time (GET)) after TLI cutoff. For this mission the RCS capability will actually exist up to about 59 hours GET for the CSM/LM combination and about 67 hours GET for the CSM only. TLI targeting will permit an acceptable earth return to be achieved using SPS or LM DPS until at least pericynthion plus two hours, if Lunar Orbit Insertion (LOI) is not performed. For this mission however, the LM OPS requirement can be met until about 20 hours after LOI.

TRANSLUNAR COAST THROUGH LUNAR ORBIT INSERTION

Within two hours after injection the Command Service Module (CSM) will separate from the S-IVB/IU and

APOLLO 15 FLIGHT PROFILE

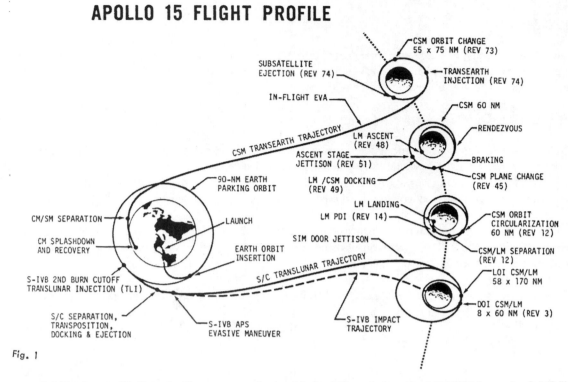

Fig. 1

spacecraft-LM adapter (SLA) and will transpose, dock with the LM, and eject the LM/CSM from the S-IVB/IU. Subsequently, the S-IVB/IU will perform an evasive maneuver to alter its circumlunar coast trajectory clear of the spacecraft trajectory.

The spent S-IVB/IU will be impacted on the lunar surface at 3° 39'S. and 7° 34.8'W. providing a stimulus for the Apollo 13 and 14 emplaced seismology experiments. The necessary delta velocity (Delta V) required to alter the S-IVB/IU circumlunar trajectory to the desired impact trajectory will be derived from dumping of residual LOX and burn(s) of the S-IVB/APS and ullage motors. The final maneuver will occur within about nine hours of liftoff. The IU will have an S-Band transponder for trajectory tracking. A frequency bias will be incorporated to insure against interference between the S-IVB/IU and LM communications during translunar coast.

Spacecraft passive thermal control will be initiated after the first midcourse correction (MCC) opportunity and will be maintained throughout the translunar-coast phase unless interrupted by subsequent MCC's and/or navigational activities. The scientific instrument module (SIM) bay door will be jettisoned shortly after the MCC-4 point, about 4.5 hours before lunar orbit insertion.

Multiple-operation covers over the SIM bay experiments and cameras will provide thermal and contamination protection whenever they are not in use.

A retrograde SPS burn will be used for lunar orbit insertion (LOI) of the docked spacecraft into a 58 X 170-NM orbit, where they will remain for approximately two revolutions.

DESCENT ORBIT INSERTION THROUGH LANDING

The descent orbit insertion (DOI) maneuver, a SPS second retrograde burn, will place the CSM/LM combination into a 60 x 8-NM orbit.

A "soft" undocking will be made during the 12th revolution, using the docking probe capture latches to reduce the imported Delta V. Spacecraft separation will be executed by the service module (SM) reaction

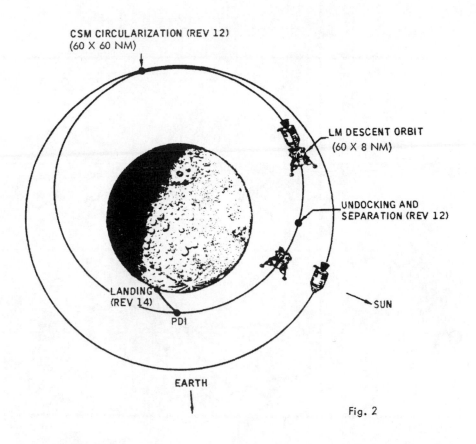

CSM CIRCULARIZATION (REV 12)
(60 X 60 NM)

LM DESCENT ORBIT
(60 X 8 NM)

UNDOCKING AND
SEPARATION (REV 12)

LANDING
(REV 14)

PDI

SUN

EARTH

Fig. 2

control system (RCS), providing Delta V of approximately 1 foot per second radially downward toward the center of the moon. The CSM will circularize its orbit to 60 NM at the end of the 12th revolution. During the 14th revolution the LM DPS will be used for powered descent, which will begin approximately at pericynthion. These events are shown in Figure 2. A lurain profile model will be available in the LM guidance computer (LGC) program to minimize unnecessary LM pitching or thrusting maneuvers. A steepened descent path of 25° will be used during the terminal portion of powered descent (from high gate) to enhance landing site visibility. The vertical descent portion of the landing phase will start at an altitude of about 200 feet at a rate of 5 feet per second, and will be terminated at touchdown on the lunar surface.

LANDING SITE (HADLEY-APENNINE REGION)

The Apennine Mountains constitute the southeastern boundary of Mare Imbrium, forming one side of a triangle-shaped, elevated highland region between Mare Imbrium, Mare Serenitatis, and Mare Vaporum. In the area of the landing site, the mountains rise up to 2.5km above the adjacent mare level.

Rima Hadley is a V-shaped lunar sinuous rille which parallels the western boundary of the Apennine Mountain front. The rille originates in an elongate depression in an area of possible volcanic domes and generally maintains a width of about 1.5km and a depth of 400 meters until it merges with a second rille approximately 100 km to the north. The origin of sinuous rilles such as Rima Hadley may be due to some type of fluid flow.

Sampling of the Apenninian material should provide very ancient rocks whose origin predates the formation and filling of the major mare basins. Examination and sampling of the rim of the Hadley Rille and associated deposits are expected to yield information on the genesis of it and other sinuous rilles. If the exposures in the rille are bedded, they will provide an excellent stratigraphic section of Imbrian material.

The planned landing point coordinates are 26°04'54"N, 3°39'30"E (Figure 3).

APOLLO 15 LANDING SITE

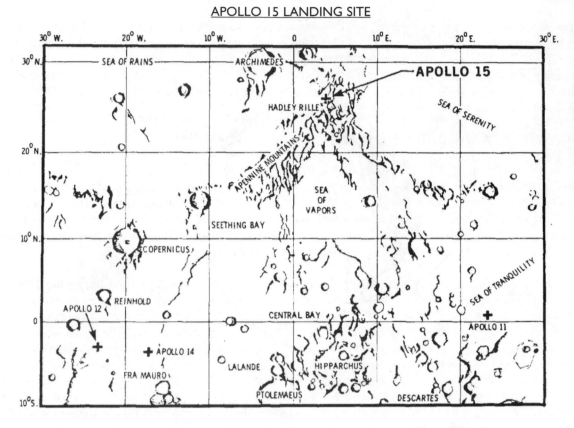

Fig. 3

LUNAR SURFACE OPERATIONS

The maximum stay time on the lunar surface is approximately 67 hours which is about double that of Apollo 14 and is a result of the addition of life support consumables in LM-10. A standup EVA (SEVA) will be performed about 1½ hours after landing with the Commander (CDR) positioned with his head above the opened upper hatch for surveying the lunar surface. The SEVA will be followed by rest-work periods which provide for 3 traverse EVA's of 7-7-6 hours respectively. The LM crew will remove their suits for each rest period and will sleep in hammocks mounted in the LM cabin.

This mission will employ the Lunar Roving Vehicle (LRV) which will carry both astronauts, experiment equipment, and independent communications systems for direct contact with the earth when out of the line-of-sight of the LM relay system. Voice communication will be continuous and color TV coverage will be provided at each major science stop (see Figure 4) where the crew will align the high gain antenna. The ground controllers will then assume control of the TV through the ground controlled television assembly (GCTA) mounted on the LRV. A TV panorama is planned at each major science stop, followed by coverage of the astronauts scientific activities.

The radius of crew operations will be constrained by the LRV capability to return the crew to the LM in the event of a Portable Life Support System (PLSS) failure or by the PLSS walkback capability in the event of an LRV failure, whichever is the most limiting at any point in the EVA. If a walking traverse must be performed, the radius of operations will be constrained by the buddy secondary life support system (BSLSS) capability to return the crew to the LM in the event of a PLSS failure.

EVA PERIODS

Approximately 1½ hours after landing the CDR will perform a 30 minute SEVA. He will stand in the LM with his head above the hatch opening to observe the lunar geographical features and photograph the surrounding area. The SEVA will assist the crew in traverse planning and in selecting a site for Apollo Lunar Surface Experiment Package (ALSEP) deployment. The crew will rest after the SEVA and before the first traverse EVA. The 3 traverses planned for Apollo 15 are designed with flexibility for selection of science stops as indicated by the shaded areas on the traverse map (Figure4).

First Eva Period

The first EVA (up to 7 hours duration) will include the following: contingency sample collection, LM inspection, LRV deployment and loading, performance of a geology traverse using the LRV, deployment and activation of the ALSEP, deployment of the laser ranging retro-reflector, and deep core sample drilling. The TV camera will be mounted on a tripod to the west of the LM early in the EVA for observation of crew activities (including LRV deployment) in the vicinity of the LM (Figure 5). The geology traverse will follow as nearly as possible the planned route shown for EVA-1 in Figure 4.

The data acquisition camera and Hasselblad cameras, using color film, will be used during the EVA to record lunar surface operations. The lunar communications relay unit (LCRU) and the ground commanded television assembly (GCTA) will be used in conjunction with LRV operations. Lunar surface samples will be documented by photography and voice description. High resolution photographic survey of rille structure and other surface features will be accomplished with the Hasselblad camera equipped with the 500 mm lens. If time does not permit filling the sample return container (SRC) with documented samples, the crew may fill the SRC with samples selected for scientific interest. Following the traverse, the crew will deploy and activate the ALSEP to the west of the LM landing point as shown in Figure 6. If time does not permit completion of all ALSEP tasks, they will be rescheduled for appropriate times in subsequent EVA's. The planned timeline for all EVA-1 activities is presented in Figure 7.

Second and Third EVA Periods

The second and third EVA's (7 and 6 hours duration respectively) will continue the extensive scientific investigation of the Hadley-Apennine region and further operational assessment of the new and expanded capability of the Apollo hardware and systems. LRV sorties are planned for exploration of the Apennine front, Hadley Rille, and other prominent features along the traverse routes as shown in Figure 4.

The major portion of the lunar geology investigation (S-059) and the soil mechanics experiment (S-200) will be conducted during the second and third EVA's and will include voice and photographic documentation of sample material as it is collected and descriptions of lurain features. The solar wind composition (S-080) will be concluded prior to termination of the third EVA and will be returned for postflight analysis. The LRV will be positioned at the end of the EVA-3 traverse to enable remote controlled color TV coverage of LM ascent, a solar eclipse on August 6, and other observations of scientific interest. The planned timelines for EVA-2 and EVA-3 activities are presented in Figures 8 and 9 respectively. Following EVA-3 closeout the crew will make preparations for ascent and rendezvous.

EVA TRAVERSES

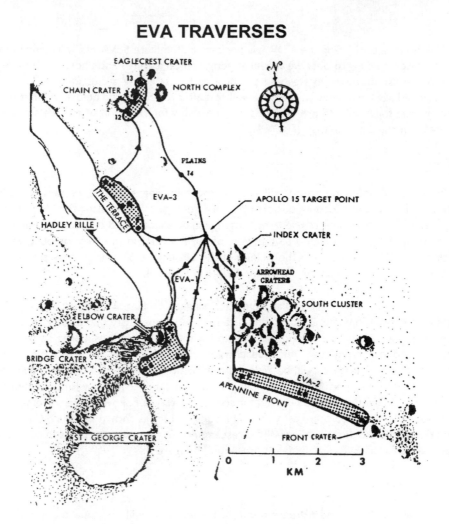

Fig. 4

NEAR LM LUNAR SURFACE ACTIVITIES

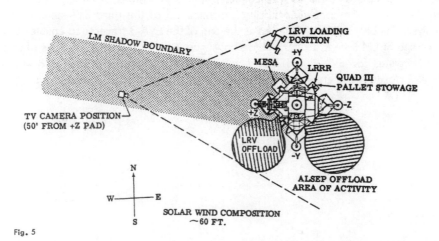

Fig. 5

ALSEP ARRAY LAYOUT

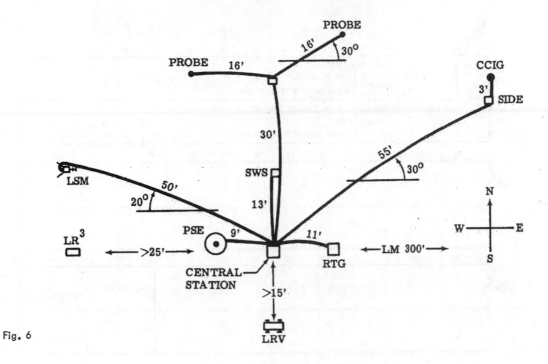

Fig. 6

EVA-1 TIMELINE

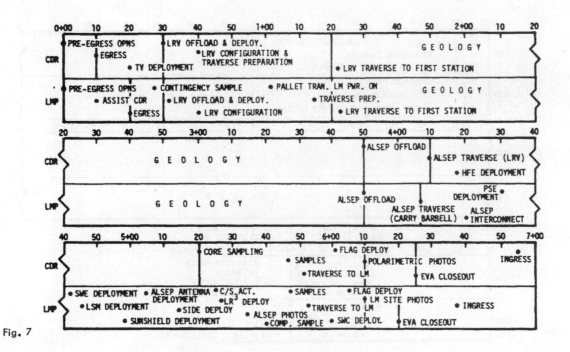

Fig. 7

EVA-2 TIMELINE

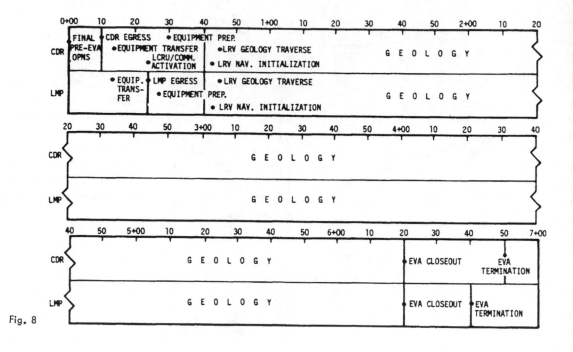

Fig. 8

EVA-3 TIMELINE

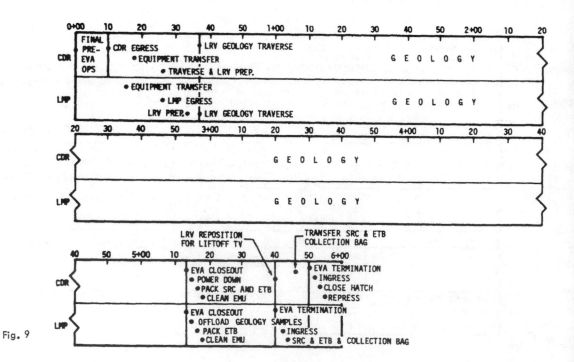

Fig. 9

LUNAR ORBIT OPERATIONS

GENERAL

The Apollo 15 Mission is the first with the modified Block II CSM configuration. An increase in cryogenic storage provides increased mission duration for the performance of both an extended lunar surface stay time and a lunar orbit science period. The new scientific instrument module (SIM) in the SM provides for the mounting of scientific experiments and for their operation in flight.

After the SIM door is jettisoned by pyrotechnic charges and until completion of lunar orbital science tasks, selected RCS thrusters will be inhibited or experiment protective covers will be closed to minimize contamination of experiment sensors during necessary RCS burns. Attitude changes for thermal control and experiment alignment with the lunar surface and deep space (and away from direct sunlight) will be made with the active RCS thrusters. Orbital science activities have been planned at appropriate times throughout the lunar phase of the mission and consist of the operation of 5 cameras (35mm Nikon, 16 mm Data Acquisition, 70 mm Hasselblad, 24 inch Panoramic and a 3 inch Mapping), a color TV camera, a laser altimeter, a gamma ray spectrometer, X-ray flourescent equipment, alpha ray particle equipment and mass spectrometer equipment.

Pre-Rendezvous Lunar Orbit Science

Orbital science operations will be conducted during the 60 x 8 NM orbits after DOI, while in the docked configuration. Orbital science operations will be stopped for the separation and circularization maneuvers performed during the 12th revolution, then restarted after CSM circularization.

The experiments timeline has been developed in conjunction with the surface timeline to provide, as nearly as possible, 16 hour work days and concurrent 8 hour CSM and LM crew sleep periods. Experiment activation cycles are designed to have minimum impact on crew work-rest cycles.

About 8 hours before rendezvous, the CSM will perform a plane change maneuver to provide the desired 60 x 60 NM coplanar orbit at the time of the LM rendezvous.

LM Ascent, Rendezvous and Jettison

After completion of lunar surface activities and ascent preparations, the LM ascent propulsion system (APS) and LM RCS will be used to launch and rendezvous with the CSM. Prior to LM liftoff, the CSM will complete the required plane change to permit a nominally coplanar rendezvous.

The direct ascent rendezvous technique initiated on Apollo 14 will be performed instead of the coelliptic rendezvous technique used on early landing missions. The lift-off window duration is about 10 seconds and is constrained to keep the perilune above 8 NM. The LM will be inserted into a 46 x 9 NM orbit so that an APS terminal phase initiation (TPI) burn can be performed approximately 45 minutes after insertion. The final braking maneuver will occur about 46 minutes later. The total time from LM liftoff to the final breaking maneuver will be about 99 minutes.

Docking will be accomplished by the CSM with RCS maneuvers. Once docked, the two LM crewmen will transfer to the CSM with lunar sample material, exposed films, and designated equipment.

The LM ascent stage will be jettisoned and subsequently deorbited to impact on the lunar surface, to provide a known stimulus for the emplaced seismic experiment. The impact will be targeted for 26° 15'N. and 1° 45'E.

Post-Rendezvous Lunar Orbit Science

After rendezvous and LM ascent stage jettison, additional scientific data will be obtained by the CSM over a two-day period. Conduct of the SIM experiments and both SM and CM photographic tasks will take advantage of the extended ground track coverage during this period.

During the second revolution before transearth injection, the CSM will perform an SPS maneuver to achieve a 55 x 75 NM orbit. Shortly thereafter, the subsatellite carried in the SIM bay will be launched northward, normal to the ecliptic plane. It is anticipated to have a lifetime of approximately 1 year.

TRANSEARTH INJECTION THROUGH LANDING

After completion of the post-rendezvous CSM orbital activities, the SPS will perform a posigrade burn to inject the CSM onto the transearth trajectory. The nominal return time will be 71.2 hours with a return inclination of 40° relative to the earth's equator.

During the transearth coast phase there will be continuous communications coverage from the time the spacecraft appears from behind the moon until shortly prior to entry. Midcourse corrections will be made, if required. A six-hour period has been allocated for the conduct of an inflight EVA, including pre- and post-EVA activities, to retrieve film cassettes from the SIM in the SM. TV, an inflight demonstration, and photographic tasks (including the solar eclipse on August 6, 1971) will be performed as scheduled in the flight plan. SIM experiments will be continued during transearth coast.

The CM will separate from the SM 15 minutes before entry interface. Earth touchdown will be in the mid-Pacific at about 295:12 GET, 12.3 days after launch. The nominal landing coordinates are 26° 07'N. and 158°W approximately 300 miles north of Hawaii. The prime recovery ship is the USS Okinawa.

POST-LANDING OPERATIONS

Flight Crew Recovery

Following splashdown, the recovery helicopter will drop swimmers and life rafts near the CM. The swimmers will install the flotation collar on the CM, attach the life raft, and pass fresh flight suits in through the hatch for the flight crew to don before leaving the CM. The crew will be transferred from the spacecraft to the recovery ship via life raft and helicopter and will return to Houston, Texas for debriefing.

Quarantine for Apollo 15 and the remaining lunar missions has been eliminated and the mobile quarantine facility will not be used. However, biological isolation garments will be available for use in the event of unexplained crew illness.

CM and Data Retrieval Operations

After flight crew pickup by helicopter, the CM will be retrieved and placed on a dolly aboard the recovery ship. Lunar samples, film, flight logs, etc., will be retrieved for shipment to the Lunar Receiving Laboratory (LRL). The spacecraft will be off-loaded from the ship at Pearl Harbor and transported to an area where deactivation of the CM propellant system will be accomplished. The CM will then be returned to contractor facilities. Flight crew debriefing operations, sample analysis, and postflight data analysis will be conducted in accordance with established schedules.

ALTERNATE MISSIONS

General

If an anomaly occurs after liftoff that would prevent the space vehicle from following its nominal flight plan, an abort or an alternate mission will be initiated. An abort will provide for acceptable flight crew and CM recovery.

An alternate mission is a modified flight plan that results from a launch vehicle, spacecraft, or support equipment anomaly that precludes accomplishment of the primary mission objectives. The purpose of the alternate mission is to provide the flight crew and flight controllers with a plan by which the greatest benefit can be gained from the flight using the remaining systems capabilities.

Alternate Missions

The two general categories of alternate missions that can be performed during the Apollo 15 Mission are (1) earth orbital and (2) lunar. Both of these categories have several variations which depend upon the nature of the anomaly leading to the alternate mission and the resulting systems status of the LM and CSM. A brief description of these alternate missions is contained in the following paragraphs.

Earth Orbit

In the event that TLI is inhibited, an earth orbit mission of approximately six and one-third days may be conducted to obtain maximum benefit from the scientific equipment aboard the CSM. Subsequent to transfer of the necessary equipment to the CM, the LM will be deorbited into the Pacific Ocean. Three SPS burns will be used to put the CSM into a 702 x 115 nm orbit where the subsatellite will be launched at approximately 35 hours GET. The high apogee will afford maximum lifetime of the subsatellite. The launching will be in the daylight with the spin rotation axis normal to the ecliptic to achieve the maximum absorption of solar energy. The gamma ray spectrometer will be employed to obtain data on the earth's magnetosphere. Two additional SPS burns will be performed to place the CSM into a 240 x 114 nm orbit with the apogee over the United States for photographic tasks using the SIM bay cameras. Camera cassettes will be retrieved by EVA on the last day of the mission. In addition, the alpha-particle spectrometer, mass spectrometer, and laser altimeter will be exercised to verify hardware operability. The x-ray fluorescence equipment will be used for partial mapping of the universe and obtaining readings of cosmic background data.

Lunar Orbit

Lunar orbit missions of the following types will be planned if spacecraft systems will enable accomplishment of orbital science objectives in the event a lunar landing is not possible.

CSM/LM

The translunar trajectory will be maintained within the DPS capability of an acceptable earth return in the event LOI is not performed. Standard LOI and TEI techniques will be used except that the DPS will be retained for TEI unless required to achieve a lunar orbit. The SPS will be capable of performing TEI on any revolution. Orbital science and photographic tasks from both the new SIM bay and from the CM will be conducted in a high-inclination, 60 NM circular orbit for about 4 days.

CSM Alone

In the event the LM is not available, the CSM will maintain a translunar trajectory within the SM RCS capability of an acceptable earth return. LOI will not be performed if the SIM bay door cannot be jettisoned. Orbital science and photographic tasks will be conducted in a high-inclination, 60 NM lunar orbit during a 4 to 6 day period.

CSM/Alone (From Landing Abort)

In the event the lunar landing is aborted, an orbital science mission will be accomplished by the CSM alone after rendezvous, docking, and LM jettison. The total orbit time will be approximately 6 days.

EXPERIMENTS, DETAILED OBJECTIVES, IN-FLIGHT DEMONSTRATIONS, AND OPERATIONAL TESTS

The technical investigations to be performed on the Apollo 15 Mission are classified as experiments, detailed objectives, or operational tests:

Experiment - A technical investigation that supports science in general or provides engineering, technological, medical or other data and experience for application to Apollo lunar exploration or other programs and is recommended by the Manned Space Flight Experiments Board (MSFEB) and assigned by the Associate Administrator for Manned Space Flight to the Apollo Program for flight.

Detailed Objective - A scientific, engineering, medical or operational investigation that provides important data and experience for use in development of hardware and/or procedures for application to Apollo missions. Orbital photographic tasks, though reviewed by the MSFEB, are not assigned as formal experiments and will be processed as CM and SM detailed objectives.

Inflight Demonstration - A technical demonstration of the capability of an apparatus and/or process to illustrate or utilize the unique conditions of space flight environment. Inflight Demonstration will be performed only on a noninterference basis with all other mission and mission related activities. Utilization performance, or completion of these demonstrations will in no way relate to mission success. (None planned for this mission)

Operational Test - A technical investigation that provides for the acquisition of technical data or evaluates operational techniques, equipment, or facilities but is not required by the objectives of the Apollo flight mission. An operational test does not affect the nominal mission timeline, adds no payload weight, and does not jeopardize the accomplishment of primary objectives, experiments, or detailed objectives.

EXPERIMENTS

The Apollo 15 Mission includes the following experiments:

Lunar Surface Experiments

Lunar surface experiments are deployed and activated or conducted by the Commander and the Lunar Module Pilot during EVA periods. Those experiments which are part of the ALSEP are so noted.

Lunar Passive Seismology (S-031) (ALSEP)

The objectives of the passive seismic experiment are to monitor lunar seismic activity and to detect meteoroid impacts, free oscillations of the moon, surface tilt (tidal deformations), and changes in the vertical component of gravitational acceleration. The experiment sensor assembly is made up of three orthogonal, long-period seismometers and one vertical, short-period seismometer. The instrument and the near-lunar surface are covered by a thermal shroud.

Lunar Tri-axis Magnetometer (S-034) (ALSEP)

The objectives of the lunar surface magnetometer experiment are to measure the magnetic field on the lunar surface to differentiate any source producing the induced lunar magnetic field, to measure the permanent magnetic moment, and to determine the moon's bulk magnetic permeability during traverse of the neutral sheet in the geomagnetic tail. The experiment has three sensors, each mounted at the end of a 90-cm long arm, which are first oriented parallel to obtain the field gradient and thereafter orthogonally to obtain total field measurements.

Medium Energy Solar Wind (S-035) (ALSEP)

The objectives of the use of the solar wind spectrometer are to determine the nature of the solar wind interactions with the moon, to relate the effects of the interactions to interpretations of the lunar magnetic field, the lunar atmosphere, and to the analysis of lunar samples, and to make inferences as to the structure of the magnetospheric tail of the earth. The measurements of the solar wind plasma is performed by seven Faraday cup sensors which collect and detect electrons and protons.

Suprathermal Ion Detector (S-036) (ALSEP)

The objectives of the suprathermal ion detector experiment are to provide information on the energy and mass spectra of positive ions close to the lunar surface and in the earth's magnetotail and magnetosheath, to provide data on plasma interaction between the solar wind and the moon, and to determine a preliminary value for electric potential of the lunar surface. The suprathermal ion detector has two positive ion detectors: a mass analyzer and a total ion detector.

Cold Cathode Ionization Gauge (S-058) (ALSEP)

The objective of the cold cathode ionization gauge experiment, which is integrated with the suprathermal ion detector, is to measure the neutral particle density of the lunar atmosphere.

Lunar Heat Flow (S-037) (ALSEP)

The objectives of the heat flow experiment are to determine the net lunar heat flux and the values of thermal parameters in the first three meters of the moon's crust.

The experiment has two sensor probes placed in bore holes drilled with the Apollo Lunar Surface Drill (ALSD).

Lunar Dust Detector (M-515)

The objectives of the dust detector experiment is to obtain data on dust accretion rates and on the thermal and radiation environment. The dust detector has three small photoelectric cells mounted on the ALSEP central station sun shield, facing the ecliptic path of the sun.

Lunar Geology Investigation (S-059)

The fundamental objective of this experiment is to provide data for use in the interpretation of the geological history of the moon in the vicinity of the landing site. The investigation will be carried out during the planned lunar surface traverses and will utilize camera systems, hand tools, core tubes, the ALSD, and sample containers. The battery powered ALSD will be used to obtain core samples to a maximum depth of 2.5 meters.

Documented Samples - Rock and soil samples representing different morphologic and petrologic features will be described, photographed, and collected in individual pre-numbered bags for return to earth. This includes comprehensive samples of coarse fragments and fine lunar soil to be collected in pre-selected areas. Documented samples are an important aspect of the experiment in that they support many sample principal investigators in addition to lunar geology. Documented samples of the Apennine front and the drill core samples have higher individual priorities than the other activities of this experiment.

Geologic Description and Special Samples - Descriptions and photographs of the field relationships of all accessible types of lunar features will be obtained. Special samples, such as the magnetic sample, will be collected and returned to earth.

Laser Ranging Retro-reflector (S-078)

The objective of the experiment is to gain knowledge of several aspects of the earth-moon system by making precise measurements of the distance from one or more earth sites to several retro-reflector arrays on the surface of the moon. Some of these aspects are: lunar size and orbit; physical librations and moments of inertia of the moon; secular acceleration of the moon's longitude which may reveal a slow decrease in the gravitational constant; geophysical information on the polar motion; and measurement of predicted continental drift rates. The retro-reflector array on Apollo 15 has 300 individually mounted, high-precision, optical corners. Aiming and alignment mechanisms are used to orient the array normal to incident laser beams directed from earth.

Solar Wind Composition (S-080)

The purpose of the solar wind composition experiment is to determine the isotopic composition of noble gases in the solar wind, at the lunar surface, by entrapment of particles in aluminum foil. A staff and yard arrangement is used to deploy the foil and maintain its plane perpendicular to the sun's rays. After return to earth, a spectrometric analysis of the particles entrapped in the foil allows quantitative determination of the helium, neon, argon, krypton, and xenon composition of the solar wind.

Soil Mechanics Experiment (S-200)

The objective of the experiment is to obtain data on the mechanical properties of the lunar soil from the surface to depths of tens of centimeters.

Data is derived from lunar module landing dynamics, flight crew observations and debriefings, examination of photographs, analysis of lunar samples, and astronaut activities using the Apollo hand tools. Experiment hardware includes an astronaut operated self-recording penetrometer.

In-flight Experiments

The in-flight experiments are conducted during earth orbit, translunar coast, lunar orbit, and transearth coast mission phases. They are conducted with the use of the command module (CM), the scientific instrument module (SIM) located in sector 1 of the service module (SM), or the subsatellite launched in lunar orbit, as noted.

Gamma-ray Spectrometer (S-160) (SIM)

The objectives of the gamma-ray spectrometer experiment are to determine the lunar surface concentration of naturally occurring radioactive elements and of major rock forming elements. This will be accomplished by the measurement of the lunar surface natural and induced gamma radiation while in orbit and by the monitoring of galactic gamma-ray flux during transearth coast.

The spectrometer detects gamma-rays and discriminates against charged particles in the energy spectrum from 0.1 to 10 mev. The instrument is encased in a cylindrical thermal shield which is deployed on a boom from the SIM for experiment operation.

X-Ray Fluorescence (S-161) (SIM)

The objective of the X-ray spectrometer experiment is to determine the concentration of major rock-forming elements in the lunar surface. This is accomplished by monitoring the fluorescent X-ray flux produced by the interaction of solar X-rays with surface material and the lunar surface X-ray albedo. The X-ray spectrometer, which is integrally packaged with the alpha-particle spectrometer, uses three sealed proportional counter detectors with different absorption filters. The direct solar X-ray flux is detected by the solar monitor, which is located 180° from the SIM in SM sector IV. An X-ray background count is performed on the lunar darkside.

Alpha-Particle Spectrometer (S-162) (SIM)

The objective of this experiment is to locate radon sources and establish gross radon evolution rates, which are functions of the natural and isotopic radioactive material concentrations in the lunar surface. This will be accomplished by measuring the lunar surface alpha-particle emissions in the energy spectrum from 4 to 9 mev.

The instrument employs ten surface barrier detectors. The spectrometer is mounted in an integral package with the X-ray spectrometer.

S-Band Transponder (SCM/LM) (S-164)

The objectives of the S-band transponder experiment are to detect variations in the lunar gravity field caused by mass concentrations and deficiencies and to establish gravitational profiles of the ground tracks of the spacecraft.

The experiment data is obtained by analysis of the S-band Doppler tracking data for the CSM and LM in lunar orbit. Minute perturbations of the spacecraft motion are correlated to mass anomalies in the lunar structure.

Mass Spectrometer (S-165) (SIM)

The objectives of the mass spectrometer experiment are to obtain data on the composition and distribution of the lunar atmosphere constituents in the mass range from 12 to 66 amu. The experiment will also be operated during transearth coast to obtain background data on spacecraft contamination.

The instrument employs ionization of constituent molecules and subsequent collection and identification by mass unit analysis. The spectrometer is deployed on a boom from the SIM during experiment operation.

Bistatic Radar (S-170) (CSM)

The objectives of the bistatic radar experiment are to obtain data on the lunar bulk electrical properties, surface roughness, and regolith depth to 10-20 meters. This experiment will determine the lunar surface Brewster angle, which is a function of the bulk dielectric constant of the lunar material.

The experiment data is obtained by analysis of bistatic radar echos reflected from the lunar surface and subsurface, in correlation with direct downlink signals. The S-band and VHF communications systems, including the VHF omni and S-band high-gain or omni antennas, are utilized for this experiment.

Subsatellite

The subsatellite is a hexagonal prism which uses a solar cell power system, an S-band communications system, and a storage memory data system. A solar sensor is provided for attitude determination. The subsatellite is launched from the SIM into lunar orbit and is spin-stabilized by three deployable, weighted arms. The following three experiments are performed by the subsatellite:

S-Band Transponder (S-164) (Subsatellite) - Similar to the S-band transponder experiment conducted with the CSM and LM, this experiment will detect variations in the lunar gravity field by analysis of S-band signals. The Doppler effect variations caused by minute perturbations of the subsatellite's orbital motions are indicative of the magnitudes and locations of mass concentrations In the moon.

Particle Shadows/Boundary Layer (S-173) (Subsatellite) - The objectives of this experiment are to monitor the electron and proton flux in three modes: interplanetary, magnetotail, and the boundary layer between the moon and the solar wind.

The instrument consists of solid state telescopes to allow detection of electrons in two energy ranges of 0-14 kev and 20-320 kev and of protons in the 0.05 - 2.0 mev range.

Subsatellite Magnetometer (S-174) - The objectives of the subsatellite magnetometer experiment are to determine the magnitude and direction of the interplanetary and earth magnetic fields in the lunar region.

The biaxial magnetometer is located on one of the three subsatellite deployable arms. This instrument is capable of measuring magnetic field intensities from 0 to 200 gammas.

Apollo Window Meteoroid (S-176) (CM)

The objective of the Apollo window meteoroid experiment is to obtain data on the cislunar meteoroid flux of mass range 10-12 grams. The returned CM windows will be analyzed for meteoroid impacts by comparison with a preflight photomicroscopic window map.

The photomicroscopic analysis will be compared with laboratory calibration velocity data to define the mass of impacting meteoroids.

UV Photography - Earth and Moon (S-177) (CM)

The objective of this experiment is to photograph the moon and the earth in one visual and three ultraviolet regions of the spectrum. The earth photographs will define correlations between UV radiation and known planetary conditions. These analyses will form analogs for use with UV photography of other planets. The lunar photographs will provide additional data on lunar surface color boundaries and fluorescent materials.

Photographs will be taken from the CM with a 70mm Hasselblad camera equipped with four interchangeable filters with different spectral response. Photographs will be taken in earth orbit, translunar coast, and lunar orbit.

Gegenschein from Lunar Orbit (S-178) (CM)

The objective of the gegenschein experiment is to photograph the Moulton point region, and analytically defined null gravity point of the earth-sun line behind the earth. These photographs will provide data on the relationship of the Moulton point and the gegenschein (an extended light source located along the earth-sun line behind the earth). These photographs may provide evidence as to whether the gegenschein is attributable to scattered sunlight from trapped dust particles at the Moulton point.

Other Experiments

Additional experiments assigned to the Apollo 15 Mission which are not a part of the lunar surface or orbital science programs are listed below.

Bone Mineral Measurement (M-078)

The objectives of the experiment are to determine the occurrence and degree of bone mineral changes in the Apollo crewmen which might result from exposure to the weightless condition, and whether exposure to short periods of 1/6 g alters these changes. At selected pre- and post-flight times, the bone mineral content of the three Apollo crewmen will be determined using X-ray absorption technique.

The radius and ulna (bones of the forearm) and os calcis (heel) are the bones selected for bone mineral content measurements.

Total Body Gamma Spectrometry (M-079)

The objectives of this experiment are to detect changes in total body potassium and total muscle mass (lean body mass), and to detect any induced radioactivity in the bodies of the crewmen. Preflight and postlaunch examination of each crew member will be performed by radiation detecting instruments in the Radiation Counting Laboratory at MSC. There are no inflight requirements for this experiment.

DETAILED OBJECTIVES

Following is a brief description of each of the launch vehicle and spacecraft detailed objectives planned for this mission.

Launch Vehicle Detailed Objectives

Impact the expended S-IVB/IU on the Lunar surface under nominal flight profile conditions.

Post-flight determination of actual S-IVB/IU point of impact within 5 km, and time of impact within one second.

Spacecraft Detailed Objectives

Collect a contingency sample for assessing the nature of the surface material at the lunar landing site in event EVA is terminated.

Evaluate Lunar Roving Vehicle operational characteristics in the lunar environment.

Demonstrate the LCRU/GCTA will adequately support extended lunar surface exploration communication requirements and obtain data on the effect of lunar dust on the system.

Assess EMU lunar surface performance, evaluate metabolic rates, crew mobility and difficulties in performing lunar surface EVA operations.

Evaluate the LM's landing performance.

Obtain SM high resolution panoramic and high quality metric lunar surface photographs and altitude data from lunar orbit to aid in the overall exploration of the moon.

Obtain CM photographs of lunar surface features of scientific interest and of low brightness astronomical and terrestrial sources.

Obtain data to determine adequate thermal conditions are maintained in the SIM bay and adjacent bays of the service module.

Inspect the SIM bay, and demonstrate and evaluate EVA procedures and hardware.

Determine the effects of SIM door jettison in a lunar environment.

Obtain data on the performance of the descent engine.

Record visual observations of farside and nearside lunar surface features and processes to complement photographs and other remote-sensed data.

Obtain more definitive information on the characteristics and causes of visual light flashes.

Inflight Demonstration

None planned for this mission.

OPERATIONAL TESTS

The following significant operational tests will be performed in conjunction with the Apollo 15 mission.

Gravity Measurement

Performance of the gravity measurement will be by ground control. Following lunar landing, the IMU and platform will remain powered up. Flight controllers will uplink the necessary commands to accomplish gravity alignments of the IMU. Subsequent to the data readouts, the crew will terminate the test by powering down the IMU. This is the only crew function required, and crew activities are not restricted by the test. If the test is not completed in the short period after landing, it may also be conducted during the powered-up pre-liftoff operations.

Acoustic Measurement

The noise levels of the Apollo 15 space vehicle during launch and the command module during entry into the atmosphere will be measured in the Atlantic launch abort area and the Pacific recovery area, respectively. The data will be used to assist in developing high-altitude, high-Mach number, accelerated flight sonic boom prediction techniques. MSC will conduct planning, scheduling, test performance, and reporting of the test results. Personnel and equipment supporting this test will be located aboard secondary recovery ships, the primary recovery ship, and at Nihoa, Hawaii.

VHF Noise Investigation

On-board audio recordings and VHF signal strengths from spacecraft telemetry will be reviewed and analyzed to attempt resolution of VHF noises and less-than-predicted communications performance experienced on previous Apollo missions. The crew will note any unusual VHF system performance, and signals will be recorded in the LM before ascent when the CSM is beyond the line of sight.

MISSION CONFIGURATION AND DIFFERENCES
MISSION HARDWARE AND SOFTWARE CONFIGURATION

The Saturn V Launch Vehicle and the Apollo Spacecraft for the Apollo 15 Mission will be operational configurations.

CONFIGURATION	DESIGNATION NUMBERS
Space Vehicle	AS-510
Launch Vehicle	SA-510
First Stage	S-IC-10
Second Stage	S-II-10
Third Stage	S-IVB-510
Instrument Unit	S-IU-510
Spacecraft-LM Adapter	SLA-19
Lunar Module	LM-10
Lunar Roving Vehicle	LRV- 1
Service Module	SM-112
Command Module	CM-112
Onboard Programs	
Command Module	Colossus 3
Lunar Module	Luminary IE
Experiments Package	Apollo 15 ALSEP
Launch Complex	LC-39A

CONFIGURATION DIFFERENCES

The following summarizes the significant configuration differences associated with the AS-510 Space Vehicle and the Apollo 15 Mission. Additional technical details on the new hardware items described below and contained in the Mission Operations Report, Apollo Supplement.

SPACECRAFT

Command/Service Module

Added third cryogenic H_2 tank with modified heating	Increased electrical power capability for extended mission duration.
Relocated third cryogenic O_2 tank isolation valve and plumbing	Eliminated potential single failure point.
Added Scientific Instrument Module (SIM) in Sector IV of Service Module	Increased in-flight science capability by addition of experiments, a subsatellite, cameras, and laser altimeter (see experiments section).
Added Scientific Data System	Provided complete scientific experiment data coverage in lunar Orbit with capability for realtime data transmission simultaneously with tape recorder playback and transmission of data recorded on the lunar far side.
Modified CM environmental control system for in-flight EVA capability	Provided for in-flight retrieval of film from SIM cameras by adding third O_2 flow restrictor; EVA control panel; and EVA umbilical with O_2, bioinstrumentation, and communications links with the EVA crewman.

Lunar Module

Enlarged descent stage propellant tanks	Provided for longer powered descent burn to permit increased LM landing weight and landing point selection.
Modified descent engine nozzle by adding a ten-inch extension with quartz liner	Increased descent engine specific impulse.
Added GOX tank, water tank, and descent stage battery	Extended lunar surface stay time from 38 to 68 hours.
Modified quadrant I for LM-LRV interface	Provided for LRV stowage and deployment to increase lunar surface mobility.

Crew Provisions and Lunar Mobility

New spacesuits for crewmen	Provided in-flight EVA capability for CMP and increased lunar surface EVA time for CDR and LMP. All suits have improved mobility. CDR and LMP suits have increased drinking water supply and 175 calorie fruit bars for each EVA.
Lunar Roving Vehicle	Provided increased lunar surface mobility for astronauts and equipment. Provided for transport and power supply for LCRU and GCTA on EVA traverses.

Launch Vehicle

S-IC

Modified LOX vent and relief valve	Additional spring increased valve closing force and improved reliability.
Increased outboard engine LOX depletion delay time	Increased payload capability approximately 500 pounds.
Removed four of the eight retro-rocket motors	Saved weight and cost and increased payload capability approximately 100 lbs.
Reorificed the F-1 engines	Increased payload capability approximately 600 pounds.

S-II

Removed four ullage motors	Eliminated single failure points and increased payload capability approximately 90 lbs.
Delayed time base 3 (S-II ignition) by one second	Maintained same S-IC/S-II stage separation as was previously achieved with S-IC retro-rockets.
Replaced LH_2 and LOX ullage pressure regulators with fixed orifices	Increased payload capability approximately 210 pounds by providing hotter ullage gases. Eliminated several single point failures
Added a G-switch disable capability	Decreased the probability of an inadvertent cutoff due to a transient signal.
Changed engine pre-cant angle from 1.30 to 0.60	Reduced probability of collision with the S-IVB stage in an engine out condition during second plane separation.

S-IV B

Added filter in J-2 engine helium pneumatic control line	Decreased probability of valve seat leakage and a possible restart problem.

IU

Added redundant +28 volt power for ST-124 stabilized platform system	Improved power supply reliability.
Modified launch tower avoidance yaw maneuver by reducing the time from command to execute	Reduced launch wind restrictions and increased assurance of clearing the tower.
Modified Command Module Computer Cutoff program to provide spacecraft computer cutoff of S-IVB TLI burn	Increased accuracy of TLI burn cutoff in event of IU platform failure.

TV AND PHOTOGRAPHIC EQUIPMENT

Standard and special purpose cameras, lenses, and film will be carried to support the objectives, experiments, and operational requirements. Table 2 lists the television and camera equipments and shows their stowage locations.

TABLE 2

TV AND PHOTOGRAPHIC EQUIPMENT

NOMENCLATURE	STOWAGE LOCATION				
	CSM AT LAUNCH	LM AT LAUNCH	CM TO LM	LM TO CM	CM AT ENTRY
TV, COLOR, ZOOM LENS (MONITOR WITH CM SYSTEM)	1	1			1

CAMERA, 35MM NIKON
  LENS - 55MM
  CASSETTE, 35MM
	1
	1
	4
					1
					1
					4

CAMERA, DATA ACQUISITION, 16MM
  LENS - 10MM
       - 18MM
       - 75MM
  FILM MAGAZINES
	1
	1
	1
	1
	10
		1
		1
					1
					1
					1
					1
					10

CAMERA, LUNAR SURFACE, 16MM BATTERY OPERATED
  LENS - 10MM
  MAGAZINES
	8	1
		1
			8	8	8

CAMERA, HASSELBLAD, 70MM ELECTRIC
  LENS - 80MM
       - 250MM
       - 105MM UV (4 BAND-PASS FILTERS)
  FILM MAGAZINES
  FILM MAGAZINE, 70MM UV
	1
	1
	1
	1
	6
	1
					1
					1
					1
					1
					6
					1

CAMERA, HASSELBLAD, 70MM LUNAR SURFACE ELECTRIC
  LENS - 60MM
       - 500MM
  FILM MAGAZINES
	13	3
		2
		1
			13	13	13

CAMERA, 24-IN. PANORAMIC (IN SIM) FILM CASSETTE (EVA TRANSFER)	1 1				1
CAMERA, 3- . MAPPING STELLAR(SIM) FILM MAGAZINE (EVA TRANSFER)	1 1				1

FLIGHT CREW DATA

PRIME CREW (Figure 10)

COMMANDER: David R. Scott (Colonel, USAF)

Space Flight Experience: Colonel Scott was one of the third group of astronauts selected by NASA in October 1963.

As Pilot for the Gemini 8 Mission, launched on March 16, 1966, Colonel Scott and Command Pilot Neil Armstrong performed the first successful docking of two vehicles in space. Gemini 8, originally scheduled to continue for three days, was terminated early due to a malfunctioning attitude thruster.

Subsequently, Colonel Scott was selected as Command Module Pilot for the Apollo 9 Mission which included lunar orbit rendezvous and docking simulations, crew transfer between CM and LM, and extravehicular activity techniques.

Colonel Scott has flown more than 251 hours in space.

COMMAND MODULE PILOT: Alfred M. Worden (Major, USAF)

Space Flight Experience: Major Worden is one of 19 astronauts selected by NASA in April 1966. He served as a member of the astronaut support crew for Apollo 9 and backup command module pilot for Apollo 12.

Worden has been on active duty since June 1955. Prior to being assigned to the Manned Spacecraft Center, he served as an instructor at the Aerospace Research Pilots School.

LUNAR MODULE PILOT: James Benson Irwin (Lieutenant Colonel, USAF)

Space Flight Experience: Lieutenant Colonel Irwin was selected by NASA in 1966. He was crew commander of Lunar Module Test Article - 8 (LTA-8). LTA-8 was used in a series of thermal vacuum tests. He also served as a member of the support crew for Apollo 10 and as backup LM pilot for Apollo 12.

Irwin has been on active duty since 1951. Previous duties included assignment as Chief of the Advanced Requirements Branch at Headquarters, Air Defense Command.

APOLLO 15 PRIME CREW   Figure. 10

BACKUP CREW

COMMANDER: Richard F. Gordon (Captain, USN)

Space Flight Experience: Captain Gordon was assigned to NASA in October, 1963. He served as the backup pilot for Gemini 8, backup CM pilot for Apollo 9 and served as CM pilot for Apollo 12, the second lunar landing mission.

Captain Gordon's total space time exceeds 315 hours.

COMMAND MODULE PILOT: Vance D. Brand (Civilian)

Space Flight Experience: Mr. Brand has served as an astronaut since April, 1966. He was a crew member for the thermal vacuum test of the prototype CM 2TV-1. He was also a member of the Apollo 8 and 13 support crews.

LUNAR MODULE PILOT: Harrison H. Schmitt, PhD (Civilian)

Space Flight Experience: Dr. Schmitt was selected as a scientist astronaut by NASA in June, 1965. He completed a 53 week course in flight training at Williams Air Force Base, Arizona. Dr. Schmitt has also been instrumental in providing Apollo flight crews with detailed instruction in lunar navigation, geology and feature recognition.

MISSION MANAGEMENT RESPONSIBILITY

TITLE	NAME	ORGANIZATION
Director, Apollo Program	Dr. Rocco A. Petrone	OMSF
Mission Director	Capt. Chester M. Lee (Ret)	OMSF
Saturn Program Manager	Mr. Richard G. Smith	MSFC
Apollo Spacecraft Program Manager	Col. James A. McDivitt	MSC
Apollo Program Manager, KSC	Mr. Robert C. Hock	KSC
Director of Launch Operations	Mr. Walter J. Kapryan	KSC
Director of Flight Operations	Mr. Sigurd A. Sjoberg	MSC
Launch Operations Manager	Mr. Paul C. Donnelly	KSC
Flight Directors	Mr. Gerald D. Griffin	MSC
	Mr. Eugene F. Kranz	MSC
	Mr. Glynn S. Lunney	MSC
	Mr. Milton L. Windler	MSC

Post Launch Mission Operation Report

No. M-933-71-15

16 August 1971

TO: A/Administrator

FROM: MA/Apollo Program Director

SUBJECT: Apollo 15 Mission (AS-510) Post Launch Mission Operation Report No. 1

The Apollo 15 Mission was successfully launched from the Kennedy Space Center on Monday, 26 July 1971 and was completed as planned, with recovery of the spacecraft and crew in the mid-Pacific Ocean recovery area on Saturday, 7 August 1971. Initial review of the mission indicates that all mission objectives were accomplished. Further detailed analysis of all data is continuing and appropriate refined results of the mission will be reported in the Manned Space Flight Centers' technical reports.

Attached is the Mission Director's Summary Report for Apollo 15 which is submitted as Post Launch Mission Operation Report No. 1. Also attached are the NASA OMSF Primary Objectives for Apollo 15. The Apollo 15 Mission has achieved all the assigned primary objectives and I judge it to be a success.

Rocco A. Petrone

APPROVAL:

Dale D. Myers
Associate Administrator for Manned Space Flight

NASA OMSF MISSION OBJECTIVES FOR APOLLO 15

PRIMARY OBJECTIVES

Perform selenological inspection, survey, and sampling of materials and surface features in a preselected area of the Hadley-Apennine region.

Emplace and activate surface experiments.

Evaluate the capability of the Apollo equipment to provide extended lunar surface stay time, increased EVA operations, and surface mobility.

Conduct in-flight experiments and photographic tasks from lunar orbit.

Rocco A. Petrone
Apollo Program Director
16 July 1971

Dale D. Myers
Associate Administrator for Manned Space Flight
Date: July 17 1971

ASSESSMENT OF APOLLO 15 MISSION

Based upon a review of the assessed performance of Apollo 15, launched 26 July 1971 and completed 7 August 1971, this mission is adjudged a success in accordance with the objectives stated above.

Rocco A. Petrone
Apollo Program Director
Date: 11 August 1971

Dale D. Myers
Associate Administrator for Manned Space Flight
Date: 16 August 1971

NATIONAL AERONAUTICS AND SPACE ADMINISTRATION
WASHINGTON, D.C. 20546

Reply to ATTN of: MAO

7 August 1971

TO: Distribution

FROM: MA/Apollo Mission Director

SUBJECT: Mission Director's Summary Report, Apollo 15

INTRODUCTION

The Apollo 15 Mission was planned as a lunar landing mission to: perform selenological inspection, survey, and sampling of materials and surface features in a preselected area of the Hadley-Apennine region of the moon; emplace and activate surface experiments; evaluate the capability of Apollo equipment to provide extended lunar surface stay time, increased EVA operations, and surface mobility; and conduct photographic tasks. Flight crew members were Commander (CDR) Col. David R. Scott (USAF), Command Module Pilot (CMP) Maj. Alfred M. Worden (USAF), and Lunar Module Pilot (LMP) Lt. Col. James B. Irwin (USAF). Significant detailed mission information is contained in Tables 1 through 13. Initial review indicates that all primary mission objectives were accomplished (reference Table 1). Table 2 lists the Apollo 15 achievements.

PRELAUNCH

The space vehicle prelaunch operations were nominal and the final countdown was exceptionally smooth.

LAUNCH AND EARTH PARKING ORBIT

The Apollo 15 space vehicle was successfully launched on time from Kennedy Space Center, Florida, at 9:34 a.m. EDT, on 26 July 1971. The S-IVB/IU/CSM/LM combination was inserted into an Earth Parking Orbit (EPO) of 91.5 x 92.5 nautical miles (NM), about 11 minutes 44 seconds after liftoff. The planned EPO was 90 NM circular.

During EPO, a navigation correction for the Translunar Injection (TLI) maneuver was uplinked to the Instrument Unit (IU), all major Command Service Module (CSM) and S-IVB systems were verified, and preparations were completed for the S-IVB engine restart for TLI. The restart was initiated at 2:50:02 GET and a nominal TLI was achieved at 2:56:03 Ground Elapsed Time (GET).

The CSM separated from the LM/S-IVB/IU at 3:22:24 GET. Onboard color television (TV) was initiated as scheduled to cover the docking of the CSM with the Lunar Module (LM). Hard docking was completed at 3:33:49 GET followed by CSM/LM ejection at 4:18:00 GET. The S-IVB Auxiliary Propulsion System (APS) evasive maneuver was performed nominally and within a few seconds of the prelaunch plan.

The first S-IVB APS burn to achieve lunar impact was initiated at 5:47:53 GET. The second S-IVB APS burn was initiated at 10:00:00 GET, about 30 minutes later than planned. The late burn provided additional tracking time to compensate for any trajectory perturbations introduced by liquid oxygen (LOX) and liquid hydrogen (LH$_2$) tanks venting. Preliminary targeting for S-IVB lunar impact was 3°39'S and 7°39'W.

The spacecraft trajectory was so near nominal that midcourse correction (MCC)-1, scheduled for 11:55:33 GET, was cancelled.

Shortly after docking, telemetry data indicated that the solenoid valve drivers in the Service Propulsion System (SPS) were on. This condition indicated an electrical short to ground in the circuitry. The crew reported the Entry Monitor System Delta V thrust light was on in the cabin. Troubleshooting appeared to isolate the problem to the Delta V thrust A switch or adjacent wiring. Additional procedures were prepared to further diagnose the problem during the SPS engine burn at MCC-2.

The MCC-2 maneuver was performed with the SPS engine at 28:40:30 GET. The burn time of 0.72 second produced a Delta V of 5.3 feet per second (fps). The maneuver was conducted with SPS bank A in order to provide better analysis of the apparent intermittent short. SPS engine checkout procedures were developed and simulations were conducted on the ground prior to passing the procedures up to the crew. The method employed for the firing isolated the intermittent short to an area of Delta V thrust switch A downstream of a necessary SPS valve function for bank A. Because power could still be applied to the valve with a downstream short, SPS bank A could be operated satisfactorily in the manual mode for subsequent firings. The redundant bank B system was nominal in all respects and could be used for automatic starting and shutdown.

At approximately 33:47:00 GET, a temporary loss of communication was experienced due to a power amplifier failure at the Goldstone wing site tracking station. Communications were restored after switching to the Goldstone prime site.

The LM crew entered the LM at 33:56:00 GET for checkout, approximately 50 minutes earlier than scheduled. LM communications checks were performed between 34:21:00 and 34:45:00 GET. Good quality voice and data were received even though Goldstone was not yet configured correctly during the initial portion of the down-voice backup checks. Approximately 15 minutes later, the downlink carrier lock was lost for approximately a minute and a half; however, other stations that were tracking reduced the data loss to a few seconds.

TV of the CSM and LM interiors was broadcast between 34:55:00 and 35:46:00 GET. Camera operation was nominal, but the picture quality varied with the lighting of the scene observed. During checkout of the LM, the crew discovered the range/range rate exterior cover glass was broken, removing the helium barrier. Subsequent ground testing qualified the unprotected meter for use during the remainder of the mission in the spacecraft ambient atmosphere.

IVT/LM housekeeping began at 56:26:00 GET, approximately an hour and a half earlier than scheduled. The crew vacuumed the LM to remove broken glass from the damaged range/range rate meter. LM checkout was completed as planned.

MCC-3, scheduled for 56:31:00 GET, was not performed since the spacecraft was very close to the planned trajectory.

At 61:13:00 GET, during preparations for water chlorination, a water leak developed in the chlorination septum gland in the command module (CM). The leak was attributed to insufficient torque on the nut which compresses the septum washers in the gland. Procedures were read up to the crew for tightening the insert in the injector port. The leak was stopped and the water was absorbed with towels.

The CSM/LM entered the moon's sphere of influence at 63:55:20 GET.

Based on the MCC-2 burn test data, which indicated that SPS bank A could be safely operated manually, it was decided to perform all SPS maneuvers except Lunar Orbit Insertion (LOI) and Transearth Injection (TEI) using bank B only. LOI and TEI would be dual bank burns with modified procedures to permit automatic start and shutdown on bank B. The procedures to be used for these maneuvers were relayed to the crew.

MCC-4 at 73:31:14 GET was performed with the SPS engine, bank B. The burn time of 0.92 second produced a Delta V of -5. fps with no trim required since the residuals were zero.

SIM door jettison occurred at 74:06:47 GET. The LMP photographed the jettisoned door and visually observed it slowly tumbling through space away from the CSM and eventually into a heliocentric orbit.

LUNAR ORBIT INSERTION

LOI was performed using both banks of the SPS. The nominal maneuver, initiated at 78:31:46 GET, placed the CSM/LM in a 170 x 58-NM elliptical orbit around the moon. The burn time of 400.7 seconds produced a velocity change of -3000.1 fps. Bank A was shut down 32 seconds before planned cutoff to obtain performance data on bank B for future single bank burns.

S-IVB IMPACT

The S-IVB/IU impacted the lunar surface at 79:24:42 GET (4:58:43 p.m. EDT), approximately 11 minutes later than the prelaunch prediction. The impact point was 1.0°S and 11.87°W, which is 188 kilometers (km) northeast of the Apollo 14 landing site and 355 km northeast of the Apollo 12 landing site. The energy from the impact traveling through the lunar interior arrived at the Apollo 14 passive seismometer 37 seconds after impact and at the Apollo 12 seismometer 55 seconds after impact.

DESCENT ORBIT INITIATE

The DOI burn at 82:39:48 GET was nominal and the spacecraft was inserted in a 58.5 x 9.2 NM orbit. The single bank SPS burn duration was 24.5 seconds and resulted in a velocity change of -213.9 fps.

LUNAR ORBIT PRELANDING ACTIVITIES

Because the orbital decay rate was greater than anticipated, and RCS DOI trim burn of 21.2 seconds at 95:56:42 GET was executed on Revolution 10, producing a velocity change of 3.1 fps. The maneuver changed the orbit from 59.0 x 7.1 NM to 59.9 x 9.6 NM.

During the 12th lunar revolution on the far side of the moon at about 100:14:00 GET, the CSM/LM undocking and separation maneuver was initiated; however, at Acquisition of Signal (AOS), the commander reported that undocking did not occur. The crewmen and ground control decided that the probe instrumentation LM/CSM umbilical was either loose or disconnected. The CMP went into the tunnel to inspect the connection and found the umbilical plug to be loose. After reconnecting the plug and adjusting the spacecraft attitude, undocking and separation was achieved approximately 25 minutes late at 100:39:30 GET. The CSM circularization burn of 3.59 seconds was performed as planned at 101:38:58 GET and produced a Delta V of plus 68.3 fps. The SPS single bank burn was nominal with a resulting orbit of 64.7 x 58.0 NM.

POWERED DESCENT

LM powered descent was initiated at 104:30:09 GET. The descent-to-landing performance was nominal.

Touchdown occurred at 104:42:29 GET. The crew reported that they had landed at Hadley near Salyut Crater. Based on landmark bearings during SEVA and sightings from the CSM, ground crew analysis indicated the LM landing point to be about 600 meters north northwest of the planned target. The LM landing coordinates were 26°05'N and 3°39'E.

LUNAR SURFACE

The stand-up extravehicular activity (SEVA) to observe and photograph the landing site and surrounding area began at 106:42:49 GET (cabin depressurization). The CDR opened the upper hatch, stood on the ascent engine cover with his head out the hatch, and described and photographed the features of the area. SEVA termination occurred at 107:15:56 GET for a total of 33 minutes 7 seconds.

EVA-1 commenced at 119:30:10 GET at LM cabin depressurization. At 119:45:30 GET, the LMP experienced difficulty with the feedwater pressure in his Portable Life Support System (PLSS). The pressure reading was off-scale high, but this was attributed to gas bubbles in the feedwater and allowed the EVA to be continued. The CDR egressed the LM, and part of the way down the ladder he deployed the Modularized Equipment Stowage Assembly (MESA). The TV in the MESA was activated and the pictures of the CDR's remaining descent to the lunar surface were excellent. The LMP then egressed the LM to the lunar surface. While the CDR removed the TV camera from the MESA and deployed it on the tripod, the LMP collected the contingency sample.

The Lunar Roving Vehicle (LRV) was deployed with some difficulty by both astronauts. During checkout of the LRV, it was found that the LRV's front steering mechanism was inoperative. Additionally, there were no readouts on the LRV battery #2 ampere/volt meter. After minor troubleshooting of these problems, a decision was made to perform EVA-1 without the LRV front wheel steering activated. The troubleshooting determined that LRV battery #2 was carrying its share of the load.

The crew mounted the LRV and proceeded on the traverse (Figure 1) for EVA-1 at 121:44:56 GET. During the traverse, the crew obtained rock samples and photographs at the various stations. TV transmission during the stops was excellent.

At the end of the traverse, the ALSEP was deployed; however, the second coring operation for the heat flow experiment was not completed. This portion of ALSEP deployment was rescheduled to be completed during EVA-2. LM cabin repressurization terminated the EVA at 126:11:59 GET. The EVA duration was 6 hours 32 minutes 49 seconds, as compared to the 7 hours originally planned. This was occasioned by higher than anticipated O_2 usage by the CDR.

Since the LMP's PLSS was recharged 30° from the required vertical position, it was decided to vent and recharge the PLSS. A non-vertical recharge could cause gas bubbles as apparently experienced during EVA-1; the PLSS operation was nominal during EVA-2. During communications checks, the CDR reported that the LMP's PLSS antenna was broken. After powering down the EMU, the antenna was taped onto the Oxygen Purge System (OPS) and communications checks were satisfactorily completed.

EVA-2 commenced at 142:14:48 GET. The LRV was powered up, and the circuit breakers were cycled. The LRV steering was then found to be completely operational, as opposed to EVA-1 when the front steering mechanism was inoperable. The crew started the EVA-2 traverse at 143:11:00 GET. The trip included stops at Spur Crater, Dune Crater, Hadley Plains, and between Spur and Window Craters (see Figure 2). During the traverse, the crew obtained numerous samples and photographs. TV transmission was very good. Following termination of the traverse, the crew completed the heat flow experiment which was initiated on EVA- 1 and collected a core sample. The drill core stems were left at the ALSEP site for retrieval during EVA-3. The crew returned to the LM and deployed the United States flag. The sample container and film were stowed in the LM. Crew ingress followed, and EVA-2 was terminated with LM repressurization at 149:27:02 GET. The total EVA duration was 7 hours 12 minutes 14 seconds.

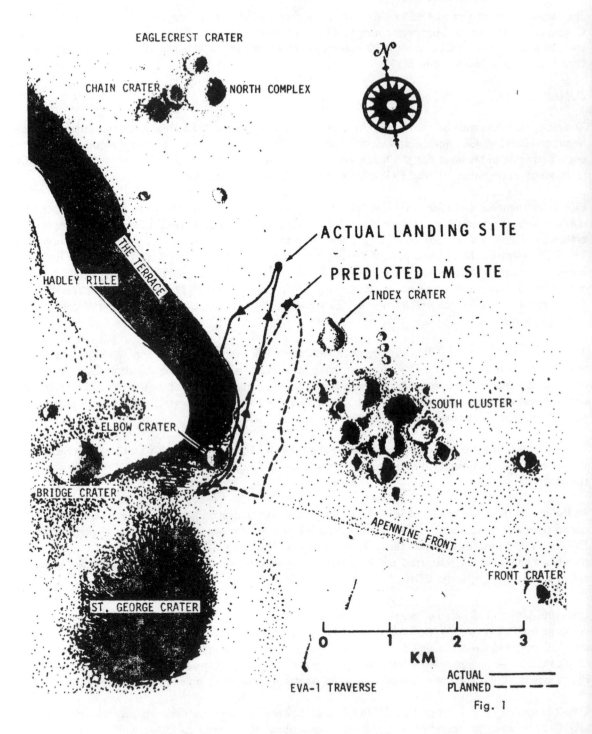

EAGLECREST CRATER

CHAIN CRATER   NORTH COMPLEX

N

ACTUAL LANDING SITE

PREDICTED LM SITE

INDEX CRATER

THE TERRACE

HADLEY RILLE

SOUTH CLUSTER

ELBOW CRATER

BRIDGE CRATER

APENNINE FRONT

FRONT CRATER

ST. GEORGE CRATER

0   1   2   3

KM

EVA-1 TRAVERSE   ACTUAL ———
                                 PLANNED – – –

Fig. 1

EVA-3 commenced at 163:18:14 GET, about 1 hour 45 minutes later than the nominal flight plan time due to cumulative changes in the surface activities timeline. The late start and the requirement to protect the nominal liftoff time required shortening the EVA. An alternate EVA plan was devised, and the traverse was made in a westerly direction from the LM to Hadley Rille (see Figure 3). The first stop was near the ALSEP site to retrieve the drill core stem samples left behind on EVA-2. Two of the sections of the drill core stem were removed and stowed in the LRV. The drill and the four remaining sections of the drill core stem could not be separated and were left for later retrieval. The remaining stops were Scarp Crater, "The Terrace" near Rim Crater, and Rim Crater. The return route was generally the same as the outbound route. Samples were obtained and documented and photographs were taken of various lunar surface features. During the sample

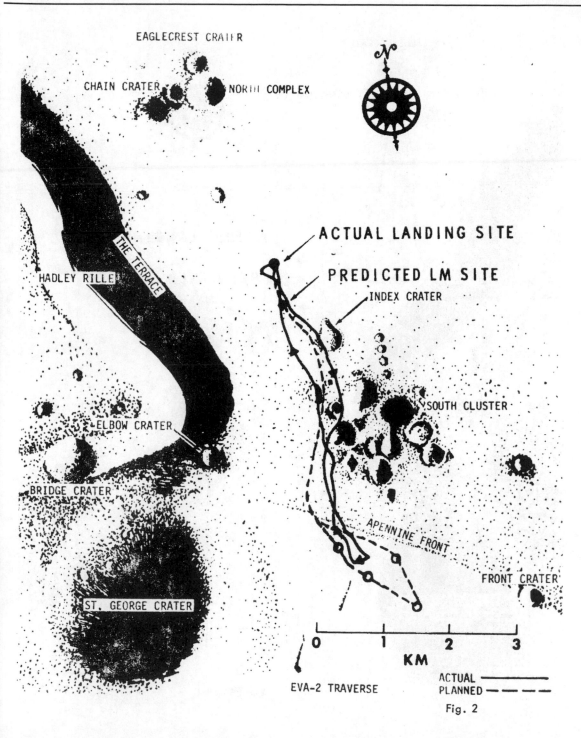

EAGLECREST CRATER

CHAIN CRATER   NORTH COMPLEX

N

THE TERRACE

HADLEY RILLE

ACTUAL LANDING SITE

PREDICTED LM SITE

INDEX CRATER

ELBOW CRATER

BRIDGE CRATER

SOUTH CLUSTER

APENNINE FRONT

ST. GEORGE CRATER

FRONT CRATER

0   1   2   3
KM

EVA-2 TRAVERSE

ACTUAL ———
PLANNED — — —

Fig. 2

collecting, the CDR tripped over a rock and fell, but experienced no difficulty in getting up.

Upon reaching the ALSEP area, the crew again attempted to disassemble the drill core stem. They managed to separate one more section, but the remaining three sections were returned still assembled.

The crew then returned to the LM, off-loaded the LRV, and stationed it for TV coverage of the LM liftoff. The CDR selected a site slightly closer to the LM than originally planned in order to take advantage of more elevated terrain for better LM liftoff TV coverage. The CDR ingressed the LM, and the EVA was terminated at 168:08:04 GET for a total duration of 4 hours 49 minutes 50 seconds.

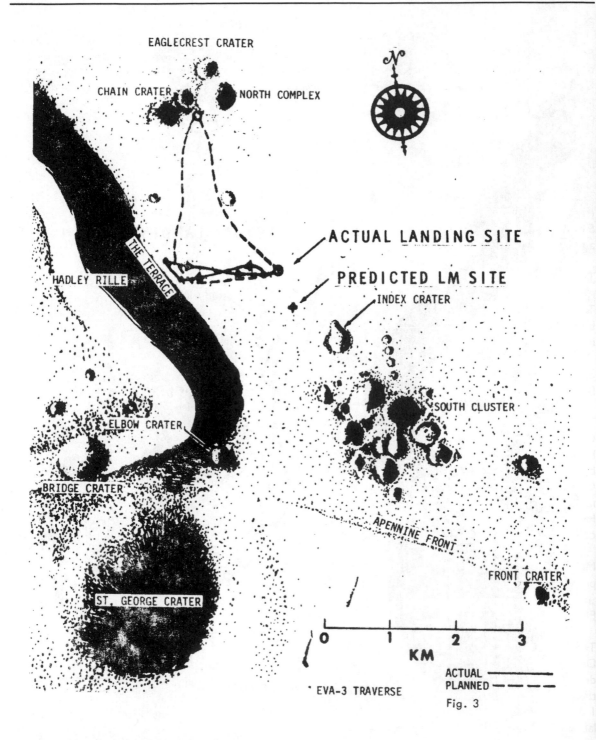

Fig. 3

ASCENT, RENDEZVOUS, AND DOCKING

Ascent stage liftoff from the lunar surface occurred on time at 171:34:22.4 GET, and the initial movement off the descent stage was televised by the Ground Controlled Television Assembly (GCTA). The ascent stage was inserted into a nominal lunar orbit and no tweak burn was required. The TPI burn was executed on time at 172:20:39 GET, with a nominal Delta V of 73 fps. The ascent stage performed a nominal braking maneuver for rendezvous with the CSM. Hard docking occurred at 173:35:47 GET. After CSM/LM docking, the CDR and LMP transferred the samples and other equipment to the CM for return to earth.

POST RENDEZVOUS

Following CDR and LMP IVT to CM, LM jettison and CSM separation were delayed one revolution in order to verify that the CSM and LM hatches were completely sealed. LM jettison occurred at 179:30:14 GET. The CSM was to perform a separation maneuver of 1 fps five minutes after LM jettison, but the delay in LM jettison caused the relative positions of the two spacecraft to be off-nominal, requiring a 2 fps Delta V posigrade burn which was accomplished at 179:50:00. The ascent stage deorbit burn (also delayed one revolution) occurred at 181:04:19. The spent LM ascent stage impacted the lunar surface at 26° 22'N and 15 'E at 181:29:36 GET, 93 km west of the Apollo 15 ALSEP site, 23.6 km from the preplanned target. The impact was recorded on the Apollo 12, 14, and 15 Passive Seismometers.

The orbit shaping maneuver for the subsatellite launch was performed during the 73rd lunar orbit revolution (rev) at 221:20:47 GET. The 3.3 second burn produced a Delta V of 66.4 fps, with a resultant orbit of 76 x 54.3 NM. At 22:39:19 GET, the subsatellite was launched into a 76.3 x 55.1 NM orbit. The launching produced a velocity of 4 fps for the subsatellite relative to the CSM.

TRANSEARTH INJECTION AND COAST

The transearth injection maneuver was performed at 223:48:45 GET. The Service Propulsion System (SPS) burn of 141.2 seconds resulted in a Delta V of 3047 fps and a flight path angle of -6.69 degrees at entry interface.

Since the spacecraft trajectory was near nominal, MCC-5 was not performed. The predicted Delta V was 0.3 fps.

The CMP performed the in-flight EVA at 241:57:57 GET to retrieve the Panoramic (Pan) and Mapping Camera film cassettes from the Scientific Instrument Module (SIM) located in the SM. Three excursions were made to the SIM bay. The film cassettes were retrieved during trips one and two. The third trip to the SIM bay was used to observe and report the general condition of the instruments; in particular, the Mapping Camera. The CMP reported no evidence of the cause for the Mapping Camera extend/retract mechanism failure in the extended position and no observable reason for the Pan Camera velocity/altitude sensor failure. He also reported the Mass Spectrometer Boom was not fully retracted. The 38 minute 12 second EVA was completed at 242:36:09 GET.

MCC-6 scheduled at 272:58:20 was canceled since the requirement was less than one fps. MCC-7 was performed at 291:56:48 GET. The 24.2 second maneuver produced a Delta V of 5.6 fps.

ENTRY AND LANDING

The CM separated from the SM at 294:44:00 GET, 15 minutes before entry interface (EI) at 400,000 ft. Drogue and main parachutes deployed normally; however, one of the three main parachutes partially closed during descent and subsequently caused a harder landing than planned. Landing occured at 295:11:53 GET in the mid-Pacific Ocean, at approximately 158°09'W longitude and 26°07'N latitude. The CM landed in a stable 1 position, about 5.5 NM from the prime recovery ship, USS Okinawa, and about 1 NM from the planned landing point.

Weather in the prime recovery area was as follows: visibility 12 miles, wind 10 knots, scattered cloud cover 2,000 ft., isolated showers, and wave height 3 feet.

ASTRONAUT RECOVERY OPERATIONS

Following CM landing, the recovery helicopter dropped swimmers who installed the flotation collar and attached the life raft. Fresh flight suits were passed through the hatch for the flight crew. The post landing ventilation fan was turned off, the CM was powered down, the crew egressed, and the CM hatch was secured.

The helicopter recovered the astronauts and flew them to the recovery ship. After landing on the recovery ship, the astronauts proceeded to the Biomed area for a series of examinations. Following the examinations, the astronauts departed the USS Okinawa the next day, were flown to Hickam Air Force Base, Hawaii, and then to Ellington Air Force Base, Texas.

COMMAND MODULE RETRIEVAL OPERATIONS

After astronaut pickup by the helicopter, the CM was retrieved and placed on a dolly aboard the recovery ship. All lunar samples, data, and equipment will be removed from the CM and subsequently returned to Ellington Air Force Base, Texas. The CM will be offloaded at San Diego where deactivation of the CM propellant system will take place.

SYSTEMS PERFORMANCE

The Saturn V stages performed nominally.

The spacecraft systems were also near nominal throughout the mission with the exception of the intermittent short circuit in the Delta V thrust switch A (downstream of a necessary SPS valve function for bank A), the CSM/LM failure to undock at 100:14:00 GET, and an increase in CM tunnel pressure subsequent to the cabin integrity check preceding LM jettison.

All anomalies were rapidly analyzed and either resolved or workaround procedures developed to permit the mission to safely continue.

All anomalies are listed in Table 9 through 13.

FLIGHT CREW PERFORMANCE

The Apollo 15 flight crew performance was excellent throughout the mission.

All information and data in this report are preliminary and subject to revision by the normal Manned Spaceflight Centers' technical reports.

C.M. Lee

IN-FLIGHT SCIENCE

The first Command and Service Module Orbital Science Payload commenced operation after successful door jettison and Lunar Orbit Insertion. Although there was a problem with the Panoramic Camera V/H Sensor and the Laser Altimeter, all the major photography objectives using the service module cameras were achieved. All of the other orbital experiments operated as designed. The subsatellite was successfully placed in lunar orbit, and its experiments all operated as planned.

EXPERIMENTS

Gamma-Ray Spectrometer

The Gamma-Ray Spectrometer performs a remote compositional survey of the upper 30 cm of the lunar surface by detecting the gamma rays emitted during the radioactive decay of the naturally occurring radioisotopes (40_k, 238_u, 232_{Th}, and their daughter products) and of the radioisotopes produced by cosmic ray bombardment of lunar surface materials (O, Mg, Al, Si, Fe). The instrument measures the number of gamma rays emitted from the lunar surface in each of 512 energy increments between 0.10 and 10 Mev. The instrument is mounted on a 25-ft boom to remove it from the gamma ray background caused by radioactive materials on the CSM and by secondary gamma rays due to cosmic ray bombardment of the CSM.

The instrument performed particularly well both in lunar orbit and during Transearth Coast (TEC). All commands to the instrument via the crew were implemented, including gain changes and charged particle rejections.

During operating periods prior to CSM/LM unlocking, gamma rays from the ALSEP Radioisotope Thermoelectric Generator (RTG) dominated the gamma ray spectrum and several RTG gamma ray line spectra were identified. After unlocking a significant decrease in the gamma ray background was observed at boom extension. Fifty-two hours of prime gamma ray data (minimum background configuration) were obtained during the lunar orbital phase and the data is 100% useful. Although detailed analysis of the data is required before the chemical composition of a particular region can be identified, on the basis of real-time data, several lunar features have been identified which have an above average Th concentration. In addition, the total gamma ray activity was found to be slightly higher on the backside than on the frontside.

The spectrometer was also operated for 50 hours during TEC to obtain background data necessary for the detailed analysis of the lunar data, to perform a galactic gamma ray survey, and to examine particular galactic gamma ray sources. Significant gamma ray activity was observed at this time, but the real time data displayed in the Mission Control Center was insufficient to identify the line spectra.

The instrument was also operated in lunar orbit for 58 hours in various non-minimum gamma ray background configurations. In addition to providing significant lunar gamma ray information, these data will determine the actual effect of the various background sources on the prime data and may allow a relaxation of the constraints on instrument operation on Apollo 16.

X-Ray Spectrometer

The X-ray Spectrometer performs a compositional survey of the topmost lunar surface by detecting secondary fluorescent X-rays emitted by the constituent elements when they are bombarded by solar X-rays.

The instrument performed exceptionally well both in lunar orbit where over 100 hours of data were acquired and during TEC where 50 hours of galactic data including seven discrete X-ray sources were acquired.

During TEC the recently discovered X-ray Pulsar CX-1 was observed continuously for 30 minutes, the first time a galactic X-ray source has been observed this long continously. Simultaneously, with the Apollo observations of CX-1, the source was also monitored by the Soviet observatory in the Crimea (Crimean Astrophysical Observatory) with a 100-inch optical telescope. The Apollo X-ray data and the Soviet visual observations will be used to derive a model consistent with both sets of data.

The data acquired is 100% useful, Since the count rate is higher than predicted, the data can be compiled in shorter time intervals permitting compositional maps with improved spatial resolution.

Preliminary analysis of the data acquired on rev 16 between 95°E and 40°E shows a definite correlation of Al concentration and Mg/Al ratios with lunar features. Over both Mare Smithii and Mare Crisium a depletion of Al and an enhancement of Mg was found over that observed in the adjacent highlands.

Alpha-Particle Spectrometer

The Alpha-Particle Spectrometer seeks to locate cracks or fissures in the lunar surface by detecting the alpha particles emitted by the decay of two of the isotopes of the inert radioactive gas radon. In addition to providing data on radon, the instrument also corroborates the data from the Gamma-Ray Spectrometer by detecting alpha particles emitted during the decay of U, Th, and their daughter products.

The instrument was operated for over 100 hours in lunar orbit and about 90-95% of the data are useful. Two of ten detectors were intermittently noisy at high temperatures, but this noise can be removed by later detailed analysis.

Preliminary analysis of the data indicates that the rate of radon evolution on the moon is at least one thousand times less than that on the earth.

Background data were also acquired for 50 hours during TEC.

Mass Spectrometer

The Mass Spectrometer (MS) measured the composition and density of neutral molecules present in the lunar atmosphere at 60 NM. Due to outgassing and venting from the CSM, the MS is mounted on a 24-ft, boom.

Operation of the instrument was nominal, but an apparent problem with the boom retraction mechanism resulted in the decision not to extend the MS during one scheduled data acquisition period, and the crew failed to turn the instrument on during a second scheduled data acquisition period. However, the subsequent experiment timeline was modified to include additional MS operating periods, and over 46 hours of useful data were acquired including 7 hours of background data. In addition, the instrument was operated for 50 hours during TEC with the boom extended various distances from the CSM to determine spacecraft venting and outgassing levels. A preliminary analysis of the data indicates the presence of several constituents which may be native to the lunar atmosphere. Argon mass 40, varies in constituents from approximately 2×10^5 particles/CM^3 on the dark side to about 6×10^5 particles/CM^3 in sunlight, while Argon 36 varies similarly from 6×10^4 to 2×10^5 particles/CM^3. Short intense bursts of several gases including CO_2 as well as masses 36 and 56 have been observed on the backside, but detailed analysis is required before these can be attributed to either the CSM or to lunar phenomena.

S-Band Transponder

S-band Doppler tracking of the CSM during inactive periods and of the LM during unpowered descent occurred over the three largest known lunar gravitational anomalies (Mare Imbrium, Mare Crisium, and Mare Serenitatus).

Preliminary data from Apollo 15 at an altitude of 15 km corroborates the mascon data obtained from Lunar Orbiter from an altitude of 200 km over these features. The gravitational profile has been compared with the Laser Altimeter data acquired over these same features, and some interesting correlations have been observed. However, further reduction of the data is required in order to remove the topographic effect and to determine the size and depth of the mascon.

The S-band Transponder onboard the Subsatellite (described below) will be tracked for approximately one year as the altitude decays from its present orbit and eventually impacts the lunar surface. Repeated overflights of the regions between 28°N and 28°S at these varying altitudes will permit an accurate detailed gravitational profile of the frontside between these latitudes and also permit an estimate of the gross gravitational anomalies on the backside.

Subsatellite Magnetometer and Charged Particles Detectors

The Subsatellite was successfully deployed from the CSM into a 76.3 x 55.1 NM lunar orbit with an inclination of -28.7°. Scientific data with the particle detectors and magnetometer is being obtained.

The particle experiment uses five curved plate particle detectors and two solid state telescopes to study the boundary layer of the solar wind plasma interaction with the moon and with the earth's geomagnetic tail. This interaction region extends outward from the lunar surface to 100 km and is characterized by the properties of the plasma as well as those of the moon. These data will yield information on the external plasma, the lunar interior, the lunar surface electrical charge, and the lunar ionosphere.

The magnetic field experiment employs a biaxial fluxgate magnetometer to measure the magnetic field at orbital altitudes. In addition to providing data in the interaction of the solar wind with the moon, the magnetometer data will be correlated with surface magnetic field measurements made by the Apollo 12 and 15 ALSEP/LSM's. These simultaneous measurements will lead to a determination of the electrical conductivity of the deep interior.

Bistatic Radar Experiment

The Bistatic Radar Experiment was conducted using the onboard CSM S-band and VHF communications systems. Two complete dual-frequency frontside passes were conducted on revs 17 and 28, respectively. In the dual-frequency mode, the S-band high gain antenna and the VHF omni antenna were used, and a spacecraft maneuver was performed to maintain the proper geometry between the HGA, the lunar surface, and the earth. The S-band data was received by the 210' Goldstone antenna and the VHF by the 150' antenna at Stanford University. A VHF only bistatic radar operational period, consisting of six complete frontside passes, was conducted during the crew sleep period from 180 hours GET to 193 hours GET. During the VHF only mode the CSM remains in the SIM down attitude, and data collection by the SIM experiments was not impacted in any way.

Although considerable processing of the received signal is required before the experimental data can be analyzed, a first look indicated strong received signals with good data potential for determination of the bulk dielectric constant and near surface roughness along the spacecraft track.

The spacecraft ground track during both the dual-frequency and VHF-only portions of the bistatic radar experiment intersected the ground tracks of the Apollo 14 bistatic radar experiment. This should permit a cross correlation of the Apollo 14 and 15 bistatic data with a common reference point.

UV Photography — Earth and Moon

The UV photography experiment was conducted from the CM using a 70 mm Hasselblad camera with a 105mm UV transmitting lens and four spectral filters centered at 2600 Å, 3250 Å, 3750 Å, and 5000 Å respectively. The filters were sequentially rotated over the camera lens and the film exposed. The photographs were taken through the CM right hand side window which is of a special double pane quartz construction.

The purpose of the UV photography of earth is to determine if there is a correlation between the observed UV radiation and known meteorological conditions. If a correlation is shown to exist, it will then be possible to extrapolate by planetary analogs to the atmospheres of Mars and Venus.

Photographs of the earth were taken in each of the 4 spectral regions, from earth orbit, from 1/4, 1/2 and 3/4 of the lunar distance during TLC, from lunar orbit, and from 2/3, 1/2, and 1/4 of the lunar distance during TEC.

UV photographs of the moon were also taken and will be used to extend the earth-based calorimetric work and to search for lunar UV fluorescence.

Gegenschein Photography

The Gegenschein experiment used a 35 mm Nikon camera in the CM to take 12 exposures (~90 sec each) of a region encompassing the Moulton Point and the antisolar point of the earth. Since the Gegenschein is a very faint glow, it is necessary to have a minimum of scattered light in the vicinity of the camera. The Gegenschein photography conducted on Apollo 15 took advantage of the CSM operating in the darkest region of the universe accessible to man — the "double umbra" behind the moon where the CSM was shielded from both sunlight and earthshine.

In addition to providing a dark photographic point, the unique geometry offered by the "double umbra" permitted photography of the region surrounding the Moulton Point from 15° off the earth-sun line. Analysis of these photographs then will determine whether a relationship exists between the Gegenschein and the Moulton Point (i.e., whether the Gegenschein is due to dust particles trapped at the Moulton Point or whether the observed light comes from the cosmic dust of the zodiacal light).

CM Window Meteoroid

The experiment takes advantage of the CM windows as meteoroid collectors for particles with masses $\geq 10^{12}$ gms and will use the data to investigate the degradation of surfaces in the space environment due to such bombardment. The windows are examined optically postmission to accurately locate all impacts. On the basis of the crater diameter, the meteoroid material may also be present in the impact crater diameter, the meteoroid mass can be calculated. Sufficient meteoroid material may also be present in the impact crater to permit an analysis of the meteoroid composition.

CSM Photographic Tasks

SM Camera System

Panoramic Camera - The objective for the panoramic camera was to obtain high resolution approximately 2 meters) photography for all areas overflown by the spacecraft in daylight. Priorities for coverage were:

> Landing site, pre-and post-EVA
> Several areas considered as possible candidates for Apollo 17 landing site
> LM impact point
> Near terminator areas
> General coverage

Telemetry from the first camera pass on rev 4 indicated that the V/H automatically resets to a nominal 60-NM altitude. For the remainder of the mission the sensor ocillated between off scale and nominal. It is expected that 80% of the photography will be high quality and 20% degraded. All critical areas have been photographed with good pictures.

Mapping Camera — The objective of the mapping camera was to obtain cartographic quality photography for all areas overflown by the spacecraft in daylight. To assist in data reduction, a stellar photograph and a laser altitude measurement were to be made in synchronism with each mapping camera photograph: Mapping camera operation was desired on all pan camera passes and on selected dark side passes where the laser altimeter was operating. Mapping camera functioning was nominal throughout the mission, and it is expected that the associated stellar camera was operating, although there is no telemetry data for that system to confirm its operation. On rev 38 the laser altimeter ceased to operate. All subsequent dark side passes with the mapping camera for altitude data (no mapping imagery) were deleted from the flight plan.

On rev 50, the mapping camera was turned off during the pan camera pass over the landing site to check the remote possibility that pan camera malfunctioning might be related to mapping camera operation. There was no effect on pan camera operation, and deletion of the mapping camera pass resulted in an insignificant loss in coverage.

Changes in planned photo passes occasioned by the delay in LM jettison caused a small decrease in sidelap between rev 50 and rev 60. This will cause a slight weakness in data reduction in this area.

Loss of laser altimeter data will also cause a decrease in accuracy of the lunar control network established from the mapping camera photographs. The objective was ±15 meters, and this may be reduced to approximately ±30 meters.

In spite of the above losses, all major objectives were met.

Laser Altimeter — The objectives for the laser altimeter were to provide an altitude measurement in synchronism with each mapping camera exposure on the light side, and to provide independent altitude measurements on the dark side to permit correlation of topographic profiles with gravity anomalies.

Altimeter operation was nominal through rev 24. On rev 27 it was noted that some altitude words were clearly in error, and this situation became progressively worse on revs 33, 34 and 35. The errors were associated with a rise in cavity temperature. The mapping camera altimeter was left extended, but not operating, for the dark side pass on rev 38, it was determined that the altimeter had failed completely.

The altimeter was deleted from all subsequent dark side passes, but it was operated with all mapping camera passes.

On rev 63 an attempt was made to correct the altimeter operation by a switch operation routine conducted by the CMP, but it was not successful. Only about half of the total planned altimeter data was obtained. Although correlation of topography with gravity can be done on the light side using elevation data obtained from eventual reduction of the mapping photographs, the ability to do this on the dark side will be reduced to those areas where valid telemetry was obtained.

Command Module Photography

Lunar surface photography from the Command Module was planned to complement the SM photography. This included:

Oblique photography of special targets using the electric Hasselblad camera with the 80 mm and 250 mm lenses using both BW and CEX film. All scheduled targets were taken as planned except for the following cases:

- Target 25 was scheduled on rev 16 and was taken on rev 33. The delay was caused by VHF check on REV 16.

- Target 14 and 12 which were scheduled on revs 58 and 59 respectively were deleted due to the delay in LM jettison and subsequent shift of the rest cycle.

Near-terminator photography using high speed BW film. Two targets scheduled on rev 58 (farside and nearside) were deleted for same reason; near-terminator photography was taken on rev 63 (farside only).

Earthside photography on rev 34 was completed as planned.

Visual Observations from Lunar Orbit

The objective of "visual observation from lunar orbit" was implemented for the first time on Apollo 15. The CMP was asked to make and record observations of special lunar surface areas and processes. Emphasis was placed on characteristics which are hard to record on film and could be delineated by the eye, such as subtle color differences between surface units. All of the scheduled targets were accomplished, and the CMP relayed to the ground results of his careful and geologically significant observations such as:

The discovery of fields of cinder cones made by volcanic eruptions on the southeast rim of Mare Serenitatis (Littrow area) and southwest rim of the some Mare basin (Sulpicius Gallos area).

The delineation of a landslide or rock glacier on the northwest rim of the crater Tsiolkovsky on the lunar farside.

Interpretation of the ray-excluded zone around the crater Proculus on the west rim on Mare Crisium is due to the presence of a fault system at the west rim of the crater.

The finding of layers on the interior walls of several craters which were interpreted as volcanic collapse creaters of "calderas" in the maria.

CM Astronomical Photography

Several tasks of astronomical photography were carried out using very high speed black and white film in the 35 mm Nikon, the 70 mm Hasselblad, and the 16 mm Data Acquisition Cameras as appropriate. These included three sets of photographs of the solar corona, one set each of zodiacal light and lunar libration point L_4, a set of the moon as it entered full eclipse and another set as it exited, and four sets of star field photos using the sextant. Although data quality cannot be assessed until the recovered film has been processed, all operational procedures were carried out as planned, and crew reports on film usage indicated the expected values.

The Apollo flights provide two unique conditions necessary for the performance of some of these tasks - the long earth-moon baseline for parallax shift in studying concentrations of dust-like micrometeoroids reflecting sunlight compared to the background star field and also the extremely dark conditions of the moon's double umbra. The eclipse photography is simply a chance opportunity to study reddening of the lunar disk as a known reflectance target for spectral scattering and transmittance effects of the earth's atmosphere. Stellar field sextant photos will indicate the future usefulness of that system for comet and stellar photography during translunar and transearth coast periods.

Visual Light Flash Phenomenon Test

The Apollo 15 visual light flash phenomenon test was successfully completed as scheduled.

The test consisted of three separate observation periods of approximately one hour duration for each period. The first session was conducted during translunar coast (51:37-52:33 GET), the second during lunar orbit (197:00-198:00 GET), and the final session during transearth coast (264:35-365:35 GET).

During the observation sessions, crew members wore eyeshields (blindfolds) to prevent light from entering their eyes, and reported by voice communications to ground each time a light event was perceived. This report was followed by comments pertaining to the characteristics of the individual light flashes. These descriptive comments were recorded on CM tape in lieu of real-time voice communication, for subsequent playback to the Mission Control Center.

Voice data obtained have been tabulated and are being evaluated. Preliminary assessment of the frequency data reported by crew members reveal the average frequency of occurrence of light flashes to range from a high of about one light flash event every two minutes (reported during translunar coast) to a low of about one light flash event every seven minutes (reported during transearth coast). Final assessment of the data stored on tapes and reporting of results will be made available to the Principal Investigators.

SURFACE SCIENCE

The first Apollo 15 surface science event was the impact of the S-IVB stage at 79:24:41 GET. The impact point was 1.0°S and 11.87°W, approximately 188 km northeast of the Apollo 14 site and 355 km east of the Apollo 12 site. The seismometers at both sites recorded the impact, and preliminary analysis indicates that the lunar subsurface east of the instruments is similar to the subsurface to the west and south where previous impacts have taken place. The distance of the impact from the seismometers will facilitate analysis of the subsurface to depths of 50-100 km. Previous impacts extended our subsurface knowledge to a depth of approximately 30 km.

The debris and particle cloud created by the impact was recorded by the Apollo 12 Solar Wind Spectrometer, and Suprathermal Ion Detector. At the Apollo 14 site the Charged Particle Lunar Environment Experiment, Suprathermal Ion Detector, and the Cold Cathode Ionization Gauge detected the event. Analysis of the data is now continuing.

Approximately two hours after landing, CDR Scott commenced the stand-up EVA (SEVA). Bearings were taken of known lunar features to assist in the landing point determination. A photographic panorama of twenty-two frames of 70 mm photos was completed as well as a number of 500 mm photos. A detailed verbal description was made of both distant and near field objects. Trafficability for the LRV was considered to be good, and a usable deployment site for the ALSEP was felt to be present to the west. Rock fragments larger than 8 inches were not observed near the LM. The surface appeared smooth and rounded with many more 8-10 meter craters present than anticipated. Lineaments were seen on the slopes of Hadley Delta, but no flows or landslides could be seen. One dark black fragment 6-8 inches long was observed; all other fragments were light colored.

The crew commenced the 1st EVA traverse aboard the LRV at 121:44:56 GET. Traverse route was direct to Station 1 (Elbow Crater), omitting the first check point. Crew observations of the craters, rock fragments, and lurain conditions were made while the LRV was underway. Hadley Rille rim was encountered north of Station 1 and followed toward the south until arrival at Elbow Crater. Documented samples and a radial sample were collected. The traverse continued south to Station 2, on the flank of St. George Crater. Documented samples, a comprehensive sample, and 500 mm photos were taken. From Station 2 the crew returned directly to the LM, omitting Station 3. ALSEP deployment occupied the remainder of the EVA. All experiments were commanded on before EVA termination at 126:11:59 GET.

To provide adequate geological exploration at the Apennine front and accommodate ALSEP closeout activities at the end of EVA-2, changes were made to the nominal plan. The traverse commenced at 143:10, proceeding south to the front. Descriptions were made of the secondary crater cluster as the crew drove by and observations made of the front. The crew continued up the front toward Spur Crater and noted that rocks were more abundant here than on the flank of St. George and were especially abundant on the rim of Spur. A stop was made east of Spur (Station A). Documented samples, some of which were breccias, single core, special environmental sample, and soil samples were collected. A panorama and 500 mm photos were taken. The soil was described as more powdery than seen earlier. A trench was also dug and sampled. The crew then turned southwest to sample a large (3M on a side) boulder (Station B). The boulder was layered and consisted in part of breccia with a thick, green colored layer. Communications became noisy at this station, and crew voice was relayed through the LM which was within sight but approximately 5 km away. Spur Crater was sampled next, and the crew believed they collected a large fragment of anorthositic rock. Light green to gray rocks and breccias were also collected and documented. A comprehensive sample was also taken. From Spur the crew returned to Dune Crater, one of the secondary craters, and sampled the west rim. A comprehensive sample and documented samples were collected. Returning to the LM the crew completed deployment of the Heat Flow Experiment and ALSEP documentation. Station 8 activities were carried out near the ALSEP. A trench was dug and sampled, and 6 penetrometer readings made. At the end of the EVA, a 7' 4" core was drilled but not extracted from the ground.

The beginning of the crew sleep period was delayed between EVA-2 and EVA-3, necessitating a shortening of the final EVA. The crew started the 3rd EVA at 163:18:14 GET and extracted the core sample. The crew then proceeded westward to the rille. Documented samples, a comprehensive sample and bedrock samples were collected. Pans and 500 mm pictures were also taken. North complex stations were omitted, and the crew returned directly to the LM. The core was separated and stowed. The Solar Wind Composition Experiment, deployed at the end of the 1st EVA, was retrieved after 41 hours 8 minutes of exposure. The LRV was parked east of the LM for liftoff and post liftoff TV pictures. Samples and camera magazines were transferred to the LM, and the 3rd EVA terminated at 168:08:04- GET. Total samples collected during all three EVA's is estimated to be between 170-180 pounds.

After LM jettison the LM was deorbited and impacted west of the Apollo 15 landing site at 26.3° N and

9.25°E. All three seismometers recorded the impact, which was 93 km west of Apollo 15, 1057 km north of Apollo 14, and 1144 km north of Apollo 12 (see Figure 4; also Experiment Description).

TV panoramics were taken throughout all the surface activities.

On August 4 a final TV panorama was taken of the Hadley Apennine area. The higher sun angle revealed lineations on the front which were previously poorly defined or unobserved.

Apollo Lunar Science Experiments

The Apollo Lunar Science Experiments (ALSEP) package was deployed on EVA's-1 and -2 and the antenna on the central station was properly aligned. The Radioactive Thermoelectric Generator (RTG) continues to supply 74 watts of electric power and the downlink telemetry signal strength is -136 db.

The Passive Seismic Experiment

The Passive Seismic Experiment (PSE) was leveled on command to 0.2" arc, and the sensor reached equilibrium at 126°F. The Apollo 15 PSE, along with Apollo 12 and 14, constitute a network of widely spaced stations simultaneously recording seismic signals. The S-IVB/IV impact was recorded by the Apollo 12 and 14 PSEs, and the Apollo 15 LM impact was recorded by all three instruments. These five new data points, plus the earlier data, indicate that a major increase in velocity could occur at a depth of about 25 km. Such an increase at this depth would indicate a change of composition in the lunar material. A depth of 25 km is not considered sufficient to produce a change in velocity with change in pressure alone. The higher velocity indicated by these new data is on the order of 7.5 km/sec which is equivalent to velocities in the earth's mantle.

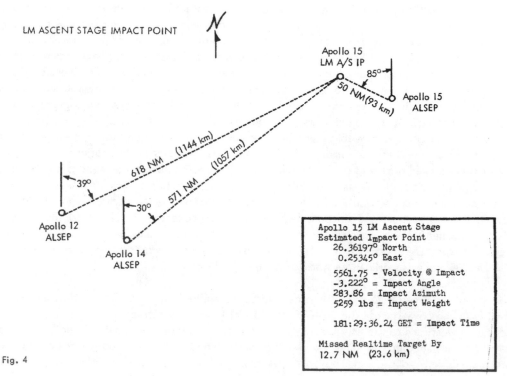

Fig. 4

The recording of the Apollo 15 LM impact at great distances (see Figure 4) will be valuable in calibrating long-range characteristics of the moon.

Seismic signals generated by the LRV during EVA-2 and -3 traverses vary smoothly in amplitude according to the distances between LRV and PSE. These data will help us understand the physical properties of near surface lunar materials to depths of 1-2 km.

The Lunar Surface Magnetometer

The Lunar Surface Magnetometer (LSM) completed its one-time site survey sequence. The site survey data are now being analyzed. The internal (remanent) magnetic field at the Apollo 15 site appears to be much lower than at Apollo 12 and Apollo 14 sites. The Principal Investigator suggests that this difference may be associated with the proximity of Apollo 15 site to one of the mascons. The Apollo 15 LSM, along with the Apollo 12 LSM, gives us two magnetometers functioning on the moon. In addition, data is being recorded simultaneously with the magnetometer on the Apollo 15 subsatellite. From analysis of this data it will be possible to determine the interior electrical conductivity and calculate the temperature profile to the center of the moon.

The Solar Wind Spectrometer

The Solar Wind Spectrometer (SWS) began recording science data shortly after LM liftoff. The Apollo 12 SWS, as well as the Apollo 15, are now both functioning nominally. Since the moon is now in the magnetospheric tail of the earth, no solar wind data is being recorded. Science data will begin August 9, 1971, When the moon comes back into interplanetary space (see Figure 5).

The Suprathermal Ion Detector Experiment and Cold Cathode Galuge Experiment

The Suprathermal Ion Detector Experiment (SIDE), which measures the lunar ionosphere, and Cold Cathode Gauge Experiment (CCGE), which measures the pressures of the ambient lunar atmosphere, both were deployed and turned on for an initial checkout. After 28 minutes, they were commanded to standby, i.e., instruments are on, but high voltage turned off. The instruments were commanded "ON" to record the LM cabin depress and LM impact. They are now in standby until lunar sunset August 13, 1971. Three SIDE's are operating simultaneously (Apollo 12, 14, 15) and two Cold Cathode Gauges (Apollo 14, 15) (see Figure 5).

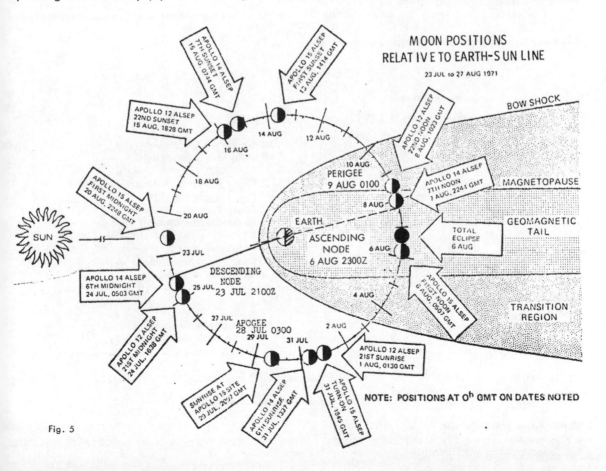

Fig. 5

Heat Flow Experiment

The two heat flow probes are deployed in the lunar subsurface. Their deployment was not nominal due to the shallow boreholes drilled. The investigators are confident that, despite the shallow depth of the probes, valid measurements of the net heat flow from the moon can be obtained over relatively long periods of time. The net heat flow will be based on thermal gradient and thermal conductivity measurements. Thermal gradient measurements are taken from both the bridge sensors and ring sensors on the probe giving a total of eight data points on each reading. A thermal conductivity measurement will be run within the 45-day period of real-time support. Initial data shows a temperature drop of over 100°C in the first 80cm (32 in.) and a slight increase (a few hundredths of a degree) in the next meter. This indicates that the lunar material has extremely low thermal conductivity. During the four-hour period of the eclipse the thermocouple on the lunar surface registered a drop from +87° C to −128° C and then returned to +87° C. This data will help determine the thermal conductivity of lunar surface materials.

Laser Ranging Retroflector

The Laser Ranging Retroflector (LR³) was acquired by McDonald Observatory subsequent to LM liftoff. Good quality signals are now being received from all three LR³'s (Apollo 11, 14, and 15).

TABLE I
APOLLO 15 OBJECTIVES AND EXPERIMENTS PRIMARY OBJECTIVES

The following were the NASA OMSF Apollo 15 Primary Objectives:

Perform selenological inspection, survey, and sampling of materials and surface features in a preselected area of the Hadley-Apennine region.
Emplace and activate surface experiments.
Evaluate the capability of the Apollo equipment to provide extended lunar surface stay time, increased EVA operations, and surface mobility.
Conduct in-flight experiments and photographic tasks from lunar orbit.

APPROVED EXPERIMENTS

The following experiments were performed:

Apollo Lunar Surface Experiments Package (ALSEP)

S-031 Lunar Passive Seismology
S-034 Lunar Tri-Axis Magnetometer
S-035 Medium Energy Solar Wind
S-036 Suprathermal Ion Detector
S-158 Cold Cathode Ionization Gauge
S-037 Lunar Heat Flow
M-515 Lunar Dust Detector

Lunar Surface

S-059 Lunar Geology Investigation
S-078 Laser Ranging Retro-Reflector
S-080 Solar Wind Composition
S-200 Soil Mechanics

In-Flight

S-760 Gamma-Ray Spectrometer (SIM)
S-161 X-Ray Fluorescence (SIM)
S-162 Alpha-Particle Spectrometer (SIM)
S-164 S-band Transponder (CSM/LM) (Subsatellite)
S-165 Mass Spectrometer (SIM)
S-170 Bistatic Radar (CSM)
S-173 Particle Shadows/Boundary Layer (Subsatellite)
S-177 UV Photography - Earth and Moon (CM)
S-178 Gegenschein from Lunar Orbit (CM)
Visual Observations from Lunar Orbit
Visual Light Flash Phenomena

Other

M-078 Bone Mineral Measurement
M-079 Total Body Gamma Spectrometry

DETAILED OBJECTIVES

The below-listed detailed objectives were assigned to and accomplished on the Apollo 15 Mission:

Contingency Sample Collection
LRV Evaluation
EVA Communications with LCRU/GCTA

EMU Assessment on Lunar Surface
LM Landing Effects Evaluation
SM Photographic Tasks
CM Photographic Tasks
SIM Thermal Data
SIM Bay Inspection during EVA
SIM Door Jettison Evaluation
Visual Observations from Lunar Orbit
Visual Light Flash Phenomena
Impact S-IVB on Lunar Surface
Postflight Determination of S-IVB Impact Point

SUMMARY

Fulfillment of the Primary Objectives qualifies Apollo 15 as a successful mission. The Experiments and Detailed Objectives which supported and expanded the scientific and technological return of this mission were successfully accomplished.

TABLE 2
APOLLO 15 ACHIEVEMENTS

Fourth Manned Lunar Landing
Largest Payload Placed in Earth Orbit (309,330 lbs.)
Largest Payload Placed in Lunar Orbit (74,522 lbs.)
First SIM Bay Flown and Operated on an Apollo Spacecraft
First LM Descent Using Steepened 25° Approach
Longest Lunar Surface Stay Time (66 Hours 55 Minutes)
Longest Lunar Surface EVA (18 Hours 34 Minutes)
Longest Distance Traversed on Lunar Surface (27.90 KM on LRV Odometer)
First Use of Lunar Roving Vehicle (Manned)
First Use of a Lunar Surface Navigation Device
First Use of Lunar Communications Relay Unit for Direct Voice, EMU Telemetry, and TV From Distant Traverse Stations Without LM Relay
First Use of Ground Controlled Remote Operation of TV Camera on the Moon
First Subsatellite Launched in Lunar Orbit
First EVA From CM During Transearth Coast
First Standup EVA with Astronaut's Head Positioned Above the Opened Upper Hatch (Lunar Surface)
Largest Amount of Lunar Samples Returned to Earth (Approximately 180 lbs.)
Longest Lunar Orbit Time (74 Orbits)

TABLE 3

APOLLO 15 POWERED FLIGHT SEQUENCE OF EVENTS

EVENT	PRELAUNCH PLANNED (GET) HR:MIN:SEC	ACTUAL (GET) HR:MIN:SEC
Guidance Reference Release	-17.3	-17.5
Liftoff Signal (TB-1)	0	0
Pitch and Roll Start	11.0	11.0
Roll Complete	23.0	23.0
S-IC Center Engine Cutoff (TB-2)	2:15.5	2:15.5
Begin Tilt Arrest	2:35.2	2:35.2
S-IC Outboard Engine Cutoff (TB-3)	2:38.4	2:39.0
S-IC/S-II Separation	2:40.1	2:40.7
S-II Ignition (Command)	2:40.8	2:41.8
S-II Second Plane Separation	3:10.1	2:10.7
S-II Center Engine Cutoff	7:58.4	7:59.0
S-II Outboard Engine Cutoff (TB-4)	9:09.0	9:08.5
S-II/S-IVB Separation	9:10.0	9:09.5
S-IVB Ignition	9:10.1	9:09.6
S-IVB Cutoff (TB-5)	11:38.6	11:34.3
Insertion	11:48.4	11:44.1
Begin Restart Preps (TB-6)	2:40:17.9	2:40:24.2
Second S-IVB Ignition	2:49:55.9	2:50:02.6
Second S-IVB Cutoff (TB-7)	2:55:52.3	2:55:53.3
Translunar Injection	2:56:02.1	2:56:03.1

Prelaunch planned times are based on MSFC Launch Vehicle Operational Trajectory.

TABLE 4

APOLLO 15 MISSION SEQUENCE OF EVENTS

EVENT	PLANNED (GET) HR:MIN:SEC	ACTUAL (GET) HR:MIN:SEC
Liftoff 00:09:34.6 EDT, July 26	00:00:00	00:00:00.6
Earth Parking Orbit Insertion	00:11:48.4	00:11:44.1
Second S-IVB Ignition	02:49:55.9	02:50:02.6
Translunar Injection	02:56:02.1	02:56:03.1
CSM/S-IVB Separation, SLA Panel Jettison	03:20:54	03:22:24
CSM/LM Docking	03:30:54	03:33:49.5
Spacecraft Ejection From S-IVB	04:16:00	04:18:00
S-IVB APS Evasive Maneuver	04:39:01	04:39:38
Midcourse Correction-1	11:55:53.9	Not Performed
Midcourse Correction-2	30:55:53.9	28:40:30
Midcourse Correction-3	56:31:14.7	Not Performed
Midcourse Correction-4	73:31:14.7	73:31:14
SIM Door Jettison	74:01:14.7	74:06:47
Lunar Orbit Insertion (Ignition)	78:31:15	78:31:45.9
S-IVB Impact	79:13:26	79:24:41.5
Descent Orbit Insertion (Ignition)	82:39:32	82:39:48.3
CSM/LM Undocking	100:13:56	100:39:30
CSM Separation	100:13:56	100:39:30
CSM Circularization	101:34:55	101:38:58
Powered Descent Initiate	104:28:55	104:30:09
LM Lunar Landing	104:40:57	104:42:29
Begin SEVA Cabin Depress	106:10:00	106:42:49
Terminate SEVA Cabin Repress	106:47:00	107:15:56
Begin EVA-1 Cabin Depress	119:50:00	119:39:10
Terminate EVA-1 Cabin Repress	126:50:00	126:11:59
Begin EVA-2 Cabin Depress	141:10:00	142:14:48
Terminate EVA-2 Cabin Repress	148:10:00	149:27:02
Begin EVA-3 Cabin Depress	161:50:00	163:18:14
CSM Plane Change (LOPC)	165:12:51	165:11:32
Terminate EVA-3 Cabin Repress	167:50:00	168:08:04
LM Liftoff	171:37:24	171:37:22
LM Tweak Burn	171:47:39	Not Performed
Terminal Phase Initiate Maneuver	172:29:39	172:29:39
LM/CSM Docking	173:30:00	173:35:47
LM Jettison	177:20:45	179:30:14
CSM Separation	177:25:45	179:50:00
Ascent Stage Deorbit	179:08:26	181:04:19
Ascent Stage Lunar Impact	179:31:41	181:29:36
Shaping	221:25:52	221:20:47
Subsatellite Launch	222:36:13	222:39:19
Transearth Injection	223:46:06	223:48:45
Midcourse Correction-5	240:48:24	Not Performed
CMP EVA Depress	242:00:00	241:57:57
CMP EVA Repress	242:48:00	242:36:09
Midcourse Correction-6	272:58:20	Not Performed
Midcourse Correction-7	291:58:20	291:56:48
CM/SM Separation	294:43:20	294:44:00
Entry Interface (400,000 ft)	294:58:20	294:58:54
Landing	295:11:46	295:11:53

TABLE 5

APOLLO 15 TRANSLUNAR MANEUVER SUMMARY

DATE: 7 August 1971

MANEUVER	GROUND ELAPSED TIME (GET) AT IGNITION (HR:MIN:SEC:)			BURN TIME (SECONDS)			VELOCITY CHANGE (FEET PER SECOND - FPS)			GET OF CLOSEST APPROACH — HT (NM) CLOSEST APPROACH		
	PRE-LAUNCH PLAN	REAL-TIME PLAN	ACTUAL	PRE-LAUNCH PLAN	REAL-TIME PLAN	ACTUAL	PRE-LAUNCH PLAN	REAL-TIME PLAN	ACTUAL	PRE-LAUNCH PLAN	REAL-TIME PLAN	ACTUAL
TLI* (S-IVB)	02:49:58	02:50:00.6	02:50:00.6	356	353.4	353.4	10 421	10414.7	10414.7	78:35:00.5 / 68.0	78:35:02 / 79	78:31:21 / 139
CSM SEP	03:20:54	03:10:51	03:22:24	3	-	-	0.5			78:35:00.5 / 68	---	78:31:21 / 139
CSM DOCK	03:30:54	03:33:54	03:33:49	NA	-	-	NA	-	-	78:35:00.5 / 68	---	78:31:21 / 139
LM EJT	04:16:00	04:16:00	04:18:00	3	3	4.6	0.3	0.3	0.5	78:35:00.5 / 68	---	78:31:20 / 126
S-IVB EVASIVE	04:39:01	04:39:38	04:39:38	80.2	5.1	5.1	10.1	9.6	9.7	79:14:354 / 0	78:28:26 / 0	78:28:55 / 0
MCC-1 (SPS)	11:55:539	-	NP	0	-	NP	0	-	NP	78:35:00.5 / 68	---	NP
MCC-2 (SPS)	30:55:539	28:40:00	28:40:30	0	0.85	0.72	0	4.8	5.3	78:35:00.5 / 68	78:35:01 / 68	78:35:17 / 63
MCC-3	56:31:147	-	NP	0	-	NP	0	-	NP	78:35:00.5 / 68	---	NP
MCC-4	73:31:147	73:31:14	73:31:14	0	0.92	0.92	0	5.4	5.4	78:35:00.5 / 68	78:35:06 / 68	78:35:06 / 68
SIM DOOR JETT	74:01:147	74:01:00	74:06:47	NA			NA			NA	---	---

NP - Not Performed

* - S-IVB Ignition

DATE: 7 August 1971

TABLE 6
APOLLO 15 LUNAR ORBIT SUMMARY

MANEUVER	GROUND ELAPSED TIME (GET) AT IGNITION (HR:MIN:SEC:)			BURN TIME (SECONDS)			VELOCITY CHANGE (FEET PER SECOND - FPS)			RESULTING APOLUNE/PERILUNE (N. MI.)		
	PRE-LAUNCH PLAN	REAL-TIME PLAN	ACTUAL	PRE-LAUNCH PLAN	REAL-TIME PLAN	ACTUAL	PRE-LAUNCH PLAN	REAL-TIME PLAN	ACTUAL	PRE-LAUNCH PLAN	REAL-TIME PLAN	ACTUAL
LOI	78:31:15	78:31:45.9	78:31:45.9	392.0	400.7	400.7	3000	3000.1	3000.1	170/58.3	169.6/58.4	170.1/57.7
S-IVB IMPACT	79:13:26	79:22:57	79:24:42	NA			NA			NA		
DOI	82:39:32	82:39:48.3	82:39:48.3	22.9	24.5	24.5	207.6	213.9	213.9	58.4/9.6	58.4/9.2	58.5/9.2
UNDOCKING	100:13:56	100:38:00	100:39:30	NA			NA			NA		
CSM SEP	100:13:56	100:38:00	100:39:30	3.3	6.54	7.2	1.0	1.0	1.1	59.8/8.4	60.8/8.9	60.9/9.0
CSM CIRC	101:34:55	101:38:58	101:38:58	3.9	3.59	3.59	70.8	68.3	68.3	64.7/54.3	64.9/54.3	64.7/53.
PDI	104:28:55	104:30:08	104:30:09	722.1	718.6	740.0	6697.6	6694	6694	0		
LANDING	104:40:57	104:42:07	104:42:29	NA			NA			NA		
CSM LOPC	165:12:51	165:11:32	165:11:32	16.5	18.1	18.1	308.6	330.9	330.9	59.8/59.2	64.5/53.2	64.5/53.
ASCENT	171:37:24	171:37:22	171:37:22	435.2	436.7	436.7	6055.5	6059	6059	45.6/9	42.5/9.0	42.5/9.0
TWEAK	171:47:39	NP		0.0	NP		0.0	NP		45.6/9		NP
TPI	172:29:39	172:29:39	172:29:39	2.6	---	---	73.7	74.2	72.7	61.5/43.9	64.4/39.9	64.4/38.7
DOCKING	173:30:00	---	173:35:47	NA			NA			---	---	64.1/53.8
LM JETT	177:20:45	177:20:33	179:30:14	NA			NA			NA		
CSM SEP	177:25:45	177:25:33	179:50:00	6.4	6.3	12.6	1.0	1.0	2.0	59.8/53.6	63.7/53.4	66.2/52.6
ASC DEORB	179:05:48	179:06:22	181:04:19	85.2	86.5	86.5	201.2	200.3	200.3	NA		
ASC IMPACT	179:31:41	181:29:23	181:29:35	NA			NA			NA		
SHAPING	221:25:52	221:20:47	221:20:47	3.4	3.3	3.3	64.2	66.4	66.4	77.6/575	76.1/54.3	76/54.3
SAT JETT	222:36:13	222:39:27	222:39:19	NA			NA			77.3/57.7	76.5/55.1	76.3/55.

TABLE 7

APOLLO 15 TRANSEARTH MANEUVER SUMMARY

DATE 7 August 1971

MANEUVERS	GROUND ELAPSED TIME (GET) AT IGNITION (HR:MIN:SEC:)			BURN TIME (SECONDS)			VELOCITY CHANGE (FEET PER SECOND - FPS)			GET ENTRY INTERFACE (EI) / VELOCITY (FPS) AT EI / FLIGHT PATH ANGLE AT EI		
	PRE-LAUNCH PLAN	REAL-TIME PLAN	ACTUAL	PRE-LAUNCH PLAN	REAL-TIME PLAN	ACTUAL	PRE-LAUNCH PLAN	REAL-TIME PLAN	ACTUAL	PRE-LAUNCH PLAN	REAL-TIME PLAN	ACTUAL
TEI (SPS)	223:46:06	223:48:45	223:48:45	137.8	141.2	141.2	3049.7	3046.8	3047.0	294:58:20 / 36 097.2 / -6.5	294:58:05 / 36097.2 / -6.5	294:57:45 / 36097. / -6.69
MCC-5	240:48:24	NP		0.0	NP		0.0	NP		294:58:20 / 36 097.2 / -6.5	---	---
MCC-6	272:58:20	NP		0.0	NP		0.0	NP		294:58:20 / 36 097.2 / -6.5	N/P	---
MCC-7	291:58:20	291:56:48	291:56:48	0.0	24.2	24.2	0.0	5.6	5.6	294:58:20 / 36 097.2 / -6.5	294:58:55 / 36096.4 / -6.49	294:58:55 / 36096.4 / -6.49
CM/SM SEP	294:43:20	294:41:55	294:44:00	NA			NA			NA / NA / NA		
ENTRY	294:58:20	294:58:55	294:58:54	NA			NA			294:58:20 / 36 097.2 / -6.5	294:58:55 / 36096.4 / -6.49	294:58:55 / 36096.4 / -6.49
SPLASH	295:11:46	295:12:23	295:11:53	NA			NA			NA / NA / NA		

NP--Not Performed

TABLE 8

APOLLO 15 CONSUMABLES SUMMARY
END OF MISSION

DATE: 7 August 1971

CONSUMABLE		LAUNCH LOAD	FLIGHT PLANNED REMAINING	ACTUAL REMAINING
CM RCS PROP (POUNDS)	U	208.6	122.0	122.0
SM RCS PROP (POUNDS)	D	1220.0	427.0	396.0
SPS PROP (POUNDS)	TK	40497.0	1379.0	1509.0
SM HYDROGEN (POUNDS)	U	82.1	22.7	20.9
SM OXYGEN (POUNDS)	U	944.5	380.0	391.0
LM RCS PROP (POUNDS)	U	532.7	119.1**	190.0**
LM DPS PROP (POUNDS)	U	19412.0	627.3*	1087.0*
LM APS PROP (POUNDS)	U	5175.2	203.8**	216.0**
LM A/S OXYGEN (POUNDS)	T	4.8	3.7**	4.6**
LM D/S OXYGEN (POUNDS)	T	87.1	46.5*	34.6*
LM A/S WATER (POUNDS)	T	85.0	23.5**	73.8**
LM D/S WATER (POUNDS)	T	423.0	27.2*	36.0*
LM A/S BATTERIES (AMP-HOURS)	T	592.0	181.0**	137.0**
LM D/S BATTERIES (AMP-HOURS)	T	2075.0	327.0*	595.8*

D -	DELIVERABLE QUANTITY
U -	USABLE QUANTITY
TK -	TANK QUANTITY
T -	TOTAL QUANTITY

* At LM Ascent Stage Liftoff
** At LM Ascent Stage Impact
N/A Not Available

TABLE 9
SA-510 LAUNCH VEHICLE DISCREPANCY SUMMARY

IU state vector error limits exceeded during EPO. Open
Intermittent talk back on S-IVB LH$_2$ prevalve closed position during second start preparation. Open
Venting from S-IVB/IV after APS-1 burn. Open
Reported S-IVB propulsion disturbances at time base-5 and time base-6. Open

TABLE 10
COMMAND/SERVICE MODULE 112 DISCREPANCY SUMMARY

Leakage observed at water panel injection port. Open
Service module reaction control system quads B and D isolation valves closed sometime during boost and closed again at S-IVB/spacecraft separation. Open
Service propulsion system thrust light on entry monitor system came on. Open
Circuit breaker 33 panel 226 (integral lighting) found to be open at GET 33:34 (AC main bus B under voltage alarm indicated problem). Open
At approximately 81:25 GET the battery relay bus read 13.66 volts (CC0232) instead of 32.00 volts. Open
Tunnel pressure increased subsequent to integrity check preceding LM jettison. Open
AC-2/main bus B glitches from GET 195:33 to GET 201:41 occurred 11 times and were later identified as nominal electrical signals for vacuum cleaner operation. Closed
Mission timer on panel 1 inoperative and stopped at 124:47:37 GET. Reset digits to zero at 125:52:26 GET and timer started. Open
Tape recorder tape deterioration during transearth coast on front portion of tape. Open
One main parachute partially closed after deploy. Open

TABLE 11
LUNAR MODULE 10 DISCREPANCY SUMMARY

Broken glass on range/range rate meter (tapemeter). Open
Variations in pump Delta P (GF2021) reading while operating on pump #1 just after cabin depressurization for second extravehicular activity (142:18 GET). Open
Bacteria filter for lunar module water gun is broken. Open
AGS fail alarm at 180:55 GET. Open

TABLE 12
LUNAR ROVING VEHICLE DISCREPANCY SUMMARY

Difficulty removing deployment saddle from lunar roving vehicle following deployment. Closed
No volts or amps on rover battery 2 readout. Open
No rover front steering (EVA-1). Cycling breaker and switch at start of second extravehicular activity restored front steering. Open
Automatic closing of one of the rover battery thermal covers did not operate properly. Closed

TABLE 13
APOLLO 15 CREW/EXPERIMENT EQUIPMENT DISCREPANCY SUMMARY

Panoramic camera velocity/altitude sensor erratic. Open
Suit water separator speed decreased to below master alarm level (800 rpm). Open
Lightweight headset failed: Fell apart when removed from stowage the first time. Open
Tone warning on lunar module pilot (intermittently). The warning was caused by air compressed in the water line which resulted during water charging operations. Closed
Drill stem stuck to drill (would not release). Open
Mass spectrometer boom talkback indicates half gray on "extended" and full barber pole on "retract." Open
Lunar module pilot's oxygen purge system antenna broken. Open
Lunar module pilot's and commander's retractable tethers failed. Open
Lunar surface 16mm magazine inoperative. Open
Sample return container handle would not latch. Open
Elevation control of TV control unit erratic. Open
Unable to separate drill core tube sections. Open
Failed 70mm camera. Open
Suit integrity check unsuccessful on initial attempt prior to LM jettison. Open
Laser altimeter not providing altitude data. Open
Lunar surface TV LCRU FM downlink lost at 211:30:06 GET. Open
Water in suit hose during lunar orbit. Open
Mapping camera did not retract at 228:00 GET. Open

ABBREVIATIONS AND ACRONYMS

AGS	Abort Guidance System
ALSEP	Apollo Lunar Surface Experiments Package
AOS	Acquisition of Signal
APS	Ascent Propulsion System (LM)
APS	Auxiliary Propulsion System (S-IVB)
ARIA	Apollo Range Instrumentation Aircraft
AS	Apollo/Saturn
BIG	Biological Isolation Garment
BSLSS	Buddy Secondary Life Support System
CCATS	Communications, Command, and Telemetry System
CCGE	Cold Cathode Gauge Experiment
CDR	Commander
CPLEE	Charged Particle Lunar Environment Experiment
CM	Command Module
CMP	Command Module Pilot
CSI	Concentric Sequence Initiation
CSM	Command/Service Module
DAC	Data Acquisition Camera
DDAS	Digital Data Acquisition System
DOD	Department of Defense
DOI	Descent Orbit Insertion
DPS	Descent Propulsion System
DSKY	Display and Keyboard Assembly
ECS	Environmental Control System
EI	Entry Interface
EMU	Extravehicular Mobility Unit
EPO	Earth Parking Orbit
EST	Eastern Standard Time
ETB	Equipment Transfer Bag
EVA	Extravehicular Activity
FM	Frequency Modulation
fps	Feet Per Second
FDAI	Flight Director Attitude Indicator
FTP	Fixed Throttle Position
GCTA	Ground Commanded Television
GET	Ground Elapsed Time
GNCS	Guidance, Navigation, and Control System (CSM)
GSFC	Goddard Space Flight Center
HBR	High Bit Rate
HFE	Heat Flow Experiment
HTC	Hand Tool Carrier
IMU	Inertial Measurement Unit
IU	Instrument Unit
IVT	Intravehicular Transfer
KSC	Kennedy Space Center
LBR	Low Bit Rate
LCC	Launch Control Center
LCRU	Lunar Communications Relay Unit
LDMK	Landmark
LEC	Lunar Equipment Conveyor
LES	Launch Escape System
LET	Launch Escape Tower
LGC	LM Guidance Computer
LH_2	Liquid Hydrogen
LiOH	Lithium Hydroxide
LM	Lunar Module
LMP	Lunar Module Pilot
LO	I Lunar Orbit Insertion
LOS	Loss of Signal
LOX	Liquid Oxygen
LPO	Lunar Parking Orbit
LR	Landing Radar

LRL	Lunar Receiving Laboratory
LRRR (LR³)	Laser Ranging Retro-Reflector
LSM	Lunar Surface Magnetometer
LV	Launch Vehicle
MCC	Midcourse Correction
MCC	Mission Control Center
MESA	Modularized Equipment Stowage Assembly
MHz	Megahertz
MOCR	Mission Operations Control Room
MOR	Mission Operations Report
MPL	Mid-Pacific Line
MSC	Manned Spacecraft Center
MSFC	Marshall Space Flight Center
MSFEB	Manned Space Flight Evaluation Board
MSFN	Manned Space Flight Network
NASCOM	NASA Communications Network
NM	Nautical Mile
OMSF	Office of Manned Space Flight
OPS	Oxygen Purge System
ORDEAL	Orbital Rate Display Earth and Lunar
PCM	Pulse Code Modulation
PDI	Powered Descent Initiation
PGA	Pressure Garment Assembly
PGNCS	Primary Guidance, Navigation, and Control System (LM)
PLSS	Portable Life Support System
PSE	Passive Seismic Experiment
PTC	Passive Thermal Control
QUAD	Quadrant
RCS	Reaction Control System
RR	Rendezvous Radar
RLS	Radius Landing Site
RTCC	Real-Time Computer Complex
RTG	Radioisotope Thermoelectric Generator
S/C	Spacecraft
SEA	Sun Elevation Angle
SEVA	Stand-up EVA
S-IC	Saturn V First Stage
S-II	Saturn V Second Stage
S-IVB	Saturn V Third Stage
SIDE	Suprathermal Ion Detector Experiment
SIM	Scientific Instrument Module
SLA	Spacecraft-LM Adapter
SM	Service Module
SPS	Service Propulsion System
SRC	Sample Return Container
SSB	Single Side Band
SSR	Staff Support Room
SV	Space Vehicle
SWC	Solar Wind Composition Experiment
TD&E	Transposition, Docking and LM Ejection
TEC	Transearth Coast
TEI	Transearth Injection
TFI	Time From Ignition
TLC	Translunar Coast
TLI	Translunar Injection
TLM	Telemetry
TPF	Terminal Phase Finalization
TPI	Terminal Phase Initiation
T-time	Countdown Time (referenced to liftoff time)
TV	Television
USB	Unified S-Band
USN	United States Navy
USAF	United States Air Force
VAN	Vanguard
VHF	Very High Frequency
Delta V	Differential Velocity

MSC-04561

NATIONAL AERONAUTICS AND SPACE ADMINISTRATION

APOLLO 15
TECHNICAL
CREW DEBRIEFING
(U)

AUGUST 14, 1971

PREPARED BY
TRAINING OFFICE
CREW TRAINING AND SIMULATION DIVISION

MANNED SPACECRAFT CENTER
HOUSTON, TEXAS

1.0 SUITING AND INGRESS

SCOTT — Starting out with the suiting and ingress, there were no problems. The suiting was on schedule, I think a little ahead of time. Ingress was nominal, and the cabin closeout looked good.

2.0 STATUS CHECKS AND COUNTDOWN

SCOTT — The communications were good. The countdown was smooth; I think we were probably 20 or 30 minutes ahead all the way. Can't think of any anomalies during the count. I thought the EDS checks went particularly well.

IRWIN — I guess I was surprised that the hydrogen flow indicator on fuel cell 2 was out. I hadn't been briefed on that before the flight.

SCOTT — Yes, that's right. That was a surprise. Nobody told us that. As soon as Jim called it, Skip came back and said, "that's right, it's out." Like you should have known it, I guess. Controls and displays were okay.

3.0 POWERED FLIGHT

SCOTT — My evaluation, compared to Apollo 9, was that the lift-off itself was softer and quieter. When the tiedowns went, we could feel definite motion, but it didn't seem like as much as it was on Apollo 9. The noise was relatively low-level, and none of us had any trouble with the comm at all. We had vibrations within the S-IC which were just about the same frequency as the noise you hear standing on the ground. You hear the reverberations from the engines or the S-IC vibrations were about the same frequency, low amplitude - just something you could feel. Going through max q was noisy, but we still had good comm. And it didn't seem to me that that was as loud as it was on Apollo 9 either. I could hear Jim call cabin pressure relieving very clearly. You could hear pretty well all the way through there, too.

IRWIN — Yes, I thought the comm was excellent.

SCOTT — The staging was as we expected, I guess. It was what I'd call violent when the S-IC shuts down and everything uncoils there, and that was almost identical to Apollo 9. It was really just a big bang. We saw the fireball come up to the BPC; I saw it in my left side window. I saw the fireball out the front window, too. Right after, or just prior to, the S-II ignition, there was a lateral motion, attitude-wise, in the vehicle. Sometime in this staging sequence, we got a slight yaw. The S-II was very smooth, and all the way through the S-II burn, we had a very slight - I'd guess in talking about it - we figured 10- to 12-cycle-per-second vibration, something in that range, low-amplitude, something you could just feel, but it was continuous all the way through. There was no pogo, no change in the oscillation.

S71-4036S

Tower jet was smooth and came away very cleanly. We didn't notice the PU shift; when we went through it, I couldn't feel anything. Could you, Jim? I remember on Apollo 9, we also didn't feel the PU shift, but I guess other crews have felt it.

The S-II to S-IVB staging was about a quarter to a fifth the force of the S-IC staging. It was again a positive kind of feeling, but it wasn't a violent crash like we felt on the S-IC, I didn't think. We had the same light 10- to 12-cps vibration on the S-IVB all the way into orbit. The shutdown was smooth.

All the sequences throughout the launch were nominal and as expected. All the lights worked good; controls and displays were good, comfortable.

IRWIN — The noise and the vibration were less than I was expecting; it was much less. I was impressed about the lateral vibration on launch. It was much greater, of course, on the S-IC than it was on the S-II. Just a shaking, back and forth, lateral vibration all the way through the launch. It was a pretty smooth ride.

4.0 EARTH ORBIT AND SYSTEMS CHECKOUT

IRWIN — The insertion parameters are here. I can't read them; my eyes are dilated.

SCOTT — Okay. 93.2, 88.9, and I think the ground eventually had us in a 93 circular. But that's in the ball park. The postinsertion systems configuration and checks went very smoothly. I don't think we had any problems at all. We took our time. We spent about 10 minutes or so just looking at the scenery after we cleaned everything up with the gimbal motors and all.

IRWIN — We had one secondary propulsion barber pole.

SCOTT — Yes, that's right.

IRWIN — Insertion B, B secondary, right?

SCOTT — Yes, we set that and it went great. I had it written in this one too. RCS-B secondary isolation valve barber pole, cycle to gray. It didn't come on at insertion. It came on at some other point.

IRWIN — Yes, we noticed it when we did the check.

SCOTT — I have a note in here when you did the fuel cell purge check. You confirmed, there wasn't any H_2 flow. Of course, we knew that.

IRWIN — We didn't get the MASTER ALARM on the H_2.

SCOTT — I have a note here. At about 53 minutes, we noted that the D primary and secondary RCS isolation valves were barber pole, and cycled them to gray. And with all that going on, I made an SM RCS minimum impulse check, just to make sure that the RCS was working okay. I did that at 01:00 g.e.t., and it worked fine. It was night and we could see the flashes. So I was fairly well convinced it was okay, that there wasn't any problem with it then. I don't remember what event would have triggered those barber poles unless somebody hit a switch, and nobody remembered hitting a switch. We talked about it, how did that thing get barber poled? When we noticed it, Al and I had been down in the LEB getting the helmet bags or something.

71-H-1125

WORDEN — Okay, you've already given your comments. I don't really have anything more to add, other than the fact that I guess the shaking of the S-IC was a little bit more than I expected. More lateral shaking, a little more vibration than I expected right at lift-off. When we got away from the tower and got away, maybe from some ground effects, whatever it was, it smoothed down. After being briefed several times of what to expect at separation, it didn't seem as violent as I was really expecting it to be.

SCOTT — Which one?

WORDEN — The first one.

SCOTT — I thought you agreed that it was pretty violent.

WORDEN — It was pretty violent, but I guess I was expecting something even more than that.

IRWIN — You had us so well briefed, Dave, that we were expecting it.

71-H-855

WORDEN — The guidance in the CMC was just dead-on, like what we looked at in simulation. I could almost repeat the numbers verbatim, because we had seen them so many times in simulation. It was just absolutely perfect, dead-on. The Z-torquing angle that we got after we got insertion was about half, as I recall, and that's just about what the first P52 showed, right in that ball park. We have the numbers written there somewhere, but the guidance was right-on, super. We had no problems at all with the alignment. In fact, that was generally true with all the alignments. The first alignment went very smoothly. Tracking it in ORB RATE was no problem. The Z-torquing angle came up about the same as they had called up.

71-H-1182

SCOTT — UV photography.

IRWIN — I don't know that we had a color mag out at that time. I think we just had a UV mag out.

WORDEN — I recall now, we did discuss that in flight. I think we decided that the color mag would be nice if we could get the same area that we had taken the UV pictures of. But we couldn't do that because of the time. It wasn't valuable taking a color of some spot other than where we had taken the UV.

SCOTT — The attachment of the ORDEAL to the spacecraft was very loose. I just couldn't believe it; the thing was really rattling. Somebody ought to check that.

IRWIN — The ORDEAL itself worked just fine.

SCOTT — We had about a 1 hour check there. We did the attitude reference check at 01:28 g.e.t., so that was an hour and 20 minutes or so; and it drifted 2 degrees in pitch, 1 degree in roll, and about 1 degree in yaw. So, it was good confirmation on the SCS. Optics cover jettison. Did you see any debris?

WORDEN — Didn't see any debris. Didn't see anything through the optics when they went. All I heard was a slight thumping noise when the covers came off. That was it, never saw anything in the optics.

SCOTT — Okay, COAS looked good, horizon check looked good, the S-IVB was driving very smoothly, ORDEAL was tracking right on. The whole launch vehicle was just super. Unstowage went as planned. Comm was good. TLI preps were nominal. Subjective reaction to weightlessness. I guess we might go into that one. I had fullness of head as I expected to have. I had no other sensation whatsoever. On Apollo 9 I had felt some tendency not to want to move my head, but in this case I felt completely at ease. I noticed in looking around, that I felt quite well adapted immediately upon getting into orbit. I think that probably had to do with all the flying we did prior to the flight, the acrobatics and everything in the T-38. That's the one thing that I did different from Apollo 9. I really believe that was a help, because that was the only thing that was different. I felt much better this time than I had on Apollo 9.

WORDEN — I had the same thing, a little fullness in the head. But I never at any time noticed any problems with equilibrium, sensation of spinning, or any problems with moving my head. The thought crossed my mind at the time that it was probably a result of zero-g flight. I was ready to move right away, get down in the LEB and get on with that part of it. Dave kept telling me to slow down a little bit. I think we both came to the conclusion that there wasn't really any reaction. We weren't getting any reaction out of it. We could proceed on normally after a few minutes.

SCOTT — Yes, that's right. How did you feel, Jim?

IRWIN — Well, I definitely had a fullness of head that persisted for 3 days. I had just a slight amount of vertigo. I didn't want to move my head very fast or move very fast in any direction. That was more pronounced, of course, once we got inserted. That feeling gradually subsided, but I still had a slight amount of vertigo, even after 3 days.

WORDEN — I really felt like we were right at home when we got into orbit. I really felt very comfortable in the environment. Maybe that's part of it too. If you feel comfortable with that kind of environment, that may help you adapt more to it:

IRWIN — I just didn't want to move very fast, but not nauseous.

SCOTT — That's the way I felt on Apollo 9. I just didn't want to go fast. It might just be the time of year, as far as anybody knows. But there were no problems. As far as any other anomalies, I can't think of anything else prior to TLI. We were well ahead of the checklist all the way. We had plenty of time to look out the window and watch the scenery. We took in a couple of looks at the sunrise and the earth airglow and everything. I think the time line was well organized.

WORDEN — As a matter of fact, I thought we had a lot more time in flight to just look out the windows, see the Earth and see what was going on, etc., than we ever had in simulation. The time line seemed to work out so much better, for some reason, that we really had additional time, and it just flowed so smoothly that we didn't miss anything in the checklist.

SCOTT — I'll go through the notes we wrote down in the Flight plan on TLI. The time base 6 events were on time. The one thing that I noticed, which was a fair surprise, was that the helium repress was very slow compared to the CMS. In the simulator, helium repress goes very rapidly, and pressure on the oxidizer tank comes right up. In this case, it came up very slowly. It was almost an imperceptible beginning. I called the ground and questioned them on it. They called back and said it was a normal repress. I think we ought to have the simulator people take a look at that. It was a little bit of concern even though we had the ambient bottle if we didn't have the repress. But in the back of my mind, I was wondering what was wrong, and nothing was wrong. One minute after ignition on the S-IVB we had PU shift, which we hadn't been aware of and which we weren't expecting. We did feel a very noticeable change in thrust, and that hadn't been discussed preflight. It was something, I guess, we just missed along the way. It seemed strange to me that we didn't have it in our checklist or time line. We had the PU shift for launch, and I think it might be a nice thing to stick in the TLI timeline also, just so you'll know it's going to happen. It's no big deal.

SCOTT — The new procedures that Mike Wash worked out for the TLI, putting that automatic and manual together, were really good. The ORDEAL setup was just right. The numbers came out just right, and ORDEAL track was right on zero until the last minute when the guidance starts trimming things out. Had we been required to fly a manual TLI, the ORDEAL would have been excellent because it really worked well.

WORDEN — That procedure is nice, too, because it is easy to keep up with. It is sequenced in the checklist so that there are plenty of check points in there so you can get everything squared away. It really works best.

SCOTT — Yes, if you ever had to step into a manual TLI, you could do it about any place and wouldn't be behind. I think you did a good job on that. I guess we all felt that same low amplitude 10 or 12 cps vibration all the way through S-IVB burn, just like we did during the launch. And we got a call from the ground on, it seems to me, a 3-second-early shutdown.

WORDEN — One further thing on TLI. I guess we wrote them down in the Flight Plan, but the residuals on the CMC at the end of TLI were very close to zero. We wrote them down. The CMC kept very good track of the TLI burn. I can't read them.

SCOTT — Oh, you have bad eyes?

WORDEN — Yes, I still have bad eyes.

SCOTT — Yes, it looks like we had 2 seconds before cutoff. You gave all these to the ground, didn't you?

IRWIN — Right.

SCOTT — That's all recorded. V_c was plus 145 and V_I was 35 599. You got a 35 614 written over here.

IRWIN — Yes.

WORDEN — That's probably what it was.

IRWIN — That's right.

SCOTT — DELTA-VC was minus 14.9, which meant the EMS was tracking well, too.

WORDEN — That's right. Everything was working.

SCOTT — We got everything squared away in time to do the T&D. Al, why don't you comment on the T&D? I think that went pretty smoothly.

WORDEN — Yes, as far as I'm concerned, the transposition and docking was just as nominal as it could be. We came off the S-IVB and did the SCS turnaround and then trimmed the final maneuver with the G&N. I guess I started translating in a little bit more slowly than would have been - may have been more comfortable if I had translated a little bit faster. And everything was nominal inside the spacecraft at the time. The only thing I noticed about T&D was that the different reaction you get from the spacecraft as opposed to the simulator. The reaction you get from the spacecraft is very positive. You put a little bit of thrust in translation, and you get it right away. You can see the rates right away, which is something you don't always see in the simulator. Outside of that, I thought the T&D was pretty nominal. We went right on in; docking was no problem. It went very smoothly; all the latches worked except one.

SCOTT — The procedures, coming off the S-IVB and turning around, put us in a very good relative position when we got around. It was just nicely positioned as to distance from the S-IVB. We weren't too far away and we weren't too close, just very comfortable. The high-gain antenna worked right away. We cranked it up, turned it on, and went to the values set in the flight plan; and it worked fine. Formation flight was no problem; transposition was no problem. In docking, you had to give a little squirt on the plus-X to get the capture latches engaged.

WORDEN — Yes, that's right. We came in the first time, and I could feel the probe contacting the drogue. We just sat there, and it just seemed to at least slow down any forward rate. When I felt that the closing rate had reached a minimum, I gave it a little squirt of X, and it went right in from there.

SCOTT — Yes, I think there is the tendency to go in a little too slowly. On a dock, that could be compensated by a little plus-X when you got there.

WORDEN — Yes.

SCOTT — Once you gave it the plus-X, I was watching the talkbacks, and they flipped right in the barber pole. We retracted and cinched right on down and heard a good bang on the latch.

WORDEN — That's very positive. Not only can you hear it, but you can feel it, too, when those latches go. You really know you are there.

IRWIN — Number 3 was the one that wasn't latched. It took two strokes.

SCOTT — Yes, that's right. All of them were locked up tight except number 3. It took two strokes to lock it. Could you see anything hanging from the LM?

WORDEN — I didn't see a thing.

SCOTT — You mentioned the handling characteristics. Sunlight and CSM docking lights must have been okay.

WORDEN — We didn't need the docking light. Everything was illuminated by the Sun. We didn't have any problems with shadow; no problem seeing the docking target. It was very clearly illuminated. We didn't use the docking lights.

SCOTT — I guess we've got to go in and get the tunnel all configured, and that went according to plan. The extraction was a pretty good bang, wasn't it?

WORDEN — Yes.

IRWIN — Yes. It was more than I had expected.

SCOTT — That's a pretty good thump when it goes off - those springs pushing out - but there was no question that we had had extraction. You could see the S-IVB going away, couldn't you? Or could you?

WORDEN — No, I couldn't. Jim could. I couldn't see out my window. I watched the EMS, and when we first separated, the EMS counted up as I expected. When we turned around, the thing kind of backed down again. We started out at 100, and it went up to about 125 or 126. When we turned around, it was down to 99.2, 99.3, or something like that. So, the EMS was affected by the turnaround. As a matter of fact, during the whole TD procedure, I had the EMS set up and had the accelerometers turned on. I was in DELTA-V and normal, but I really didn't rely at all on the EMS for any indication of DELTA-V. I used strictly time on plus-X thrusters and only looked at the EMS as a backup. In fact, I don't even recall looking at it more than maybe once or twice during the T & D.

SCOTT — Yes, I think that's a good procedure, too, especially when you check and make sure you have all your isolation valves open.

WORDEN — Yes.

SCOTT — You could be pretty well sure that it is going to get you the DELTA-V you want. The pyros going off made pretty positive sounds. RCS, retraction: I guess we've hit all those. We got the camera on right at the docking point. I guess we didn't take any pictures of the ejection because there's nothing you can take pictures of.

WORDEN — That's right.

SCOTT — The next thing is attitude control and stability during and immediately after the SEP ejection. Did you notice anything?

WORDEN — Attitude control was very good. There was no problem with attitude control. We did the whole thing in G&N. We came off the S-IVB with the LM and did some minus X. After we got off we did our VERB 49 maneuver into the S-IVB viewing attitude. It took us some time to get around to that attitude because we had low rates loaded in the DAP. I thought that maneuver went very smoothly. No problem at all with the maneuver. As soon as we saw the S-IVB, we called the ground and told them that we had the S-IVB in sight and for them to align and do their yaw maneuver.

SCOTT — Then we gave them a GO, also, for their basic burn very soon thereafter. We were good and clear. There wasn't any problem with that in the basic burn. The basic burn was a very slow, low-thrust maneuver. We could see some of the propellant coming out. There was a very fine mist if you looked very carefully, and the S-IVB moved very slowly along it's plus X-axis. I rather expected a burn there - some sort of impulsive DELTA-V - but it was a very slow thing. It wouldn't be any problem getting out of its way if you were in its way. We have talked about SEP and evasive maneuvers. Our S-Band comm was good all the way.

71-H-1225

WORDEN — I thought it was superb comm the whole way.

SCOTT — Nothing there on the S-IVB yaw and evasive burns or on the S-IVB closeout. Workload and time lines. I thought that was a very well planned sequence of events. We were busy.

WORDEN — I don't think we were ever overloaded during that time.

SCOTT — That takes us up through all the S-IVB activities.

WORDEN — I don't recall now whether we did one P52 or two P52s before we went to PTC REFSMMAT

6.0 TRANSLUNAR COAST

SCOTT — We did two. I'd like to comment on one thing here in the Flight Plan. Every time you had two P52s due there was only one box to fill in the numbers, and I thought it would be handy to have two boxes.

WORDEN — Yes. The second one is a starting point for the next series of drift checks.

SCOTT — They have stars.

WORDEN — I couldn't agree with you more, but I'm so conditioned to writing down the gyro torquing angles in any P52 that I write them down even if there isn't a box in there. I'm sure it might be helpful to go ahead and write them down. As a matter of fact, it does give an indication of performance of the IMU because it tells how the coarse align is working; how accurately you're getting a coarse align.

SCOTT — They have a place in here for shaft angle and trunnion angle for the stars. Do you know why?

WORDEN — Yes. You go to SCS narrow deadband and do your first P52. Then you take shaft and trunnion angles on those two stars. You do your second P52 with an option 1, and then if you have any problem locating the star, you go to the shaft and trunnion angles because you're still there at the attitude. It's just a warm feeling kind of thing. You should see those same shaft and trunnion angles come up on the second P52 within the deadband of the SCS.

SCOTT — Okay. Well, the torquing angles looked good. I felt we had a pretty good platform at that stage. Let's see, OPTICS CALIBRATION.

WORDEN — We ran all the P23s the same way. We did an automatic maneuver to the optics calibration activity and then an optics calibration, which is the first part of the P23 series. We had no problems doing the optics calibration, and I don't recall now what the exact numbers were on the calibration, but they were within 3/1000ths of being zero. Then we did a VERB 49 maneuver to the P23 attitude so that we made sure we had a good view from the LM and that the subsequent star horizon sightings would be done at approximately the same attitude and the same roll angle. I thought P23s went very well. I used minimum impulse to control attitude while I was doing the P23s. With the LM on, minimum impulse was very slow in correcting any attitude errors that we had, but it was very positive, and there were certainly no problems with that particular mode of operation for P23s. Outside of recalibrating myself on which horizon I wanted to mark on, the P23s were quite straightforward.

Incidentally, I might add here, we haven't yet found out what the results of the P23s were. I felt that one of the biggest helps I had in doing the P23s was in flying the really accurate simulator at MIT, where they have very accurate calibration and good slides and they can tell you within tenths of kilometers what altitude you're marking on. I spent a good session with Ivan Johnson (MIT) doing nothing but P23s - all the way through the flight, right from translunar coast back into transearth coast. I thought the P23s in flight were very close to what we practiced in the simulator. I really had a fairly warm feeling about that P23 system.

SCOTT — I think the updates you were giving there, particularly on the way home, sort of verify that.

WORDEN — The command module simulator (CMS) is just not set up to do P23s accurately. There's just no way you can build that kind of procedure into that part of the simulator. The simulator is good for procedures, in that case; but, if you really want a fine calibration on the altitude, on the horizons you're looking at, and on your eyeball and what you're doing, then the MIT simulator is the place to go, really. That's really a must. I was really impressed with that whole thing.

SCOTT — Yes. I really think you could have got us home.

WORDEN — Yes. They kept telling us all the way home that our vector was as good as theirs.

SCOTT — The next thing is SYSTEMS ANOMALIES. I guess the first one we noticed was the SPS THRUST light being on.

WORDEN — I think I noticed that SPS light right after T&D.

SCOTT — Yes.

WORDEN — We were just sitting with nothing going on. The EMS was turned on, and I looked up and saw the SPS light on. That was fairly close after T&D.

71-H-1228

SCOTT — We went through the procedures, I think, which are already documented by the ground, as they recommended, and found that there was a short in the switch. I think pounding on the panel turned the light on for us, and finally, manipulating the switch gave us a constant light out. First time we tried it, I guess, we moved the switch up and the light went out. Then I played with the switch some and got familiar enough with the short that I could put the switch in the mid-position on the lower portion of the foreskirt and hold the light out. It would be interesting to see the inside of one of those switches some day to see if you could identify the contacts they're making. I think the ground did a fine job of coming up with procedures to evaluate the switch, and when we got down to checking the engine out, I thought the procedures they recommended were real good. They were simple, easy for us to change, easy for us to work through, and, in general, I thought that was a real fine bit of support they were showing on the ground.

WORDEN — That was superb, I thought. Like you say, the procedures were simple. It was very easy to make changes in the checklist, and the changes that they made were straightforward. It was a pretty straightforward systems problem, I think, at that point. The procedures that they recommended worked well for both the dual bank and single bank. You could just forget part of the procedure. All the burns that I did in lunar orbit, the short burns, I did on one bank only, so we would avoid any problem with that bank that had the short in it. The procedure was almost the same, and it worked out very nicely. The only change was in one extra circuit breaker that was out, the MAIN A pilot valve, in leaving the normal A DELTA-V thrust switch on.

SCOTT — I think we had a good, warm feeling that we had two complete SPS systems going into lunar orbit.

WORDEN — Sure did.

SCOTT — Let's see, what else on the systems on the way out?

IRWIN — Well, we had the AC problem on the circuit breaker, with the circuit breaker popping.

SCOTT — Yes, I guess that popped, the lights went out, and we left it out.

WORDEN — And we taped the rheostats.

SCOTT — And, I think the only consequence of that was, as you mentioned, it sort of changed your pattern of operation down in the LEB because you didn't have the timer down there.

WORDEN — We didn't have the mission timer down there, and the other thing that I missed in the LEB, which I found, I stubbed my toe on a few times, was that all of the program lights in the DSKY were out. The only thing we had was caution and warning panel down there. That was operational, but all of the status lights on the DSKY were out. They went with that circuit breaker. The backlighting in the EMS scroll was out. That was also off of that circuit breaker. I don't know if you want to comment on the backlighting on the EMS.

SCOTT — It wasn't any problem.

WORDEN — I could see it all the way down. We didn't need the backlighting.

IRWIN — The water system.

WORDEN — Yes. The leak in the water system.

SCOTT — Yes. I guess we can go through that one here. Was it the third night out when you were chlorinating?

IRWIN — We were just getting ready to chlorinate. We had just taken the cap off when the leak appeared. I guess it had been leaking before that. But I don't know; it's hard to tell.

SCOTT — It was found, subsequently, that Karl Henize had experienced the same type leak prelaunch, which is another thing we hadn't heard about. It would have been nice to have known that there was some sort of expected problem in the chlorination and how to take care of it before the problem occurred because I think all three of us were a little concerned when that thing started leaking at the rate it was leaking. It was really pouring out. I guess we'd just got the dump turned on to suck the stuff out. We were going to dump it overboard during the PTC when the ground came up with the procedure to tighten down that little valve in there. It was a very simple procedure, and had we known that it might occur, we could have taken care of it and saved a lot of anxiety and a lot of wet towels here.

WORDEN — Unfortunately, we looked for a break in the line. It's the chlorination part in panel 352, and that thing comes out of the panel, straight out, and then it angles; it has about a 45-degree angle on the vent and the first thought I had was that we had cracked that tube right there.

IRWIN — You know, if we'd suspected that they'd had a problem before, we could have left the injector screwed on there so that it wouldn't have messed up that valve. Left it on there all the time.

SCOTT — I don't think it was an injector problem, Jim.

IRWIN — Well, I think you aggravate the situation by taking the injector valve off and on each time you chlorinate. You loosen that little valve.

SCOTT — In looking at that little valve, it's a rubber seated valve. It seems to me that that wouldn't be a bad thing to take along as a spare because if you ever tore up that rubber seal there, you'd be in trouble. It was obvious that there's no way to check that off, because we closed everything. Al got out the systems book and we closed everything, and it turns out there's only one check valve between the potable tank and that valve there. And if you lost that check valve and the thing started leaking, it'd be all over. So I think, in summary, it would be good to know about those things, even if they do happen just before you go. If somebody could just pass the word and say, "Hey, we had a little trouble with the water system." And if you take a tool, B3 or whatever it was, and tighten it down, it would be just fine. But there were a few anxious moments there mopping up.

IRWIN — One other humorous note. You know, you were yelling for towels and I couldn't get into that compartment that had all the dry towels.

SLAYTON — You couldn't get into the compartment?

SCOTT — It was stuck and we got into it a little later.

WORDEN — Yes, we had a funny with the latch on that compartment, and once we got it open, it worked fine the rest of the time. I guess it was jammed. Something was jammed underneath that latch. You know, those are those radial latches, and the one under the head-end of the couch was jammed.

SCOTT — Those compartments are too big and they're not partitioned. Once you open one-half of one door, well, everything in there comes floating out unless it's tied down. To try and restrap things when you get through with them is a pretty good job because those straps aren't that easy to use. I think it would be very helpful if somebody could partition the various sections within the compartment with Beta cloth and a snap, or something, because we were forever and a day opening one of those things up to get one item out. Everything else came out and it was just floating, and, as you know, everything floats up plus-X-wise and you just have to leap on the whole compartment with all your arms and legs to hold everything down while you search for the one item you're wanting. They've grown to such large size, it's almost like having a whole aft bulkhead in one compartment. It was sort of a nagging problem all the way through. It just took that much more time.

WORDEN — Yes, it did. Those straps are very good at holding things down, but they're really designed to strap everything down pre launch. They're not very well designed for use in flight. They have a very difficult little button fastener in them and the straps themselves have a rubberized feel to them that makes it hard to cinch those things down and get that little button into the loop. I agree with Dave. I think with compartments that large and with so many small pieces that we're fooling with inside the spacecraft, there ought to be some smaller compartments.

SCOTT — Every time you opened a compartment, everything just jumped right out at you.

IRWIN — Oh, yes. The screws on the restraint system came loose.

WORDEN — Yes. That was during the lunar orbit, but that's a good thing to mention so that we don't forget it.

IRWIN — It's surprising in that it came out of the center seat and the right seat, both on the right side.

WORDEN — On the right side, yes. The lower lap belt restraint attach point on the center seat and on the right seat came loose. The small bolts that hold them to the attach point and the nuts all came loose and just floated around. The lap belts came loose. I thought it was kind of funny in lunar orbit. When I took the center couch out, I noticed the attach point on the center couch was gone. And that strap was floating free. I looked around for the little bolt that goes in there, the little screw and the nut that goes in there, and I couldn't find it. I thought to myself at the time, well, I'll just sit and wait and sooner or later it will float by. Sure enough, all four pieces to that thing floated by: Two washers, a bolt and a nut. I just grabbed them as they went by and stuck them on a piece of tape. I kept the tape in one place and then put it all back together again. But I was really surprised that those little things came loose.

SCOTT — I think it's a good idea to have restraint straps, especially if you're going to make a two-chute landing.

WORDEN — Yes.

IRWIN — We were able to repair the one in the center couch because we got all the parts. On my couch, we never did find the nut for it.

WORDEN — That's right.

IRWIN — So we ended up just taping this good one up. And it withstood that two-chute impact.

WORDEN — That little piece of gray tape.

SCOTT — Yes. Take lots of gray tape.

WORDEN — Best invention yet.

SCOTT — Next thing is the MODES OF COMMUNICATIONS, and I think all of those worked very well. Super comm all the way.

WORDEN — All the way. Fantastic tracking. We went to PTC attitude for the first time. We had arrived at PTC procedure preflight, which was different than the PTC procedures that had been used before because of the new universal tracking program. We didn't have the same kind of ORB RATE DAP that we had before. To make the thing work right we had to load 0.35 deg/sec in the rates and a half degree deadband into that, so that the thing would spin up when you first turned the DAP on. We got the attitude and used two adjacent quads to get the attitude to damp the rates. I'm trying to think now why.

We eventually ended up getting the PTC going on the third try. On the first one the rates hadn't damped properly, I guess. When we got the PTC going, it wandered off. And on the second one, because of the half-degree deadband in the DAP, as soon as you get the first firing to spin the spacecraft up in PTC, you have to take CMC mode switch and go to FREE so that you don't get another firing from the DAP, which could give you some cross-coupling. As I recall, I hesitated, I didn't go from AUTO to FREE in one motion. I hesitated in

HOLD just a second. That split second was just enough time for the thing to fire the jet, and it somehow got the roll rate screwed up because we drifted off attitude again. But I could definitely hear the jets firing when I went to HOLD. The third time, we finally got the thing started. We did everything just as per the checklist. The third try worked beautifully, and I guess it was one of the best PTCs we've seen. It worked just as advertised. And I don't think we ever had any trouble with PTC after that either. So it was just a question of getting that new procedure straightened out.

The half-degree deadband was the big thing. We used to load 30-degree deadband in there and when you first proceed on the DAP you get a forced firing, which gives 80 percent of the rate that you loaded in. With the universal tracking now, that's all been changed and it doesn't work that way. So we had to go to the half-degree deadband. And it's just a question of getting used to that program, that's all. The P23s and the P52s were just absolutely nominal as far as I'm concerned.

SCOTT — Midcourse corrections. We did that one last midcourse prior to LOI to check out the engine to make sure that bank A worked okay. It worked exactly as advertised. It was a good burn.

WORDEN — That was a little bit humorous, in a way. We talked about that burn as a checkout of the SPS engine. And the procedure was that we'd get everything set up and we'd push that circuit breaker in, and, if the engine light was off, we'd then pull the circuit breaker right away. We got all the way through that burn before it really dawned on me that that was a mid course correction. The DELTA-V we got out of the burn was exactly what they wanted on the midcourse correction. So it worked out rather nicely.

IRWIN — We did a lot of UV photography on the way out. In a way, that was good because we always maneuvered to a position where we could view the Earth or the Moon at the various stages.

WORDEN — The procedure that we established preflight to do the UV photography worked okay in flight - putting the cardboard window shade up, pulling the Lexan shield down, and all that. But it's a cumbersome procedure, and we've got to be careful about a bunch of things all at the same time. You've got to juggle the camera and the Lexan shield and the cardboard and a whole bunch of things. I guess that's really my major complaint, the Lexan shield. That thing was fine the first day, maybe the first couple of days. But that Lexan is so soft and it scratches so easily that after a couple of days it was worthless as a window to take any photography out of. I sure hope that, if we do that kind of photography on the next flight, there's a better system of protecting the interior from ultra violet than with Lexan.

IRWIN — The nominal mode was to leave the metal shield over the cardboard, leave the cardboard up all the time and leave the metal shield up. Then when we wanted to take the UV photography, just take the metal shield off. The cardboard was there. But that didn't work very well in lunar orbit.

WORDEN — That's right, because in SIM bay attitude, that particular window is the very one that you see almost all the targets out of. And you're sitting there with the Lexan shield in place, and you're trying to take pictures through it, and all you've got to do is touch it with a camera lens and it's scratched. I ended up using the Lexan shield as a shield, not necessarily in the window, but putting it between myself and the surface and taking pictures around it, so that I could get some decent pictures.

SCOTT — The television cameras seemed to work okay.

WORDEN — Yes. No problem.

SCOTT — High gain antenna performance was all right. Daylight IMU realign and star checking: You never had any trouble doing alignments. They worked out fine. You want to comment on your telescope? You didn't think you could see very much through your telescope.

WORDEN — Yes. The attenuation of the telescope. You really had to have a very bright star to see it through that telescope. You sure couldn't pick out any guide stars to any of the Apollo navigation stars, I didn't think.

SCOTT — The light loss through the telescope seemed to be considerably greater than Apollo 9. I could not see stars on the dark side of the Earth very well through the telescope; and, on the dark side of the Moon, I still couldn't see stars very well.

WORDEN — It was absolutely amazing. You could look out the window and the sky was just bright, there were so many stars. You looked through the telescope and you could pick out maybe one or two major stars. That was all. A fantastic difference in the light attenuation through the windows as opposed to the telescope. It was unbelievable, really.

SCOTT — The CM/LM delta pressure seemed to work okay. When we pressurized the tunnel and did the latches, it worked as prescribed. Monitoring the delta pressure on the way out with the LM seemed to work all right. The LM and tunnel pressure were nominal.

IRWIN — Removing the tunnel hatch and the probe and the drogue was an order of magnitude easier than it's ever been in practice. It went exactly as we'd seen on the mockup over here. I thought it was a very easy operation. We put the hatch underneath the left-hand couch. We put the probe in the center couch, and lashed it down. We put the drogue underneath the left-hand couch and tied it down so that we had good clear access to the tunnel area. I thought the whole operation was very easy. No problem.

SCOTT — I thought we had a lot of odors in there.

IRWIN — I think it was probably due to all the hydrogen systems.

SCOTT — It cleared out pretty well. The spacecraft system cleared it out pretty well.

IRWIN — Not nearly as well as the LM system. Maybe that's just a function of three versus two guys. Of course, I was contributing my share too. I thought it was pretty gross in there.

SLAYTON — What we are really searching for here is some sort of burn smell up there.

WORDEN — Yes. I recall some powder smell in the tunnel.

IRWIN — I didn't smell any nitrogen up there.

WORDEN — I thought that whole tunnel operation was very clean.

SCOTT — I think if you listened very carefully you could hear the SM RCS, and to me that seemed somewhat different from Apollo 9. I thought we could hear any firing fairly well on Apollo 9.

WORDEN — On the service module RCS, I thought that the noise level of the solenoids operating on those RCS engines was much less than what we've got in the simulator.

IRWIN — Could you really hear them or just feel them? I kind of got a muffled thud when they would go off.

WORDEN — Yes, you do. It's very muffled though; it's very attenuated.

SCOTT — I'd say it was almost an order of magnitude less than we felt on Apollo 9, as I remember. Maybe that's because it's biased by a lot of simulation noise, but you sure can't hear it very well. You can see it. You can see the flashes at night, but you sure can't hear much. I think we all got the sleep as recorded. I think everybody slept fairly well. There were a couple of nights in there where I think everybody was really bedding down. That went quite well. It seems like, particularly with the SIM bay, that we really never had enough time to do our housekeeping. We were always busy trying to keep up with things. I'm not sure whether it was because of the amount of equipment on board or because we had to constantly pay attention to our SIM bay operations. But it seems like we were always pressed on the housekeeping.

We had to eat fast, had to get ready for the next thing fast, and, in general, we never had a lot of time to sit around and wait to get to the Moon; nor did we have a lot of time to sit around and wait to get home. We always had something to do. And it was mostly because the housekeeping took a fair amount of time. One thing we all commented on was that it would be better if, when you awaken, you ate first and took care of your cleanup activities before you got into the operational part of the day. To try and combine operations with eating sort of compromised both. A guy would be halfway through fixing a meal and he'd have to go turn on some SIM bay thing, which means you didn't do either one very efficiently. After waking up, you should eat, clean up and then go to work. You'd be more efficient. Is that what you were thinking of?

WORDEN — Yes, precisely.

SCOTT — There are a number of things you have to do in the spacecraft which aren't really called out in any time line. We have an eat period and then we have a rest period and vice versa. You can't go from an eat period to a rest period. There are a lot of things that have to be done, most of which are called out in the presleep checklist. You can't just go finish your dinner and in 2 minutes do the presleep checklist and go to bed. You have to have a transition period during which you chlorinate the water, change a canister, everybody take their last urination for the day, and clean things up in general. You have to have a period of time there to get ready to go to bed. Another thing that's not mentioned that I think should be passed on is cleaning the screens with tape. I'm not sure that that's ever been discussed. I know we did it on Apollo 9. The suit circuit return screen and the screens on the suit holders were constantly covered with stuff. You had to clean it at least once a day. It made a noticeable difference in cabin temp when we did that. The cabin was running sometimes up to 80 degrees. We'd clean those screens out and it would bring it back in.

IRWIN — Also, the use of the cabin fan brought the temperature down. We couldn't find that other suit umbilical screen, so we had to improvise.

SCOTT — This was on the way home.

IRWIN — In conjunction with housekeeping, I think I'd have felt a lot more comfortable if we'd taken the time before the flight to decide where we were going to put everything, where we were going to put those Gemini bags, which CSCs we were going to use for garbage, et cetera. It wouldn't have taken very long. I would have liked to physically have used the Gemini bag and cleaned it up before the flight. It was always fouling up on me, and I wasn't transferring properly. Dave said, "Well, why don't you clean it?" I tried many times and finally, after about 5 or 6 days, I got it so it would work and I could clean it. But with little things like that, I got to the point where I just didn't want to urinate if it was that much trouble.

WORDEN — There are numerous things like that that you run into for the first time on the flight that you haven't really thought about before because they're kind of low in the order of priority. Just a quick briefing on those things would be helpful, but I do think that they're the kind of things that you pick up as you go along.

IRWIN — If I'd just used the Gemini device once before the flight, I'd have gone into it prepared, knowing how to clean it and how to use it.

SCOTT — I think we did make an attempt to get oriented on how to use the urine collection bags, which seemed to work out fairly well when we were restricted to one dump a day. That operation, although again it took a lot of time to get all the hoses hooked up right, did seem to work okay. But all the different devices you have to take care of while living up there required time and familiarization. I think both of you did try the defecation device before the mission. That wasn't bad; it was a job, but it still worked. I think your recommendation to take the UCTA and the UTSs and work them over before you go would be very helpful as far as time goes. We had a lot of hydrogen in the water periodically. It didn't seem to be associated with any particular event. However, at some period after we used a lot of water, it seemed that there was a point at which the hydrogen increased in the water, and we got a lot of bubbles. And that was a problem, digesting all those bubbles. We tried the hydrogen separator and that didn't really seem to help very much. We also found that it didn't work on the food water tap; it only worked on the gun.

We tried to consume all the meals as planned and maybe a little bit more. That means that when you eat a lot you have to defecate a lot. That meant that extra time was required to take care of those little chores. Everybody stayed on an "as required schedule" and nobody would use any Lomotil. I think that was a good plan. The biomed harnesses worked well. There are a few little things we might recommend for improvement, but putting them on and taking them off was a relatively simple matter. The data that they got on the ground was good. They told us before we went that the sponges which go into the sensors were going to be somewhat larger than the diameter of the sensor. They were not. They were smaller. We stuck it in and it would float back out. That was sort of a pain. They also have the sponges in little packages of two. We have five sensors on. This means that every time you put the five sensors on you have to throw one sponge away, because they're in packs of two.

WORDEN — Yes. But it gets darn important when you're in flight and you've got a package of eight of these sponges. I think they come in packages of eight. If you only use five of them, you've got three of them that you've got to do something with.

SCOTT — Well, two of them you can put back and use again, because they're sealed in pairs.

WORDEN — That's right.

SCOTT — But the third one you've got to throw away.

WORDEN — I'd like to add something else about the biomed sensors while we're on that subject, too. I think both you and I got a reaction from the paste or whatever was used on the disk on the biomed sensors. Now,

I've still got some welts, some lesions, that I got off those biomed sensors.

SCOTT — I'm not sure it was the paste or just the pressure of the sensor being on that same spot on my skin.

WORDEN — Well, that could be. I'm just wondering if maybe there isn't something that could be looked into to see if there's a different kind of adhesive or something that would alleviate that problem.

SLAYTON — Did they test you for allergy to that paste preflight as they are supposed to?

SCOTT — Yes.

WORDEN — Yes.

SCOTT — We wore it all preflight.

WORDEN — Maybe you're right, Dave.

SCOTT — We tried to put them exactly in the same spot on launch morning. Dr. Teegen painted a circle on us where they had put those things. I redrew mine every once in a while, and it was very helpful. You just stuck them right where the mark was. And it was real easy to locate them there. But I think that doing it over and over sort of made you a little sensitive in that area. Al and I both have little rings there where the thing sort of cut in. But it was far better than wearing them all the time. I know that we could hardly wait to get ours off after we came up off the surface, because it was really getting irritable. I think that it was very beneficial to be able to take those things off and let your skin dry out. I guess the next one I had here was our little Exergym. As we were on the way to the Moon, we were using the thing and trying to use it correctly, but it looked like the rope was wearing. There was quite a bit of fraying on the rope. So we decided that, instead of having everybody work out twice a day, to let Al have it twice a day and Jim and I would do something else on the other time. That way Al would have the benefit of the thing.

WORDEN — Well, I don't know what it's going to boil down to. The Exergym is good for keeping some muscle tone, but I found that there was just no way I could get a heart rate established and keep it going. There was just no way I could do that. So I finally decided on a combination of two exercises. I used the Exergym a little bit, just to keep my shoulders and arms toned, and I ran in place. I took the center couch out and wailed away with my legs, just like running in place as a matter of fact. I didn't say anything to the ground, but the doctors watching the biomeds called up and said, "Hey, you must be exercising. We can see your heart rate going up." And they kept me advised as to what my heart rate was. It worked out very nicely, I thought, because they could tell you that you're up to 130, going up to 140. Then I would exercise a little bit harder, and true, even though I wasn't exerting any pressure on anything, just moving the mass of your legs around really gets your heart going. I'm really convinced that that's the way to exercise in flight; get that kind of motion going and keep it going not let up on it at all. I did that for 15 to 20 minutes at a time. I just ran in place as hard as I could. As a matter of fact, I thought I'd strained some muscles that I had never used before because I was just free wheeling my legs and wasn't exerting any pressure on anything. I really thought that was a useful exercise, and as far as cardiovascular was concerned, I thought that was a much better exercise than the Exergym. I used the Exergym just for muscle tone. I think it's good for that. It's a good thing. You can pull against that and it's almost like doing situps.

SCOTT — That's the best way to really keep yourself up to snuff up there, especially for people in the command module. They really don't have anything else. I guess our recommendation would be to get that small ergometer and put that onboard, because that's the only way you're going to get a dynamic exercise.

WORDEN — I found out that with the center couch out, there's just almost the right amount of room. In fact, the same thing could be done up in the tunnel area. You don't need a whole lot of space. In fact, that particular exercise doesn't take as much space total as does using the Exergym.

IRWIN — We strained against the struts, against the bulkhead, and against the straps; this was kind of an isometric form of exercise. I think it's almost as good as the Exergym.

SCOTT — You can put your feet on the bulkhead and hold onto the seat struts and do deep kneebends against your arms.

WORDEN — In fact, Dave had the spacecraft moving all over every time he'd exercise. I'd sit there and watch the rates jumping up and down. He was really moving us around.

IRWIN — As far as comfort is concerned, I think that after about 1 day out, we all took off the inflight coveralls. We got down to the CWGs and were very comfortable in CWGs until we got to the Moon. In fact, around the Moon, it was even warmer. It was almost too warm to wear the inflight coveralls, and it didn't really cool off until we got back to about 1 day out from the Earth.

SCOTT — It seemed to be warm in the spacecraft. And I think we felt warm around the Moon, so we all just wore CWGs. It would have been too warm to wear the coveralls. On the way back, it really cooled off; the last day in particular. We were pretty chilly when we woke up the last morning in our coveralls, CWGs, and in the sleeping bag. We were down to about 65 on the cabin temp. But on the way out, in the vicinity of the Moon, we were running 75 to 78 in the cabin temp; I think that it was a fairly warm environment. It would have been nice to have been a little cooler. That was with the cabin fan on and everything.

IRWIN — And, in conjunction with that, it might be useful to have some pockets on the CWG if you get down to that mode. Then, you can have your pencils where you can get to them.

SCOTT — That's a good point.

IRWIN — Also, under anomalies, we ought to mention the problem with the cabin fan and whatever that loose object that was in there and the fact that we couldn't cycle it freely. We were afraid to turn it off because it might not start up again.

SCOTT — And there was some piece of metal somewhere in the cabin fan. This jumps down a little farther along the line, but after lunar orbit docking and attempting to get the cabin cleaned out with all the lunar dirt and everything, Al heard a couple of clicks in the fan and then it picked up whatever it was and really started to groan. And then, there was a low-frequency, very hard vibration. We turned the cabin fan off and on several times and finally got whatever it was to lodge into a corner somewhere and the fan ran cleaner; but whenever we subsequently turned the fan off, we'd have to go through a couple of cycles to get the foreign object to relodge in a corner to get a clean run on the fan. There was something in there.

WORDEN — It was funny. You could hear the fan running free - I mean running as it would normally - and then you could hear a ping just as if you'd thrown a piece of metal into the fan and it got picked up by one of the blades. Then you'd hear it rattle around in there and then you'd hear this groaning sound where it obviously caught in there.

IRWIN — Well, I'm wondering if the filter for the cabin fan shouldn't be a smaller mesh to prevent an object from getting in there to interfere with the fan operation.

SCOTT — You know, we recommended that for Apollo 9; that they put a screen in there so you couldn't have something drift down into the blades. And the blades are wide open to the cabin. If you have a loose nut or bolt or anything floating around when the cabin fan is not running, it can float right down into the blades. You ought to have some protection in there, or put the cabin fan filter up right away to keep it clean. You really need the cabin fan after a docking.

IRWIN — Of course, we don't know whether that object came in through the outlet or the inlet. It could have come in the inlet. That's a pretty wide mesh.

SCOTT — Anything else on housekeeping? The last thing in this section is the SIM door jettison. In general, that was a very light bump. You could hardly feel it. I think there's no need to suit up in the future. However, I think the suiting operation was a good exercise, because it gave us a chance to run through the descent time line. What we had planned to do was to wake up and suit up in the order in which we do it during the descent day to make sure we didn't have any problems and to see what the time line looked like. And I think we learned a few things in doing that suitup in zero-g, which Jim and I had never done in the 7LB. We learned a few things and I think it helped us on descent day to get a little ahead of the time line. But as far as a requirement, would you both agree that I don't think there's any requirement to be suited to blow the SIM door? It's just not that big a shock. It was the lightest pyro charge by far of anything we had. Jim took photos of it.

7.0 LOI, DOI, LUNAR MODULE CHECKOUT

SCOTT — We discussed the procedures which were unique to the SPS, and I suppose we should talk about the LOI first. That was a pretty novel burn. It all worked out pretty much as we had seen in the simulator. The only surprising things about the burn were the residuals, all of which were zero. At that point, we were convinced we had a pretty good guidance system. No trim. It was a very smooth burn. Everything worked as advertised. Do you have anything else on that?

WORDEN — No, I was impressed with the smoothness with which the engine came on and the smoothness with which the guidance worked. There were no abrupt changes; the gimbal motors were very smooth. It did jump around very, very slightly, but there were no big oscillations. We were right on trim when the burn started. The procedures worked fine. At 5 seconds after ignition, I pushed the circuit breaker in, and we got the second bank on. I could see the chamber pressure come up an estimated 3 percent when the second bank came on. It gave us a positive indication of the bank coming on. We pulled the circuit breaker 10 seconds before cutoff, and it shut down right on time. Dave was ready on the switches to shut down at the burn time plus 10 seconds. The burn was terminated automatically and, like Dave says, no residuals.

IRWIN — I think you got that circuit breaker at 6 minutes.

SCOTT — That was the 6-minute call. That's right.

IRWIN — As for the PUGS operation, the unbalance was in normal and stayed constant at about minus 200 until crossover. After crossover, it started to increase out of the green band, so I had to give it a decrease and brought it back to normal. It looked like it needed the decrease position and was left in the decrease position for the remainder of the burn.

SCOTT — That put us in a pretty nice lunar orbit. We enjoyed the scenery and had plenty of time to get ready for DOI, and I think that's a great idea. You do the first rev to take a look at what is there. The time line was very smooth with no problems, and we got ready for the DOI. The DOI was again a nominal burn. We shut down on time manually, but the G&N beat us to it. I guess, Al, you could see I put my hands on the switches and timed it. When the time ran out, I put the switches down. Al could watch the PC, and I guess he saw it. Why don't you just say what you saw?

WORDEN — Well, I heard Jim counting down. I knew Dave was ready to throw the switches, and just as his hand started to move, the PC dropped off. So the automatic shutdown and Dave's shutdown were almost simultaneous; except the automatic shutdown was, I think, just before his. It was perfect.

SCOTT — One of the things I might mention that is different from the simulator is that we always use PC in the simulator for our cue to start. In the real world, a physiological cue is far better. I mean you don't have to look at the PC to know when to start the watch. When it comes on, it is on; and you know darn well the engine is on. When it is off, it is off; and you know it. That is something you cannot possibly simulate, but that is something to be aware of.

IRWIN — I might make a comment in that connection, Dave. The valve indicator actually opened about one-half second before I got any physical sensation that they were burning. I would see it move, and a fraction of a second later, I could feel the light off.

SCOTT — I think you mentioned that during the burn. After LOI, you mentioned it. I took a stop watch along because we were timing it in tenths of seconds, as I said was necessary for DOI. I think that unless you really have a double failure, you can't get in trouble on DOI. I don't think that is the problem that some people might have thought it was some time ago. That is a good solid method of getting into a low orbit. Very little chance I think of getting into a bail-out situation. Sounds SPS. I don't think we heard anything other than the force of the engine coming on. You couldn't hear anything.

95-13004

IRWIN — During DOI, it was left in the decrease position. For such a short burn, of course, it wouldn't stabilize anyway; so it was just left in a decrease position. I guess that after that burn, it was put back in the normal position and was left there for the other burns.

SCOTT — Gravitational effects on spacecraft attitude. I'm not sure we ever noted any because we were always in some prescribed ORB rate or the SIM bay attitude.

WORDEN — That is correct. We never really went out of an ORB rate maneuver. We were either straight heads down, straight heads up, or in SIM bay attitude. I guess that was true continuously throughout the lunar orbit. I don't think we ever went out of that particular attitude except for the rendezvous. My hat is off to the people who designed the DAP because that just worked so smoothly it was almost unreal. It was so smooth all the way around you never noticed the thruster firings. We stayed right in the ORB rate attitude all the time.

SCOTT — The confidence factor in the RCS goes up by orders of magnitude every day to the point where I think some of the training we do on RCS failures might be superfluous, because everybody powers down and goes to sleep. I guess my confidence factor on those jets is 100 percent. I don't think we ever worry about a jet failing on or failing off because you could hardly do the mission and have to worry about that. We were running through the ORB rate during the sleep, and they weren't bothering anybody.

WORDEN — That's right. As a matter of fact, during the sleep periods, I don't recall hearing a thruster fire or any maneuvers at all.

SCOTT — I don't either.

WORDEN — It was just as quiet as it could be the whole time.

IRWIN — It would be interesting to compare notes with the doctors. I thought I was getting a fairly good night's sleep; but I talked to the doctors this morning, and they said that wasn't necessarily true. It might be because I normally move around when I sleep (change positions), and they might interpret that as loss of sleep or loss of rest.

SCOTT — SIM experiment prep was standard procedure.

IRWIN — It was cook book.

SCOTT — Communications were excellent. PGA donning: We set up a plan on the LOI day to try out our sequence of suiting for PDI day. As a result, we changed our minds on PDI day to make it a little bit more efficient.

WORDEN — You two put your suits on and then went into the LM to zip them up.

SCOTT — Because it's a lot easier zipping up the 7 lb suits in the LM and it gave us a chance to do the tunnel work shirtsleeve. We helped you (CMP) get your suit on. It is worthwhile to run through suit donning because the first day we did it we had you (CMP) put your suit on. Then we put our (CDR & LMP) suits on in the command module, and it is hard to zip them up in the command module. That was a sort of chore. Jim suggested we suit up and go to the LM before zipping them up. That made it a lot easier. We recommend cleaning the tunnel out or putting the suits on unzipped, cleaning the tunnel out, and then the CDR and LMP transferring to the LM to do their suit zip. It would be a good idea to have a little trial run one of the days on the way out.

WORDEN — As to the time line, that works out much better, too, because while you were over there putting suits on and zipping them up, that gave me a chance to put my suit on which is done in parallel rather than sequentially.

SCOTT — In general, there were no problems in donning the suits. Tunnel mechanics went very smoothly, as Al previously stated. We did it the same way on PEI day as to where we put the equipment. The hatch and the drogue went in the left couch, and the probe went up at the head of the center couch. IVT to the LM was straightforward. I guess we didn't observe the condition of the CSM thermal coating.

IRWIN — LM status checks. The first thing we noticed when we got in the LM was the fact that the glass was broken on the tape meter. That initiated a requirement to clean up as much glass as possible. We transferred over the vacuum cleaner and started cleaning it up. That was the only anomaly we noted on housekeeping day. I guess we might have gone through the comm checks too fast for the ground because they asked us to go in again so they could look at the battery operation.

SCOTT — We will have to ask them, but I think they just wanted to get another data point on the battery. We found that we could run through the time line much faster on the comm checks than was allocated in the checklist. So, I guess that we didn't have the batteries on as long as they expected us to because it just didn't take that long to make the checkout. There is another factor we might mention. Having the LM housekeeping day moved up a day, or the day after TLI, gave them a chance to do all that testing on the tape meter. That gave me a warm feeling to know that they checked the thing out and it would work with a broken outer pane of glass. I think it is a good idea to go take a look at the LM early and analyze your problems and get a good handle on them before you get too far down the road. Then if they do want to take another look at batteries, the second housekeeping day is no problem. It is nice to go back to the LM and take another look around anyway. We got another chance to clean up some more glass. We did find a number of pieces on the second day. I think we got most of the glass cleaned up; don't you, Jim?

IRWIN — Yes, I think the use of the vacuum cleaner from the CM was probably just as effective or more effective than the LM cabin thing.

SCOTT — The comm check worked well. Transfer of equipment worked well. I am glad we had that preflight training exercise to get all that equipment transfer laid out. That went rather smoothly. Housekeeping was nominal, and the power transfer back and forth from the CM to the LM worked as prescribed.

IRWIN — There was one thing that we did not do, and that was to take the water bags out of their stowage bags. We left them stowed. We couldn't see any reason to cut them open and take them out of the stowage bags.

SCOTT — As we went through activation, we were always 10 to 15 minutes ahead of schedule. The power transfer was nominal.

0 LUNAR MODULE
ACTIVATION
THROUGH
SEPARATION

WORDEN — Tunnel closeout was just as per the decal in the tunnel. Just follow that down step by step.

I guess there was an anomaly that happened during maneuvering to an undocking attitude. I checked things off on the Flight Plan as we went. We went right down the line on the Flight Plan and the checklist. We released the docking latch, put the suit on, and did a suit circuit integrity check. We installed the hatch, got a LM/CM DELTA-P, and went right on down the time line. We did a VERB 49 maneuver to the undocking attitude and the SEP attitude, went into P41 SCS and the whole thing. We went through the undocking checklist and got the probe circuit breakers in. I guess the major thing is that everything was nominal, except when I went to RELEASE on the probe EXTEND/RELEASE switch; nothing happened. Nothing.

I rechecked the circuit breakers and hit the EXTEND switch again, but nothing happened. At that point, there wasn't anything I could check inside. The only two things that you've got are the circuit breakers and the switch. So, I figured that there had to be something back in the tunnel. I went back and pressurized the tunnel. I looked in the tunnel and there was nothing there that was out of order. So I thought I'd go ahead and check those connectors again. I pulled the connectors off and put them back on. I figured that if that wasn't it then we had a serious problem. I put the hatch back in, depressurized the tunnel, and went through the checklist again, depressurizing the tunnel. We got a new attitude from the ground, which was the local vertical attitude. That time it worked fine. That's really a mystery to me.

SCOTT — We got a couple of good calls from the ground on that, one when we came around the corner. I called and told them we had not had ACCEPT and that you were in the tunnel checking the umbilicals. Right away they came back and said they had no TM on the program, which gave them the indication that there was something loose on the umbilicals, and that was, of course, the last thing to check. Soon after you checked everything they reported getting their TM, so that was a pretty good confirmation that that was the problem. Then, I thought another good call was immediately or very soon after. They came up and said no problem on the time, that we had 40 minutes to get the SEP done and just go to the local vertical attitude, somewhere around there, which was a big help to us. Jim and I were trying to plan ahead to make sure we didn't get too far behind the time line and get hooked into having to delay PDI rev. We were trying to plan our next series of events for a late separation. It was nice to have that call, to know that we had 40 minutes to get things squared away and move on.

WORDEN — The MCC-H came up with an attitude after you'd requested that they give us the time and an attitude. We went to that attitude and we were there 4 or 5 minutes before the time. It worked out fine.

SCOTT — I thought that was a very good recovery for an off-nominal situation.

SLAYTON — It sounds like you guys were ahead of it, though, by the time you came around the corner.

WORDEN — Yes, that's right. There was only one way to go.

SCOTT — You check the switches and the circuit breakers, and the next thing you have to do is go into the tunnel.

WORDEN — Anyway, we got undocked and from there on it went pretty well, except that the undocking was too late to do that low altitude P24. So we skipped that.

SCOTT — You didn't do any formation flight; you went to the SEP maneuver.

WORDEN — That's right.

SCOTT — You gave us a good call on gear down.

WORDEN — Right.

SCOTT — Which was nice to hear. I guess, you didn't see anything hanging from the LM that looked funny?

WORDEN — No. The LM was clean.

SCOTT — You didn't have any calibrations, did you?

WORDEN — No. As a matter of fact, that was all supporting the things that you were doing. We got the pads for the P24, which we did in the next rev.

SCOTT — Okay. You gave us a good call on your transponder, and I think that's a good sequence of events. We checked out the radar in the LM, and right afterwards Al gave us a call on the transponder. He checked his transponder right away. I think that's a good series, because that gave us a warm feeling about that whole system. I like the way he did that.

WORDEN — The circularization burn was exciting, but it was perfectly nominal. They had updated the short burn constants for the engine characteristics. And the circularization burn was done on Bank B only, because of the problem we had with the SPS.

WORDEN — I went to attitude. All the star checks worked fine. The burn was done on time, and the residuals were 00 and minus 0.5, which is a no-trim kind of maneuver. So that was perfect. It put us in a 65.2 by 54.8, and in fact, the circularization burn was absolutely nominal. It was a very nice burn, very smooth, and is sure a difference when you get the LM off. You can really feel that mother go. It's really quite impressive.

IRWIN — All the fuel cell purges were nominal.

WORDEN — We always got a MASTER CAUTION on fuel cell 3, and that was about all.

IRWIN — We normally do on the H_2. We weren't getting it toward the end.

WORDEN — No, we weren't, as a matter of fact.

IRWIN — It cured itself for some reason. Early in the flight, when we purged the O_2 on fuel cell 3 we'd get the MASTER ALARM, which we should not have. Then, toward the end of the flight, that did not occur.

SCOTT — Okay. We'll go through the LM side of the undocking and SEP. I guess on the separation maneuver we got about a tenth of a foot per second. Wasn't that what it was in P47?

IRWIN — I didn't write it down. Before we get there we ought to talk about the anomalies we had or surprises during the activation.

SCOTT — Okay. Why don't you go through the stuff up to SEP.

IRWIN — The first one was when you brought up the computer. We got a PROGRAM ALARM on the 1105, which was something I don't think we'd ever seen before.

SCOTT — Yes. We had a lot of up link/down link too fast in the LGC. We got that several times, 1103s and 1105s, and I don't really know why.

IRWIN — That's a good point. But it's an inconsequential alarm, I think. We mentioned it to the ground and they never seemed to say anything. As I mentioned before, we were always 10 or 15 minutes ahead of time. I don't think we were ever really rushed there. Your alignment went real well. You didn't have any trouble seeing the stars.

SCOTT — Yes, I did too. I had a tough time seeing Dabih. Dabih was a good star as far as position goes, but it was a very difficult star to see as far as alignment goes. If you can pick out bright stars, it'll sure help you. I guess the message there is, even if you don't have a NAV star, I think I'd ensure I had a good bright star for those alignments through the AOT. But the alignment went very well. The P57 docked is a very practical technique. You get a good alignment out of it, and subsequent drift checks showed that we had a good platform. I think that's the way to go, rather than a docked IMU align.

IRWIN — The next surprise, and it was probably our biggest surprise of the activation, was the pressure integrity check. When we obviously did not have integrity, we tried going to the secondary canister and still didn't have any integrity. We decided to press on through it and do the rate check, which we did. Then, later on I guess it was about 10 minutes before undocking, we came back and redid the pressure integrity check. Of course, we cycled through it right from the start; and this time, it worked out great. I think we had a 1/10th drop in 1 minute.

SCOTT — I guess on the first one, we had something like 1 psi drop in 1 minute, didn't we?

IRWIN — Yes. It was obviously something open, and I don't know whether the valve was just not seating properly or just what it was.

SCOTT — Okay, you cycled the valve back there several times. We both fiddled with the détente and had a good détente in it, but couldn't come up with an answer.

IRWIN — Well, I guess the ground never came back to us with anymore words on it either.

SCOTT — There was a question in my mind as to what the mission rule was at that time. I guess the mission rule was to undock and press on, which we were going to do had we not gotten a good check. But it was a good thing we started that check a little early. It gave us a chance to come around and do it again. The message is to get ahead and stay ahead; that's why we stayed 10 to 15 minutes ahead. Every time we got to a point in the time line where we could do something, we went ahead and did it, even though it was a little early. We got a tenth of a foot per second on the undocking, and I trimmed that out on P47. The comm was good. On the PGNS activation self-test. You mentioned the ALARM. Everything else went well. How about going through the AGS, Jim.

IRWIN — We didn't do AGS, of course, until after we had undocked. That was about an hour later. AGS was unpowered until later.

SCOTT — We'll pick up the AGS activation in sequence. The landing gear deployment was positive, and Al gave us a good check when we undocked. The DAP loads were fine. Rendezvous radar and landing radar checked out as per checklist. The tape meter worked just like it was supposed to, even though the glass was broken. The next thing is landing site photography. As I remembered, you took some pictures as we went over the landing site. Incidentally, they came out very well, and so did the pictures of the command module.

IRWIN — How about the sequence camera? We had the sequence camera on there, too. I guess we did have moisture on the LM windows, and we had to turn the window heaters on. I think it was fortuitous that Al had delayed the undocking until our windows cleared, so we could get good pictures of that operation. They were cleared up at just about that time.

SCOTT — Well, I think Al wanted pictures of himself. He wouldn't have gotten them if we had undocked on time. He would have gotten a bunch of fog. I guess I'd like to go back to REV 10 and discuss something that's not in the debriefing.

IRWIN — You mean looking at the landing sites through the sextant?

SCOTT — One of the questions on the landing site was general terrain relative to boulders, debris, and craters. A couple of months before the flight, we had worked out a plan whereby we do the low altitude landmark tracking technique without the spacecraft rate drive in order to take a look at the landmark through the sextant. There had been some question as to whether or not we could see anything. I took a look on REV 10, and it was as we had expected based on previous flights and fidelity of the optics. I could see the landing site very well. I could see Index Crater and the rille very well. I determined that there was no problem relative to boulders and debris, and it looked pretty smooth and flat.

It was a comforting feeling to know that we wouldn't have a rockpile to land in. If the advertised resolution of 3 feet was correct, we had no problem with boulders on the order of 3 feet and above. This was subsequently verified when we got there. It was a nice thing to have behind us in the way of validating the surface at the landing site because of the poor resolution of photography we had from Orbiter. The technique worked very well. It was easy to track in Inertial Attitude. I think you found that during your J-1 track, also, didn't you Al? I think you could have done your landmark tracking without a spacecraft rate drive.

WORDEN — I almost feel that way, yes. The J-1 tracking was really easy. With a high rate in the optics, it was fairly easy to track if it was off track. You have some roll in there, so that you're not coming through zero on the track.

SCOTT — But the optics are very easy to control, have very positive response, and once you lock on the target, you can stay right on it.

WORDEN — That's right.

SCOTT — That gets us down to lunar landmark recognition. Of course, the landing site at Hadley was particularly unique, when relative to landmarks. When I looked at it through the optics, I could recognize the craters that lead into Index and Index Crater quite well, even though there didn't seem to be as many shadows, crater shadows, as I had expected. MSFN Relay seemed to work all right.

9.0 SEPARATION THROUGH LM TOUCHDOWN

WORDEN — I want to say something about the VHF tracking. Of course, we didn't do any optics tracking prior to touchdown, but we did check out the VHF against the rendezvous radar. I think I reset the VHF three times, and it came up each time with half the value that the rendezvous radar had in it. This made me wonder at the time how good the VHF was operating, and it subsequently turned out that it was operating just fine. I don't know what caused the difference in the range between the rendezvous radar and the VHF at that close range, because it was 0.79 mile, or something less than a mile, I think.

SCOTT — Yes, we had on the NOUN 78. In the LGC, we had 0.78; you had 0.4; and the tape meter had 0.78,

Of course, the NOUN 78 is just a tape meter readout, but you did have, for some strange reason, just half value.

WORDEN — I reset that thing three times, and I think it came up with the same value each time.

SCOTT — I might add, in the IPA, we could tell when you were resetting. It was audible, so we tried to observe a no-comm silence period while you were getting your reset. There's no question there that you reset. Okay, you did your circularization burn, and the next thing is SIM BAY EXPERIMENTS DEPLOYMENT.

WORDEN — Before we talk about the SIM bay, I guess I should talk about the landmark tracking because that was the next thing that came up. After the circularization burn was the landing mark tracking from higher altitude. That went as planned, and everything worked out fine on it. I tracked the landmark with the telescope, and it was no problem. It was very easy to track the landmark. There was some concern about the shallowness of the particular crater that we were using, but it was pretty clearly visible when I checked it out.

I guess my only comment about the landmark tracking is that I would have felt much warmer about the landmark tracking if I had done it with the sextant, rather than with the telescope. The telescope presents a pretty large field of view, and you're trying to track a very small object down there. Apparently the numbers don't show that to be true - that there is a great deal of difference between the two. I think my own personal feelings would have been that I would have felt much better about it if I had done it with a sextant, because then I know I'm really on the target.

SCOTT — You could have locked up on that target easy, I think.

WORDEN — Easy with a sextant. In fact, I did it on subsequent revs when I came by in an attitude that would allow me to get the optics on it and track it all the way through. I guess the results of the landmark tracking were satisfactory for the ground to go ahead and update your descent.

SCOTT — Okay, why don't you hold the SIM bay information until we get past the landing.

WORDEN — The SIM bay stuff is sort of separate from the landing, and perhaps we should go all the way through that first.

SCOTT — We probably should stick with our Flight Plan, rather than the debriefing guide, at this point. We did a DOI trim burn, and we might go back in history a little bit on that one. We expected a long time ago, I think during the data priority meetings, to see some orbital perturbations out of plane due to the mascons. We had them put in the Flight Plan an extra period of time to do a DOI trim burn, if it is required. Prior to descent day (after the DOI, just before going to bed) in order to plan the next morning, we asked what was the probability of doing the DOI trim burn, and they said very remote. The next morning, they called up and said we were going to have to do a DOI trim burn. Fortunately, we had the time allocated in the Flight Plan to do it.

IRWIN — It seems to me that the perilune had degraded quite a bit during the night.

SCOTT — When we got up the next morning, they told us that PDI was going to be 33,000 plus or minus 9000 feet, which meant that we could be down to 24,000 feet at PDI. They were still thinking about a DOI. Well, that was a cue to me that we were definitely going to do a DOI trim burn, and I expected a 6-foot-per-second burn, which we prepared to do. That threw a glitch into our thinking that morning, because we planned that morning to try and get everything done early. Because we had run through the suited exercise good once before, we got ahead of the game there, and we were able to get that burn done. I think it should be included in the Flight Plan if there's any question at all about it. If there's a 10-percent probability that you're going to have to do it, you should probably stick it in there. You will need the time to get in the LM and get it cranked up.

IRWIN — The DOI trim turned out to be 3.1 feet, which we did with the RCS.

SCOTT — Theoretically, that was to bring us up to 50,000 feet at PDI. I guess we'll have a couple of words to say about that when we get around to our altitude check. Okay, anything we missed along the way, Jim?

IRWIN — We can start after undocking and just go through the time line, I guess.

SCOTT — We'll pick up with the LM from undocking to PDI. A general comment before we begin is that I thought the coordination of the two time lines in the two vehicles went very well. I think we always knew where the command module was, and the command module always knew where we were and in what sequence of events we were engaged at the time. I think it was a very comfortable time line, and we had plenty of time even to eat during that period in the LM and to take care of all the systems things on time, although we undocked a little late. The thing we ran was the DPS throttle check, and when we ran it through the first time, the DECA POWER circuit breaker was open. We have no explanation for that. I've checked it according to the circuit breaker list. I checked the row of circuit breakers, and then I counted the number open and compared it with the numbers we had there; it checked out. I don't know whether the circuit breaker opened during the check or if I missed it.

IRWIN — It should have been closed before undocking.

SCOTT — It should have been closed. The ground called us and said they didn't see anything and how about closing it, which we did. Then the DPS throttle check worked out fine. The window heaters, which we had turned on to take the condensation off the windows, seemed to work quite well. It got both windows cleaned up, and we never had to use them again.

Concerning the approach to the landing site, I think we missed this one because we undocked late. We took movies of that because Al was right below us and we were just about over the landing site. So, we have a combination of undocking plus landing site pictures. Rendezvous radar checkout went as prescribed. We mentioned the numbers already compared to the VHF. The alignment used the same two stars, Dabih and Alpheratz, again and used a regular P52 rather than P57. Again, Dabih was very difficult to see. The star angle difference was five zeros and the P52 was four zeros and a one. It was relatively easy to do. In my estimation, if we could convince ourselves that the P57 was as accurate as a P52, it would be an easier way to do an alignment because you don't have to maneuver the spacecraft; although, the P52 is not that difficult. The torquing angles were small. The LPD calibration worked out well. I had to turn the lights to approximately the same intensity as in the simulator, and the star was about the same as in the simulator. We used Nunki and it fell right on 40 degrees.

IRWIN — The AGS activation worked out real smooth. I think we got all the entries in before we had LOS. The ground can probably confirm that. I think there was no doubt that we had all the entries in, and we had the K-factor before we came upon LOS. The timing worked out real well on that. Then we maneuvered to the AOS CAL attitude plenty early, so the rates were damped when we got around to the AOS CAL.

SCOTT — One of the things that happened in our planning was a little confusion as to attitudes. The AGS CAL attitude is rather arbitrary within certain limits, and we originally chose an attitude which would enable us to view the command module during the circularization burn. Unfortunately, we were using the wrong reset tapes at the Cape, which we discovered about a month before flight. When we got the correct reset tapes, we felt it was too late to make a change in attitudes at that time. Therefore, we didn't get to see the circ burn, which is only a "matter of interest" kind of thing, but it would be nice to watch. If you choose your AGS CAL attitude correctly, you can see the command module do that. The overriding factor is getting to the attitude early so that you have about 20 or 30 minutes to let the spacecraft damp. Then when you do the AGS CAL, you have very low rates, which we did. The comm worked fine; configuration for LOS and AOS was fine.

IRWIN — The AGS CAL worked out real well. I read the values to the ground, and I don't think there was any problem at all on the calibration.

SCOTT — The AGS looked good; DPS pressurization and checkout went well. The landing radar checkout went well. We came by and made an altitude check as we went over the landmark, over the landing site, and we got something like 8 or 9. Didn't you write that number down? We did two of them as we went by and we called it down to the ground. So, they have the data.

IRWIN — I think it was 9 seconds.

SCOTT — We read it down to the ground. I guess it's just a warm feeling. I'm not sure what you can do with the data anyway. We just ran it. Landing site observation worked out well.

IRWIN — We might make a comment. Even after the 40-minute delay in undocking, we still picked up on the checklist and were right on the time line in very short order.

SCOTT — Which indicates it's a comfortable time line.

IRWIN — Yes.

SCOTT — The maneuver to the landmark LPD altitude check was done in AGS, and that was when we found, during the simulations, that we had no checkout of the AGS part of the landing, which we felt would have been a good idea. And I checked it out in both attitude hold and pulse and it worked very well. Both control systems were very stable and positive.

IRWIN — The next event was another alignment on the same two stars. We got four zeros and a one on the star angle difference, and the torquing angles were quite small again (0.010, 0.023, and 0.030, which gave us an indication that we had a good platform. The COAS calibration was approximately half a degree up. The star was about half a degree above the center of the COAS, and that looked pretty good. We ran the P63 ignition algorithm test, and the ground seemed satisfied with that.

SCOTT — Pre-PDI ECS looked okay. We went to the switch list and came around the corner for PDI. Then we started getting a few surprises. The first thing we got was a PIPA bias update right after we came around the corner, wasn't it? Do you have those erasable loads in there?

IRWIN — Yes, I wrote them down. Let me see if I can find them.

SCOTT — Up to this point, we felt we had a pretty good platform, and soon after AOS on the PDI rev, the ground called up two erasable quantities for PIPA bias updates, which we loaded manually. I was a little surprised to see that. We went through the procedures into the PDT as per the checklist, and everything seemed to be working just right.

IRWIN — 1454 and 1452.

SCOTT — Then, as I remember, at PDT minus 2 minutes, or something like that, we got another PIPA bias update, an erasable load, which we loaded.

IRWIN — I had a surprise here on the loading 231. I had never done that in training, and apparently hadn't interpreted the time line properly. They called me on loading 231.

SCOTT — Yes. I remember you even asked them about that.

IRWIN — That was a surprise for me. I would have thought someone would have caught it during the training period.

SCOTT — We got into PDT, the ignition part. Everything went as planned. We got a good ignition, good throttle-up, and were on our way to 2 minutes, and we got a call for a NOUN 69 or 169.

IRWIN — What do you remember it as?

SCOTT — Minus 2100.

IRWIN — I wrote down minus 1600. That wasn't definitely a minus.

SCOTT — Yes. It was an uprange load. Well, we loaded it real time, as they called it, and the ground verified it was the right number. We entered it and proceeded on down to 3 minutes, at which time I yawed around to zero. Very shortly thereafter we got the altitude and velocity lights out on the landing radar. We had a good DELTA-H. The ground confirmed it was good and around 2500, as I remember. We saw a DELTA-H on the order of 2000 almost all the way down. But, it accepted the updates quite well, and I think we noted that there was a fair difference in the PGNS altitude and the nominal, wasn't there, on the way down? Like about 3000 feet?

IRWIN — I think you're right.

SCOTT — It almost agreed with the DELTA-H we were seeing on the DSKY. We got throttled down a couple of seconds early, as I remember. I evaluated manual control with the PGNS MODE CONTROL switch in ATTITUDE HOLD. All I did was check roll, pitch, and yaw to see if we had any red flags and went back to AUTO. Everything seemed to be in order. I called up a NOUN 68 to check the time at which P64 would occur, and it was 9:23, which was nominal. Just prior to P64, two events occurred which biased my estimation of where we were going to land. The ground called and told us we were going to be 3000 feet south. Right?

WORDEN — Yes, I recall that from the command module.

SCOTT — I looked out of the window, and I could see Hadley Delta. We seemed to be floating across Hadley Delta and my impression at the time was that we were way long because I could see the mountain out of the window and we were still probably 10,000 to 11,000 feet high. I couldn't see the rille out the forward corner of the window, which you could on the simulator, out the left forward corner. So I had the feeling from the two calls that we were going to be long and south. When we pitched over, we got P64 right on time. As we pitched over and I looked out, there were very few shadows as far as craters go. I think the model gave us the impression that we could see many craters on the surface because of the shadow lines. I believe the overall problem was the enhancement of photography that was a little too high fidelity. In other words, I think they over-enhanced the photography and made themselves think the terrain had more topographic relief than it really did. When we pitched over, I couldn't convince myself that I saw Index Crater anywhere. I saw, as I remember, a couple of shadowed craters, but not nearly as many as we were accustomed to seeing. I measured my east-west displacement by my relative position to the rille, and I could see we were in fairly good shape, relative to the rille, but we were south.

I could see the secondaries. I could see some shadowing in the areas in which the secondaries occurred. Knowing that we were 3000 feet south, which I'm sure will be discussed in the debriefing because that's not what they meant. I don't know whether you know that or not. They didn't mean 3000 feet south apparently.

They meant azimuth. They meant that we were not coming in on 91 degrees. We were coming in at some other azimuth. But my interpretation was that our landing point had been moved. I'm sure we'll get that in the debriefing, but that was a confusing call. We were south, and I redesignated immediately four clicks to the right, and then very shortly thereafter, after you called me again on the LPD numbers, I redesignated two more right and three uprange. I saw what I thought was Salyut Crater and the smaller crater to the north of Salyut, both of which are quite subdued on the model. I think, in fact, what I was seeing was Last Crater. Punch that. The Last Crater on the model is rather a sharp rim crater with shadows, and Salyut and the one north of Salyut are rather subdued. I think what I selected was a landing site relative to Last Crater rather than Salyut Crater, but it looked like Salyut and the one north of Salyut to me, and that's where I redesignated to. I'm not sure how many other redesignations I put in heading for the target as Jim called the numbers. I may have put in a couple more.

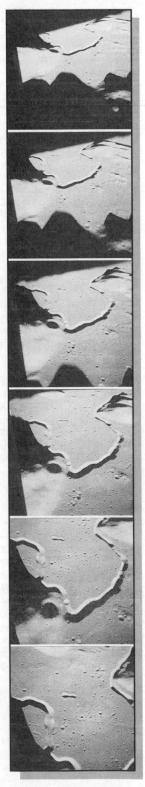

I got busy, at that time, attempting to select a point for the actual landing. I guess our preflight philosophy had been that if we were on target, we would try to land exactly on target. If we had a dispersion, we would select some point within the 1-kilometer circle which looked like a good place to land and would land as soon as possible so as not to get behind on the propellant curve. Once I realized that we were not heading for the exact landing site, and that I didn't have a good location relative to Index Crater, I picked what I thought was a reasonably smooth area and headed directly for that. We got down to 400 feet, and we had planned to switch to P66. I gave one ROD click at that time. Jim called me on the P66, which verified the ROD was working, and I went on down to 200 feet and started rounding out at 150 feet. I could see dust - just a slight bit of dust. At about 50 to 60 feet, the total view outside was obscured by dust. It was completely IFR. I came into the cockpit and flew with the instruments from there on down. I got the altitude rate and the altitude from Jim, and rounded out to 15 feet and 1 foot per second for the last portion. When Jim called a CONTACT LIGHT, I pushed the STOP button, which had been in the plan. Knowing that the extension on the engine bell was of some concern relative to ground contact, it had been my plan to shut the engine down as soon as possible after Jim had called the contact and to attempt to be at some very low descent rate, which we felt that we were at that time. The next event was the contact with the ground, which I guess was somewhat harder than the 1 foot per second. One of the sensations in the LLTV which helped me was contact on the order of 1 foot per second, which feels rather hard with a tightly sprung system like you have on either of those two vehicles. We landed in a shallow depression on the rear pad. I think the rear foot pad was in a 5- by 15-foot shallow crater. Wouldn't you say that was about the order?

IRWIN — Fifteen to 25 feet in diameter.

SCOTT — It gave us a tilt of about 10 degrees left and 10 degrees up, which was subsequently no problem. There was a rumble when we landed. I think all the equipment on board rattled. It seemed as if I could hear it all when we landed, like you would shake the vehicle. Couldn't you hear that?

IRWIN — Yes, I agree.

SCOTT — Soon thereafter, we called Houston and informed them we were on the ground.

IRWIN — The propellant was about 6 percent.

SCOTT — About 6 percent, and that gets us on the ground. A couple of general comments on the techniques. I relied on Jim's call on the altitude and altitude rate and on the LPD. I felt I had a good handle on LPD and H and H-dot all the way down. I could concentrate out the window to try and select a point. I was very surprised that the general terrain was as smooth and flat as it was, with relatively few prominent features that could be seen. There were very few craters that had any shadows at all, and very little definition. The terrain was quite hummocky. There were smooth and subtle craters everywhere, which made subsequent motion and movement on the terrain there somewhat tricky. But at the altitudes looking down as we approached the landing, it was very difficult to pick out the depressions. I did know that I was landing past the crater which I thought was the one north of

Salyut, which I believe now was probably Last Crater. I could see that I was going to land to the west of that, but as far as the other shallow depressions there and the one in which the rear pad finally rested, I couldn't see that they were really there. It looked like a relatively smooth surface.

IRWIN — I put the altitude update into the AGS at 12,000 feet, and shortly thereafter (we were probably at 11,000 feet), I put in the altitude rate update. Immediately after that, I called 367. It looked like there was probably a difference of 1 foot per second between PGNS and AGS, so I think we had a good manual update of H-dot. When we got to P64, I did not look out the window at all. I just concentrated on the systems readouts so I could give Dave as much information as he needed. Everything else should be on the tape.

SCOTT — Okay. In going through the other notes here, we were probably fairly close to zero phase, and I didn't notice any particular effects on the zero phase as we approached the landing. I don't think that contributed to the wash out or the lack of seeing shadows. It was just a subtle terrain, and the rounded features prevented any shadows from showing. I don't think that was the zero-phase effect at all. The LPD was real good. I felt we were heading toward the point for which the numbers were being read. Manual control on the vehicle was excellent. I think it was more positive than the LLTV. I'll make one general comment. I felt very comfortable flying the vehicle manually, because of the LLTV training, and there was no question in my mind that I could put it down where I wanted to. We landed exactly where I was headed. In spite of the fact that the rear pad was in a crater, that's just where I wanted to land. I think our horizontal velocities were zero lateral and I had about 1 foot per second forward to keep from backing into anything. That's exactly what I wanted. There was no tendency to overshoot in attitude or overshoot in the selection of the landing site. I think all of this is because of the time that I had to work with the LLTV. I guess I can't say enough about that training. That puts you in a situation in which you appreciate propellant margins and controllability. I think the LLTV is an excellent simulation of the vehicle. I think if you had to move from one point to another, you could do it quite well. I would recommend maintaining an altitude of at least 150 feet so you don't get into the dust problem. I think dust is going to be variable with landing sites.

10.0 LUNAR SURFACE

10.1 POSTLANDING AND SEVA

SCOTT — Everything worked as advertised. We got the venting going and and I think it vented somewhat slower than the simulator. As I remember, we had time to do a few other things before we got down to the minimum fuel and oxidizer pressure.

IRWIN — There was a little confusion on the P57, using two NOUN 88 stars. That held us up a little bit. It was just because we hadn't done it recently.

SCOTT — Yes, I think the problem was that we never got a confirmation from the ground that the erasable load in the P57 was the right thing to do. We did it, and it seemed to finally work. I guess not having worked with NOUN 88s for a long time, it took us a little while to get through it. I think we ended up fairly close to the time line in spite of that.

87-11755

IRWIN — I don't have any other comments coming into the standup EVA day.

SCOTT — Let me comment that the stars, even though new and different, we had Schedar, which was in Cassiopeia and we had Alhena which was in Orion and even though they were not mass stars, they were easily recognizable. The numbers that were called up, cursor and spiral, were very close. There was no trouble identifying the stars and ensuring that they were, in fact, the correct stars. The alignment was straightforward once we got the 188 procedures squared away. We got a .01 on the first star angle difference and five zeros on the second star angle difference. I think we finally ended up complete with the alignments at about the right time, within the checklist. The new procedures developed by MIT to perform P57s are very good and save quite a bit of time. We had an extra pad in there, based on the old techniques of having to go cursor and then spiral. Now that you can go straight through, it saves quite a bit of time, and it's a valuable improvement in the program. We did not do the 10-minute gravity exercised with the platform. It seem to me that they called us right away and said to go ahead to P6 and we wouldn't have to do that. I don't know why they canceled that; maybe they had enough data by that time anyway.

IRWIN — We were running a little bit behind time because of the delay on the P57s.

SCOTT — I wasn't watching the clock at that time. Did we end up behind on the time?

IRWIN — We were a little behind. They wanted to get us bedded down so we could get out on time the next morning.

SCOTT — Okay, we went through the switch list, did the equipment prep for the SEVA. We didn't discuss our position more than a comment on what it looked like on the way down. We were saving, I guess, trying to utilize the time so we could discuss the position of the landing site through the top hatch.

I thought the equipment prep for the SEVA was very straightforward. We went per the checklist on the SEVA prep, got the hatch open, pulled the drogue out (which was very similar to the one-sixth-g airplane exercises we had), and when I stood up in the top hatch, I found that because of the one-sixth gravity, I could support myself on my elbows without having to stand on anything, and get fairly well out of the hatch. I guess the first thing we used was the Sun compass to try to get a relative bearing on three sites. We used Benefield 305 and Mount Hadley to get bearings. And then you passed up the camera for the pans, and then the 500, and I took probably about 20 pictures with the 500 and described the general area.

My impression, looking out, was that we had good surface on which to travel with the Rover. I could see the Northern Complex almost completely, and I could see the base of the Front, and all the way up the side of Hadley Delta. There were no boulders anywhere which gave us some confidence that we could make pretty good time with the Rover, if the Rover produced for us as far as performance. There weren't any obstructions other than the many small and subtle craters. The general surface was rolling, smooth, hummocky and very much like 14. Although there were not a lot of boulders, there were a lot of small craters, which we could see were going to require some navigation. We could see that the trafficability was going to be good. I could look out to the west and see a spot that was fairly level for the ALSEP, and confirm that we did have a place we could put the ALSEP. It was not apparent that there was any place in the immediate vicinity of the LM to place the ALSEP. I couldn't tell exactly why we had the tilt on the LM. It wasn't clear that we had put the rear foot pad in the crater, which we subsequently found. I couldn't tell that from the top hatch, although I could see there were a number of shallow depressions and smaller craters in the area. But, it did look like a good place to put the ALSEP within a reasonable walking distance. I couldn't see the rille, or define the rille, but I could see the far side, Hill 305 and Bennett Hill, which looked a great deal closer than I had expected, as did the Northern Complex. Geologically, we could see there were few fragments in the area and no boulders. One apparent observation was the secondaries which had gone up the side of Hadley Delta. I think immediately it was obvious that the secondary cluster had swept up on top of the Front rather than the Front coming down on the secondaries, which gives us an age relationship.

About Hadley itself, the Swann Range, the Big Rock Mountain, and all the features to the east were still in shadow so I couldn't see anything there that I could define specifically that we had geologically. You could see the Northern Complex, however. I could see the inner walls of Pluton and they had large fragments, probably on the order of a couple of meters, on the inner walls. They probably represented 3 or 4 percent of the debris that I could see on the wall, although the inner walls seemed to be all relatively smooth, free of talus. I saw nothing on the outer wall of Pluton. It looked pretty smooth and similar to the rest of the local surface area. I could also see Icarus and Chain; a very good vantage point primarily because we landed on a topographic high. This proved to be helpful; subsequently, when we were great distances from the LM, many times we could locate it.

IRWIN — We did take color and also black and white.

SCOTT — We came back in after the SEVA, and I might say in conclusion that the SEVA was a very useful thing. It gave us a lot of confidence that we could get to the Front with the Rover and also to the rille and the Northern Complex. I felt we had all three of them pretty well in hand for traveling, in spite of the fact that it was obvious that we had not landed precisely at the preplanned point. At this time, I wasn't sure where we were located. Although I could see prominent features, I was relying on the Sun compass to give us the data for triangulation to spot our point because there was nothing in the immediate vicinity which was recognizable. I think this was general throughout the rest of the EVAs. The terrain was considerably different than we had been led to believe, because of the lack of high resolution photography. I think, in retrospect, the enhancement of the photography provided more detail than was actually there and that fooled us a little bit.

IRWIN — How about comfort while you were there, were you particularly warm without the LCG on?

SCOTT — No, I was very comfortable. As a matter of fact, I thought the cooling was fine, and there was no problem at all wearing the CWG. As a matter of fact, I thought it was more comfortable wearing the CWG. How did you feel?

IRWIN — Yes, I was plenty comfortable. I just thought maybe you would be a little bit warmer, being up in the Sun. There wasn't any sunlight at all coming in on the front panel. We were concerned about that beforehand, but there wasn't any coming in.

SCOTT — In summary, the SEVA was very easy, the procedures were simple, and there were no problems encountered. The next thing is the eat and rest period, and the suit donning and doffing. As we came back in to repress the cabin, no problems. We took our suits off, and, there again, no problems. I think training had prepared us for the doffing of the suits, and I can't remember having any trouble at getting out of the suits. Can you?

IRWIN — No, we configured them for drying; although they probably did not need drying, we decided to go ahead and do it.

SCOTT — Why don't we make a general comment for all the suit doffing while we're here since I don't think we remember having any problems getting out of the suits? I think that at the conclusion of each EVA, we configured the suits for drying. We let them go for about an hour, and more in some cases, and unplugged them and configured the ECS for sleep, which was no problem.

IRWIN — We ate and while we were eating, we did the PLSS water charge and topped it off. I frankly don't remember what the orientation of my PLSS was when we did the water recharge, but I guess it was off-vertical somewhat.

SCOTT — Maybe a little bit, but at that time, the cabin wasn't too crowded, nor was it dirty. I think there was another - maybe an advantage for sleeping. We got to sleep the first night in a clean cabin.

IRWIN — I thought we had my PLSS on the midstep.

SCOTT — I think we did when we charged it. We didn't have to do the oxygen charge. It was probably level. I guess we might relate to the problem here, since we're on it.

IRWIN — There was a subsequent problem with the water cooling in my suit, my PLSS.

SCOTT — But I think that, at that time, we didn't have it vertical when we charged it. Okay, on down to the sleep period, or the eat period. I guess we ate the meal that was provided. And I think that I'll mention, in general, that I don't think there's enough food on the LM, and I think we ate everything that was there. I think that, for the activity we have on the surface, that you need more food in the LM.

IRWIN — Particularly those food sticks would really come in handy.

SCOTT — I'd say we could take at least twice as many of those easily, because you can eat those during the preps and posts.

IRWIN — The first night's sleep on the LM was the best night's sleep I had on the total flight.

SCOTT — Yes, I slept quite well too. I was surprised that the hammock was as comfortable as it was. I think in the one-sixth g environment that those hammocks work just fine, don't you? And the suit positions were fine. I think the whole layout of the cabin was quite adequate. We have no comments or recommendations on a change on that. It worked out just fine.

In looking over the events on the surface, I think we'll go to the systems within the spacecraft in sequence and then come back and go through the events which occurred after the hatch was open. So we'll really have two categories of surface activities, one of which will be in the cabin - we'll discuss that now from end to end - the other of which will be on the surface, which we'll discuss after the cabin events. After the good night's sleep, we awakened the next morning.

IRWIN — They awakened us early because of the O_2 leak to the urine transfer device. And I think part of the problem on that was that the top seal, the double seal on it, the cork plunger there, was not completely seated after we used it the night before.

SCOTT — Had we ever been briefed on that thing?

IRWIN — No. Well, we had told John that we planned on connecting it and leaving it connected. Nothing was ever said against that.

SCOTT — Yes, but I don't think there was ever any discussion. It had the two seals on the plug and it had one valve which, I guess, we felt, prior to the flight, would be adequate for leakage prevention.

IRWIN — And it might have been, had we had the plug fully seated. Anyway, the plug wasn't fully seated, and we had leaked some oxygen. So that was the first call from the ground to check it. Well, they didn't know where the leak was, but it was pinned down quite quickly to that being the cause. So we took the urine transfer device off the hose and capped the end of that line, and that stopped the leak.

SCOTT — And I think there is a point you're inferring there. I think we'd have been better off had the ground called us once they recognized that there was some leak, even though it was in the middle of the night. I think we would have slept better on subsequent nights knowing that any small thing would be corrected immediately before it got us too far down. I guess our recommendation there would be to call the crew if the ground sees any problem which might develop into significance later on.

0.2 EVA PREP AND POST

We got up in the morning and had breakfast and proceeded with the EVA prep for EVA-1. That seemed to go, as I remember, fairly well. I don't know what the timing was on that. We might tell you here that the mission timer was turned off for power savings, and we were going on Houston's time on our watches. There really weren't too many references within the surface checklist to the Houston time, so we were not really conscious of where we were relative to the g.e.t. or relative to the timing and relied on Houston to keep us abreast of the time. We just proceeded through the checklist as expeditiously as we could. I do remember we asked them during the EVA prep when they expected us to depress. That came out on time, so we were pretty well going with the time line as planned.

IRWIN — I think I made the comment that I was glad that they had awakened us about an hour early, because we went into our first EVA very leisurely. There was plenty of time; there wasn't any rush at all. In fact, I think I made the comment that I would just as soon wake up an hour early for the subsequent EVAs to give us a little more time to think about things and get organized.

SCOTT — That's a good point, because here again, we've got a plan which says eat and rest, and we don't have all the transition things in there. They are in the checklist, but I'm not sure there is adequate time. And a rest period, I guess, we might define as not necessarily closing and opening your eyes, but as a period during which you've had no scheduled activities, wouldn't you say? If you had some little cabin things you want to take care of during your rest period, like the biosensor change or some housekeeping that has to be done, I think that can be easily included in the rest period.

Let me interject here that we're reorganizing the plan a little bit and discussing the LM activities in the cabin as one category. And then, within the surface activities, we'll subdivide that into two categories, one of which will be all the equipment that was utilized on the surface, and the other of which will be the science and the geology part. I think we can present a more organized approach in doing it that way.

So, back to the prep for EVA-1. I guess I might add to Jim's comment on having an extra hour in the morning, to go through it leisurely really helps. I think we saw this later on in the flight, too. You could be more sure of doing things right if you proceed to it leisurely, which I think we planned within the nominal training anyway. We had plenty of time during our training months to do the EVA preps leisurely.

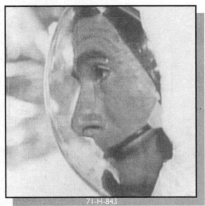
71-H-843

IRWIN — We would have had plenty of time on the subsequent EVAs if we hadn't had those problems. I might make one note there. When I unstowed my PLSS, I noticed that there was a large hunk chewed out of the antenna. About half of the width of the antenna was gone.

SCOTT — And about an inch long.

IRWIN — Yes, like somebody had taken a pair of snippers and snipped a piece out of it, right at the base, about a couple of inches from the base of the connection. We put a piece of tape around that at that weak paint, and on EVA-1, we pressed ahead.

SCOTT — They should have the antenna because we brought it back on the OPS that was in the CM. It looked like somebody really missed something in the PIA of the OPS. When Jim unstowed it, he found it right away. It was a pretty gross oversight.

The depress went all right, and then we started having some problems. hanging up in the cabin. I think that they were magnified by the one-sixth g environment because we didn't compress the suits as much in one g, and I think we both were riding a little bit higher, and a little lighter. Turnarounds within the cabin were very difficult, and my hangup problems were on the mounting lever or shaft that holds the PLSS and recharge station, in the handle of that. Jim finally figured that I was hanging up on that handle, and we put some tape on it, across the handle on subsequent EVAs, which did help. It was also hanging up on the corner of the Flight Data File, which is a sharp corner, and also on the DSEA guard, the wire cover. It's very crowded in there, and it takes a lot of time in moving about the cabin to prevent hangups, and I think we lost, overall, quite a bit of time. I wouldn't be surprised if we didn't lose a total of a half an hour. Do you remember when you were hanging up?

IRWIN — One thing was the water hose. The other was - you know, after you disconnected your umbilicals, they were not stowed as far aft as they probably should have been.

SCOTT — We got that corrected on the second one.

IRWIN — Yes, once we pushed them way back in the aft, it was all right. And another problem was when we stowed the bracket that holds the PLSS to the floor, we didn't get that pin fully secured, and that bracket did not go down flush with the floor, so the hatch would not open fully. This caused a subsequent problem for me, getting out and getting back in. Another problem that I noticed was the strap length. It's measured in one g, and I think that's a mistake. Because my controls were just too high at one-sixth g. The PLSS was just riding too high. I had a difficult time getting to the controls.

SCOTT — I thought the strap length was measured in the rig that supported the PLSS at one-sixth the weight.

IRWIN — It was on the rig, but something's different.

SCOTT — You felt your controls were a lot higher. Mine felt fine. I didn't have any trouble reaching the controls, other than when my fingers got sore. I noticed your PLSS seemed to be sitting in an angle, too, where your controls were further behind, you tilted. It looked like you had more trouble. Do you remember any of the other things you were hanging up on.

IRWIN — I can't think of any other things I was hanging up on.

SCOTT — Well, then we get down to the depress and the hatch opening. The hatch was very difficult to open partially. I guess we expected that because of the pressure on it, from previous flights. Once we got it open, it could be held without any trouble.

Got the GO for the egress, and I didn't have any particular trouble getting out. I think that's because you were guiding me as I went out. I remember you gave me a couple of "move rights" or "move lefts," something like that, and I didn't have any trouble getting out. Maybe you ought to talk about your problems getting out. I didn't realize the hatch was only partially open.

IRWIN — I guess we lacked about 40 degrees on hatch motion. I had to go a little more right-than I normally would. I think I was hanging up on the right side of the hatch. I had to ask you for guidance when I initially came through the hatch.

SCOTT — Your whole back was hanging up on the ACA mount because you were too far to the right. With the hatch only partly open, you got yourself too far over to the right. I remember when you went back in and I was going to see what was hanging up, and you were hanging up underneath the ACA mount.

IRWIN — Once the hatch was configured so it would fully open, it wasn't any problem getting in or out.

SCOTT — We'll step ahead to getting back in then. I guess you had the same problem getting in because of the same reasons. I think once we got everything in and you discovered that the hatch wasn't fully opened, why that made it a lot easier from then on. I guess I didn't have any problems getting back in, because again you were able to guide me as I came in through the door.

The cabin repress: I guess when we got back in, I noticed that things seemed to be much more crowded than I had remembered several hours before. I guess that's when we had the freedom of mobility outside. And I had a tough time getting to my water to turn it off, and I think you did too.

IRWIN — In fact, I think I asked you to get mine. It could have been a function of our hands being so doggone tired.

SCOTT — Yes, I think it probably was. But still it was very crowded and very difficult to move around. Once we got the hatch closed and repressed, why we sort of took a break right there, which I think wasn't really in the time line. But it was a good place to take a piece of our rest period. I remember we got the helmets off and stood there and talked about it for a while before we went through the rest of the function.

IRWIN — I'm trying to think of when we noticed the break in the bacteria filter.

SCOTT — I think it was right after we got our helmets and our gloves off. I think you looked down and saw it.

The water was in and out of there, the hose right at the connection where the bacteria filter joins the water hose. The bacteria filter has some plastic attachments to it. There were two little knicks about, probably a quarter of an inch long and about a quarter of an inch wide out of the side of the plastic connector. The water was flowing freely and we had no idea at that time how much water had come out, nor how long it had been flowing. There was no way to really tell. We looked at the floor, and there was a little bit of water on the floor, not much. There was no evidence of a great leakage rate, although the spacecraft was tilted. We found out subsequently it had leaked, I guess, about 25 pounds back in the aft portion of the cabin. Then we disconnected the filter; that stopped the leak. The first order of business after we got repressed was to go through the checklist and do the EVA post and try and come up with a plan on how to handle all the dirt in the cabin. We were pretty dirty. We had planned prior to the flight to take the jettison bags and step into them with the suits to keep the lower portion of the suit isolated from the rest of the cabin. Our legs from about thigh down were just about completely covered with dirt. I guess the dust brush worked fairly well. I got the most part of it, but we were still pretty dirty.

The O_2 recharge. The first thing we did was the O_2 recharge. That went as planned. We got to everything all right; even though the O_2 line was a little short at that stage, we could reach it. Then we docked the PLSSs and took off our PGAs. Then we started hunting, as I recall.

IRWIN — I had the impression we had more time there. We were moving pretty slowly. We could have easily got some of that recharge while we were eating, which we did later.

SCOTT — We could have combined some things as we did do later. We were going sort of slow, feeling our way around the cabin, trying to get settled down to some sort of system to control the dirt and stay organized. I think the jettison bags over the legs worked fairly well. I think we kept the majority of the dirt out of the cabin and kept it in the bag. We just cinched the bags up around our legs. It was no problem getting in and out of our suits with the bags on them. We took another jettison bag and stuck it up on the midstep, and I stood on that to keep my CWG clean. You stood on one of the OPSs to keep off the floor, which was pretty dirty. Now, we get down to the water charge. On my first, the water charge went as advertised. At this point, when we charged yours, we probably had to tilt it.

IRWIN — We layed the PLSS on the suits, as I recall.

SCOTT — What we were trying to do, to save a little time there, was to charge water and complete the O_2 recharge at the same time. All the connections to be made to the PLSS at about the same time would save a series operation. The high-pressure O_2 line wasn't long enough to reach the PLSS unless you tilted the PLSS and we found out later that the ground suspected some substandard water charge because of the tilted PLSS. We had to lean the PLSS over in order to get all the hoses connected to it simultaneously. I'm not sure that that was really a problem, even though we corrected it by recharging vertically later on. The stowage went as planned. We didn't have a lot to stow on the first EVA. We put bag 4 back in the box which contained the 500 millimeter on the way down, dried the suits, configured the ECS for sleep and proceeded to sleep. Can you remember anything else on that?

IRWIN — No. I thought most of that was fairly nominal.

SCOTT — Getting out of the suit again was no problem. I think the procedures established were quite adequate.

IRWIN — The hammocks are adjustable to a certain degree, and we were pitched up which meant that my hammock would be tilted back. I didn't notice any problem at all. I slept very comfortably and I think Dave did, too.

SCOTT — I was afraid I would be feeling like I was sleeping heads down with that pitch angling there, but I didn't at all. The suits were a lot fluffier than they are in one g. They compress and they were right up to the bottom of my hammock, but that didn't bother me either. As a matter of fact, it was almost like a nice little bed up there. I guess we had some concern that the hammocks were not going to provide the reasonable sleeping position; but after we had done it a few times, I think it worked all right. I think at any choke angle you can adjust those things so it will give you a good position.

IRWIN — One improvement I would suggest is extending the bottom of my hammock up to the connectors. There is a gap of about 2 feet, and my legs would dangle down at night.

SCOTT — Oh, really?

IRWIN — It wasn't any problem. I found that most of the night my legs were up and kind of resting on top of the comm panel.

SCOTT — I felt like I might put my feet on the control switches. I took a piece of the webbing we had on board, cut a hole in the bottom of the hammock, and tied the bottom up to the AOT guards so my feet wouldn't slide down onto the switches. I think you can improve the hammock by providing those two little items. You can make them wider, too. They're not quite wide enough to put your shoulders on them. Other than that, there is no problem. The night went all right; but at some point along the way in one of the nights, we got a call for Endeavour. Did you hear that one?

IRWIN — No, I didn't hear that one.

SCOTT — You didn't hear that one. Yes, the Endeavour called one night, which made me think a while about where I was. Was I on the Endeavour or the Falcon or where? I think it might behoove CAP COMM to be sure they punch up their right key when they are talking to the different spacecraft, because that can make you come out of the hammock pretty fast. When we woke up the next morning, I was surprised how clean the spacecraft was. I think most of the dust had been removed. That's right. It surely had.

IRWIN — That night, it was fairly clean, you know, when we went to sleep. I don't know how all the dust got out of there.

SCOTT — Yes, the ECS does a pretty good job of cleaning the place out. The smell was gone. When you took the helmet off, you could smell the lunar dirt. It smelled like - the nearest analogy I can think of is gunpowder. But that had all cleaned out. By the time we got up the next morning things were in pretty good shape.

The first thing that occurred, I guess, the next morning was a call from the ground about how much water we thought we had lost and to check the aft behind the engine cover. We did, and sure enough, there was a great big puddle back there. The ground suggested using a food bag and LiOH canister to get it all up, and they wanted all the water cleaned up before we depressed. That was probably a pretty good idea because immersed in a puddle of water were a couple of glycol lines and some wires.

Thereupon, we entered into another mopping operation. We took one of the large meal container bags and cut it out like a scoop, and Jim passed me the canister cans. I scooped up the water, and then we took towels and dried up the rest of it. I think we got it completely dry.

We got two full LiOH cans and locked them with their locks to make sure the tops wouldn't come out. Then we got another half can, at least, in the helmet bag which we had intended to throw out; but we subsequently found it was dripping, too, so we took the ground's suggestion and dumped it into the urine container. We had plenty of storage space after handling extra water and all the urine, too. So I would say it worked pretty well.

IRWIN — The temperature in the cabin was very comfortable. I slept in my CWG in the sleeping bag and did not use the coveralls.

SCOTT — I slept in my coveralls without a sleeping bag; so I guess we each had two layers on, and it was very comfortable. We also used our earplugs, so noise was no problem.

IRWIN — The earplugs worked very well.

SCOTT — There was some light leakage which you commented on. The stitching around the window covers provided light leakage around the main left- and right-hand windows, but it wasn't any problem. I think the final ECS configuration they came up with as a result of the chamber run was a good one. That, plus the ear plugs kept things pretty quiet. The only noise you could hear was the constant tone of the glycol console.

I guess that gets us up to breakfast on the prep for EVA 2. We were starting to run behind because of our mopping operation; however, the EVA 2 prep, went nominally. I can't think of anything off nominal up to the comm check. Can you?

IRWIN — No. It seems at this point, that we did a water recharge on my PLSS.

SCOTT — Yes, that's right.

IRWIN — At this point.

SCOTT — Yes, because of the expected tilt problem.

IRWIN — Yes, and that probably took us an additional 15 minutes to recharge the water on my PLSS.

SCOTT — I guess the suits went on without a hitch, and the banks worked all right. We kept things fairly clean. We went down through the checklist in a nominal fashion until the comm checks at which time we noticed that when Jim went to his portion of the comm check, we could not hear him. Isn't that right? Or you could not hear us.

IRWIN — I could hear. I wasn't transmitting. It was zero.

SCOTT — We took a look at the antenna again and found that it had broken off at the root right down inside the OPS; so we took a couple inches off the top, spliced it, and taped it down. That seemed to solve the problem. The ground then informed us that we would not have to have Jim's antenna up anyway because his comm was so good, so we left it down. I would say that's probably a pretty good nominal procedure to leave that antenna down if you don't really need it up.

IRWIN — I don't know why they subsequently asked for your antenna to be up because it looked like we had great comm with it down.

SCOTT — Yes, that's right. I think that would be one for the systems people to think about because it would surely prevent any possibility of knocking that antenna off somewhere along the way hooking it on the high-gain antenna on the LCRU or the LM, or something. It also saves time.

Cabin depress. I guess we got the hatch open all the way, and I did not have any trouble getting out. Did you then, with the hatch open?

IRWIN — No.

SCOTT — I guess we'll step ahead to getting back in on EVA-2. We had no problem getting in and closing the hatch. We again had trouble getting hold of the water valves. That is probably because both of our hands were hurting at the end of the EVA. It's just hard to feel with them. Anyway, I think we got them locked, but wasn't this the time that yours really didn't get turned off, and didn't we get a little bit of water in the cabin?

IRWIN — I think it happened on both EVAs. On the first one, you turned it off, and I must have bumped it on something and turned it back on.

SCOTT — Well, maybe I didn't get it off.

IRWIN — No. It happened both times, I think, because you turned it off both times, and then after we repressured, it was on again; so I must have been bumping it on something.

SCOTT — You felt the water in your suit, didn't you.

IRWIN — Yes.

SCOTT — That was a clue.

IRWIN — I felt it and heard it gurgling and running down my right leg, so there was good reason to dry the suits.

SCOTT — PLSS water, huh?

IRWIN — Yes, it was really water.

SCOTT — That's true. It would be a good idea to dry those suits. It wasn't any problem, but once you felt the water and we got the water turned off for sure, then it stopped running. I don't think we ever accumulated enough water in the cabin to even see; it was mostly in your suit.

IRWIN — I don't know. A little bit of that water on the floor there might have reduced the amount of dust on the floor. The floor was always kind of moist.

SCOTT — Yes, that's true. I might comment that lunar dust is very soluble in water. It seems to wash off very easily. I would say if you ever have a connector problem that was really stiff, you could take the water gun and spray it in and loosen it up.

IRWIN — We did not loosen the suit connections for EVA-2 but we did for EVA-3.

SCOTT — It seemed like they were still working pretty well. The connectors got covered with dust - one of mine. One of the primary problems was the LEC. On EVA-1, when I passed you the rock box on the LEC, I just got covered with dirt all down the front. The result was pretty dirty connectors. We tried to brush them off and clean them off. We found that the booties which had been placed over the PLSS connectors were good protection from dirt. A recommendation would be to put booties over all the connectors or some sort of protective device. In the old days, they had a bib to keep them clean - or for double protection, I guess. Something like that would surely prevent problems later on and would save time cleaning the connectors. They sure get dirty, and I am just not sure there is any way to prevent them from getting dirty. If you are going to go out there and do the job, you are going to get dirty. If you try to keep everything clean, you are just not going to be able to do the job on time. I think those little booties are a pretty good idea. They were no problem on the donning and doffing. EVA-post went all right. Suit doffing went all right. We made sure the PLSS was vertical when we recharged the water.

IRWIN — They did call up and ask us to go to 10 minutes. It really isn't any problem to combine the PLSS recharging with your eating. As a matter of fact, that would be a good procedure to get everything set up to do your recharging the PLSS and then let the PLSS recharge while you are eating. It would save some time.

SCOTT — It seems to me that the last night's sleep was about the same as the others. We talked over getting up early to make sure we didn't fall behind. We were going to try to awaken 45 minutes or so early, and that is exactly when the ground called us the next morning. It was just about the time we thought we ought to get up.

IRWIN — It was just about the time we thought we ought to get up.

SCOTT — They told us then Wednesday to get 7 hours for sure; and looking at the time line, we figured we needed it.

IRWIN — I guess they were concerned about whether we did not get a good night's sleep that night.

SCOTT — Yes. Didn't you feel like you had?

IRWIN — I thought I slept just as good that night as I had the night before.

SCOTT — I would question how they know you got a good night's sleep; except other than asking. The heart rate is a great idea, except do they ever measure your heart rate while you're sleeping at home?

IRWIN — No, but I guess they want to do that.

SCOTT — I think they ought to. I don't see how they can possibly correlate it, otherwise. Well, I got a good night's sleep that night, it felt like to me. When I got up the next morning, I remember asking how you slept. You said you slept fine. So, I felt like we were both well rested for that day.

IRWIN — On this last morning we didn't put water bags or food sticks in the suit, because we knew it was going to be a relatively short EVA. On previous preps we did put the food sticks and the water bags in the suit.

SCOTT — The last EVA would be short enough that we wouldn't need them. We didn't want to take the time to fill the water bags and put them in, because that was taking time away from the EVA. The prep went good, the comm went good, and I guess we got into the depress. I don't remember exactly what time it was but we were letting Houston keep track of the time. I think it would be good to have some sort of procedure for what time we could expect the various events to be occurring. We did the checklist on Houston time so we could have something to refer to. Depress and out the hatch without any problem.

IRWIN — I do not think we ever hit any circuit breakers during the operation.

SCOTT — Yes, that's right. I do not think we ever did. Everytime we checked them they were configured right. We did lubricate all the wrist rings, connectors, and helmet rings on this one, which was easy. I think that little dab of lubrication material works just fine.

IRWIN — It was easy and I think it paid off because it was very easy to make the connections.

SCOTT — We never had a problem with the zipper at all. Both zippers worked very good throughout the flight. I don't remember ever having your zipper hang up. I thought the lock box worked fine. I guess we can't think of any improvements on that. The post-EVA went well. We configured as per checklist, prepared for the equipment jettison and jettisoned all the equipment that was planned. The procedures worked well. We got into the launch pad. Can you think of anything in that period that didn't work as planned?

IRWIN — I guess we could have saved some mental activity there if we had let the ground tell us where to put the bags. They came up with a plan and I did not know that they were going to do that.

SCOTT — That is right. The preflight plan was to take the checklist we had on board, and the limitations on the weights and stow according to that checklist. After we got all stowed, the ground called up with a plan and said here is where we think you ought to stow everything. I guess we just read them our stowage from the checklist, and they accepted that.

IRWIN — It turned out to be very close to what they had, but it would have saved some time.

SCOTT — It sure would have. Once we got back in after EVA-3, they could have said stow this here, and here, and there. We would not have had to figure it out. That would have saved some time. But I thought, in general, the post EVA-3 time line went right down the money all the way. I think we were within probably 5 or 10 minutes of every event.

If you're not careful with the vertical straps going up to the cabin fitting, you can put extra stress on that PLSS mounting; also, the interface of the straps could take off the thermal cover of the PLSS.

IRWIN — The straps on the Commander's side occasionally bear down on the Y-adapter and also the PLSS hard line. We should look at it closer during the C² F². We should look strongly at some other way of securing the PLSS.

10.3 EVA-1 EQUIPMENT

SCOTT — Okay, we'll take this egress on EVA-1 to the end of EVA-3 relative to all the hardware on the surface. Okay, moving through the hatch and down the ladder was nominal. The MESA came out and went straight to the surface as we expected, there was no preplan adjustment hike and it went right on down to the ground. Jett bag and LEC went down all right. I descended to the surface and hopped out and found that the one sixth-g environment was pretty much as everybody else had said. There was no problem going down the ladder. The front footpad was only very lightly on the ground. There was only very light contact.

IRWIN — I question whether it was even in contact with the ground because it was so free to swivel.

SCOTT — Well, it was when I got out because it made an impression on the ground.

IRWIN — It might have made an impression and then it might have rocked back.

SCOTT — The pad was on the ground when I got down the first time. It was pretty solid when I stepped down because I stood on the footpad before I stood on the ground. The ETB was transferred down all right. The MESA height was easy to adjust.

SCOTT — I think it weighed some 400 pounds and there was some question before we went as to whether it would take two of us to adjust it to a reasonable level, but I had no trouble at all using the black adjustment strap and locking it in place. When I opened the blankets, I found that they had been taped together in addition to being Velcroed and that took a fair amount of time to get them open. I suggest that if we're going to tape them, then we ought to train with the tape on them. I thought the Velcro was going to be adequate, but I guess not. Jim came down and I unstowed his antenna. The TV tripod was fine and the TV camera was fine. We put the camera in the shade which nobody had mentioned prior to flight, since it was obviously looking up-Sun and the picture would be a lot better in the shade, which was somewhat closer than the preplanned location of the TV camera. Then we got ready to deploy the Rover. I noticed the first thing we noticed when checking the Rover was that the walking hinges were both loose, or disconnected. Resetting them, I found that they'd lock into position okay but it was obvious that they had been too loose, or the design needs to be improved to hold them in position. I could see why any vibration at all would shake them out of their seated positions and cause them to fall open as we found them. They reset okay, and the Rover appeared to be parallel to its mounts and the outrigger cables were taut. I deployed both tapes and when everything was ready to go, I gave Jim a call and he was ready to deploy it. Did you see anything off nominal when you pulled the handle to deploy?

IRWIN — Okay, going back a little. When I started coming out of the hatch, I hung up a little bit because the hatch wasn't fully open and Dave had to guide me out. I got down to the surface and immediately felt at home in the one-sixth-g environment because of all the good training we had on the centrifuge POGO. I immediately moved out to take the contingency sample at about the 11 o'clock position to the LM, at about 30 feet. I collected that, moved back to the LM, and immediately configured the 16-millimeter camera, which is not according to checklist, but we had talked it over and decided that we wanted to get some 16-millimeter pictures of the Rover deployment. I had no trouble making the connection. I put the correct mag on the sequence camera and mounted it on the LCRU. And then I positioned myself on the ladder to release the Rover.

SCOTT — Did you notice anything when you pulled the lanyard? It seemed normal. The Rover came out in its deployment just like we'd seen in training. I might add that it was a good thing that we'd gone through all the training we had on the deployment of the Rover because it was easy to recognize the walking hinges being open, and had we not recognized that, we probably would have had a serious problem. Anyway, the Rover came down very well with a manual deployment. And everything was nominal until we got it on the ground and attempted to disconnect the saddle and the telescoping rods from the front. I don't know why they hung up. Both pins were pulled. It finally took some pulling, picking up the Rover and pulling by both of us to get it disconnected. I didn't see why it was hanging up other than that two studs in the bottom saddle that sink into the frame on the chassis of the Rover seemed to be hanging up. Other than that I couldn't tell, could you Jim?

IRWIN — No, I'm trying to recall. We were pulling it kind of uphill. Up the slope of the crater, and whether that slope had anything to do with it, I really don't know. I guess we did modify the procedure there slightly. I was pulling on the lanyard with one hand and trying to take pictures with the other. And of course I fell down there once because I tripped backing up in that soft soil.

SCOTT — Yes, but you recovered gracefully.

IRWIN — Well, you helped me up.

SCOTT — When we finally got the thing free from the telescoping rods and the saddle, we turned it around and pointed it away from the LM so I could drive off in forward rather than in reverse. We found it was very easy to pick up and turn around. Subsequently, we moved it several times and it was easy to handle. All the pins came out and the setup went very well. Okay, the first thing that was noted in the post deployment checks was that the front steering didn't work. I cycled the switch several times and talked to the ground. We went through the various, configurations on the front steering, but to no avail. However, rear steering was available. I also noticed that the battery voltage and amp readout on battery number 2 was zero. That subsequently turned out to be an indicator problem as we did have both batteries available. The seatbelt was adjusted properly. I attached it, although it took a fair amount of effort, and I drove around behind the LM to the deployment position and found that the handling was very good even though the front steering was locked in the neutral position.

IRWIN — During this time I was attempting to take sequence camera pictures of you as you drove around the back of the LM, then I met you in front of the LM. About this time, I looked at the mag and it had apparently not moved at all. This was the first indication that we were going to have problems with the sequence camera.

SCOTT — Why don't you just hit that right now since you mentioned it? The whole sequence camera problems.

IRWIN — Okay, we had very unsatisfactory results with the sequence camera. Out of all the ones we tried on the surface, only one mag drove. I really don't know what the problem was. I suspect that it was a film loading problem because we checked the film mags when we loaded the ETB, and they seemed to be very tight. It was hard to manually advance the film in the mags. That's about all I can say, Dave.

SCOTT — Okay. And all the mags were the same way and I guess they can analyze them when we get back, but it appeared that the camera was working all right, didn't it?

IRWIN — Yes.

SCOTT — The LCRU came out of the MESA as planned. The indicator on the handle, which a support group put on, helped get it out without any problem. The LCRU mounted as easily as in training. The TCU was stuck in its mount in the MESA and I had to take the pallet out and then take the TCU off the pallet. But once I got the pallet out, it was no problem to remove it. The low gain antenna came out very well. The only factor there was the spool, about which the antenna lead was wound. I had to unwind the wire, and I recommend for future flights that we come up with some simpler method of stowing the antenna lead because it takes a fair amount of time to unravel all that wire. The high gain antenna came out nominally and was easy to mount. The unlocking of the antenna was easy, but relocking the antenna in the open position was quite difficult probably because of the new stiff antenna and the difficulty in the locking mechanism. I finally got it locked, but it took quite a bit of force. Cable connections worked well. I moved the TV camera over and mounted it on the TCU and I went through the procedures of turning the CTV power switch on and the LCRU switches. Then the ground called to say that they had no picture. So I recycled all the switches again and apparently the CTV switch was the clue to that problem, because once I recycled it the second time, the ground got the picture. The high gain antenna was pointed to Earth without any problem, although the Earth was very dim in the field of view, and I did check to make sure the filter was open. The ETB contents were stowed on the Rover as per planned with no problem. And that gets me down to the start up and the drive to the nav align site which was no problem either. It went nominally. Jim, you want to go through the loading on the back?

IRWIN — Okay, the geopallet came off very easily from the LM and surprisingly, it locked on the back of the Rover without any difficulty. Contents of the SRC-1 were transferred to the geopallet. I unstowed the equipment from the pallet. I fastened the gnomon bag to the back of Dave's seat and I guess we had a problem with the bottom of it coming loose.

SCOTT — First time I pulled the gnomon out.

IRWIN — So that requires some improvement. I attached a vise to the pallet, and there's only one way that goes off; and followed the checklist. No problems configuring the back of the Rover.

SCOTT — I might add that because of the number of different articles that the bags were so fresh and new and stiff that it took me a while to get your bag on the first time because it kept wanting to refold to its stowed position. But that was a very minor problem.

IRWIN — Okay, before that we had the first LEC transfer to the pallet, just before that.

SCOTT — Oh, yes.

IRWIN — I don't think it was any problem. I guess I was a little surprised it was as heavy.

SCOTT — Yes, you commented on it. As a matter of fact, you had to work pretty hard to haul that thing up as I remember.

IRWIN — I was surprised that it was that heavy.

SCOTT — Since we're on the LEC. I would like to say that we had - we divided the tasks up - and had to spend time cross training because there just wasn't the time available. We each had our particular thing to do on the Rover. I guess one man could have done it all with coaching from the other, but we had divided the tasks and the time line worked out well. I thought that we were both finished almost right on the money together, didn't you?

IRWIN — It did, and we kind of swapped some of the tasks there during the early part of EVA-1 because you were tied up doing some troubleshooting, and I moved out and put the geopallet on. So, we deviated from the checklist, but as it turned out when I was ready to go up the ladder for the contingency sample, we were back on schedule and it worked out real well. I think we'd done enough training so we had that flexibility.

SCOTT — In fact I thought the time lines on the surface relative to hardware loading and unloading worked out well the whole way. We were never in each others way nor was anybody ever standing around with nothing to do. Okay, I went out to the nav initialization site which was about 10 meters away, a relatively smooth place down-Sun, and gave the readout align for the nav system, and the next step was attaching the geology tools to the harness. That went pretty well as planned. The only thing that I noted was, after going up the ladder several times with other pieces of gear, I feel that the LEC is unnecessary. As a matter of fact, I think it requires time and effort that's not required. I think we can do away with that. That would be my recommendation. Do you agree?

IRWIN — Yes, as long as the Commander is willing to transfer the bags. And, of course, on subsequent EVAs I transferred a lot of bags up to the platform, too.

SCOTT — You ever have any problems?

IRWIN — No, I really didn't. I guess we had, as far as I was concerned, the worst possible problem as far as getting up on the first rung because the front strut had obviously not stroked. As far as I was concerned, the front pad was off the surface. As I initially came down and stepped on it, it was loose, and I wasn't aware of that and it tilted, pulling me back and I almost went over backwards.

SCOTT — On the first EVA?

IRWIN — Yes. So that was a surprise to me, and from then on it was a real struggle to get up to the first rung.

SCOTT — Was it really?

IRWIN — Yes. Invariably, I'd end up pulling myself up by the arms to get to the first rung, particularly if I was carrying a bag up. If I didn't have a bag, I could leap far enough to just barely get my feet on the first rung.

SCOTT — Did you have any trouble pulling yourself up?

IRWIN — No, it was just, you know, additional effort which probably raised the heart rate a little bit.

SCOTT — Well, I didn't have any problem getting up and I could get to the first rung with a leap with any bag, with a good spring. And another problem I found with the LEC was when we transferred the ETB at the end of EVA-1, the LEC line had been in the dirt and that's the dirtiest I got, I think, in the whole trip. It just spread dust all up and down the front of me as the thing went up and I guess I could have grabbed that one handle and held it, but that would have been putting an awful lot of force on you and I think that the effort expended by the guy in the cabin to hold that stuff up is not worth it. I'd recommend just taking up the bags one by one manually, putting them on the porch.

IRWIN — How about the pallet? We never transferred a pallet, I don't believe.

SCOTT — No. And I thought about that afterwards too. If you want to free your hands completely, you can have a small wrist tether with an elastic band on it just like in the command module and hook it to the wrist tether, and it wouldn't be any problem at all taking it up, with both hands free to hold on the rail. It would save a lot of time, a lot of dirt, and a lot of effort. Okay, and then we started out on the geology traverse, and I guess we should probably break down again in a subdivision within the geology traverse. The Rover operations and the geology. So maybe we ought to go from here onto the checklist to the closeout and discuss the equipment closeout and then come back to the Rover. So we did our traverse and got back to the LM for pre-ALSEP deployment and I guess we can go through the ALSEP.

IRWIN — Well, at some point along here, I think, your yo-yo failed.

SCOTT — Yes, I'm not sure exactly - oh, it was out on the first station at Elbow Crater when we started doing that radial sampling. I was holding the tongs and I looked down to see that the yo-yo string was still connected to the tongs. It had broken at its base. The remainder of the flight I kept asking myself why we had string rather than cable, but I figured it had been all worked out prior to flight. It looked to me like the string came untied or it broke right at its connecting point within the yo-yo. Then yours broke somewhere along the way. Do you remember where?

IRWIN — It was sometime during the ALSEP deployment, as I remember. Initially, I had the tool tethered there; at some point I had taken the tool off and I was looking for the yo-yo and I couldn't find it. So it was some time during the ALSEP deployment it broke off.

SCOTT — Yes, because we attempted to exchange yo-yos after EVA-1 so I could have it for the tongs. And when we went through that operation, we found that yours was no longer intact either.

IRWIN — Yes, I might make a note that we had my yo-yo on the right side because I'd hoped to use it to secure the extension handle on the scoop. That seemed to work okay for tethering that equipment and still using it while it was tethered, but it really was a problem as far as fastening the seatbelt in the Rover.

SCOTT — Okay, that puts us back at the LM ready to unload the ALSEP. The restowing of the geology equipment was rather straightforward, no problem there, as I remember. We unloaded the ALSEP packages and we had not planned to use the boot. They were tilted at the right angle, so they slid right out into our hands with no problem at all. I thought that was a slick operation. The drill came out very easily; it was easily stowed on the LRV; the LRRR came off the pallet without any problem, even though we were on a slope. I stowed everything on the Rover, and I was ready to go, I think, shortly after you were ready to go. Did you have any problem with the rest of the ALSEP? Oh, didn't you have trouble getting the UHT out of its mount?

IRWIN — Yes, but two of them were secured together, kinda locked together, which we hadn't seen. I don't know whether it was a thermal problem, or what it was. But the two UHT were kinda stuck in that bracket. We had some difficulty getting them apart.

SCOTT — I'm not sure why they were stuck. It wasn't apparent but they were really stuck for a while.

IRWIN — I also had some difficulty in getting the sequence bay doors closed; I had to cycle them, I think, three times to get them fully closed. I don't know why they were hanging up, but I did get them closed. I guess we were then ready to carry the ALSEP out. I tried to carry it in my hands and I realized that that would really tire my hands, so I ended up putting it up in the crook of my elbow and carrying it in that position out to the ALSEP site. And that was an easy task. It seems like I got out there about the same time you did.

SCOTT — Yes, we arrived there about the same time. I was surprised how easy it was to move with the ALSEP.

71-H-834

IRWIN — Did you see any swaying motion at all - of the package? It seemed like it was pretty steady.

SCOTT — It looked pretty steady to me. It looked like you were making pretty good time, but I couldn't go very fast because of all the little craters around there, subtle craters going up and down. I wanted to make sure I didn't drop anything, although the seatbelt held the LE tube and the drill very well. Okay, we're at the ALSEP site and we picked a site which was relatively level, and I think an acceptable site, although it was difficult to get completely away from some of these little craters. We parked the LRV as prescribed and proceeded into the ALSEP deployment. The major problem was the drill, so you want to go through that?

IRWIN — Okay, I attached the RTG tape and there was some question on the shorting switch reading that I was never completely clear on. I cycled the shorting switch, gave them the reading, and we pressed on. They said that wasn't really important. SIDE came off; legs deployed; I set that on the surface. We removed the carry bar, stowed that. PSE was deployed west of the central station. There was no problem there. As we remarked before, the surface soil was very soft. I spent a few minutes stomping down a place to put the stool, but it aligned very easily. Solar Wind was deployed; no problems there. It was easy to align the shadow correctly; door was open. Then the next operation was the magnetometer. It came off freely. It was taken out to its deployment site and there were no problems there. Then it was back to the central station, and I aligned it with the shadow before I started to release the sunshield. I guess the first problem that I encountered there was when I tried to release the pins that hold the rear curtain cover. When I pulled on that cord, or string, it broke and I could not release the pins. Now, there's a string that goes from one pin to the other. I tried to put the tool in there to release both pins, and the cord again broke; so I was forced to physically get down on my knees and pull them out by hand, and fortunately they came out. That could have been a real glitch because unless that's off, you can't get to the Boyd bolts on the back side of the central station. From that point on, I released all the Boyd bolts, and very surprisingly, they all released and the central station erected per checklist. I installed the antenna mast, the gimbal and I leveled it, and I guess the time to level it was about the same as what I'd been spending in training, a little more than I would like to spend. It seemed like the central station wasn't very stable, because every time I adjusted it the bubble would move back and forth, but I did get it level. I aligned it, put in the settings per checklist. Then I attempted to take the SIDE out. I had trouble with the UHT locking into the SIDE. I didn't realize this until I got it just about out to the station, and I was about to put it down when it dropped off the UHT. I hope it didn't interfere with the experiment itself. I tried to engage the UHT again and again had problems. It fell off the UHT about three times there. It was very frustrating. I don't know, maybe there was some dirt on the UHT that interfered with the engagement. We got the screen down, got the SIDE positioned, pulled the safety pin, checked its level, and aligned, reported to Houston, went back to the central station, and depressed the shorting switch. I couldn't really check amps zero because there was just too much dust on the gauge. Might make a comment here that the dust covers that were put on the various experiments, really paid off because we were in probably the worst situation that I've seen as far as dust and soil, but they kept all the Boyd bolts clear of any dust. The ground requested transmitter turn-on, and we were running out of time at about this point. Then Dave moved in and started taking the pictures. I'll end it there. Dave, you pick up.

SCOTT — Okay, the unloading of the Rover at the ALSEP site was nominal. Went to the heat flow pallet; that came out fine. Connected the central station; worked all right. When I went to remove the probe box from the pallet, the right rear Boyd bolt hung out and I had a difficult time getting that to disconnect. I finally got it disconnected, and taking a look at the two-probe parts of the box, I found that the rammer was in the left probe box and, in training, it had always been in the right. One of the things we had attempted to do was make sure that the lines were not crossed. As I remember I unraveled the lines to make sure they weren't crossed and deployed the two probes, one to the south and one to the north according to the diagram. Then I went back and removed the electrical box, and again I had a problem with the right rear Boyd bolt, the one closest to my right foot, getting it to disconnect which it finally did. Got the box off and put it down and aligned it, and it was no problem. The next operation was the drill. Drill procedures worked very well, was unstowed from the treadle and I brought over the first probe which, in this case, was the Etham probe. I proceeded to attach the drills and start the drill. When I got the first two stems in, why, it was apparent I was hitting something very hard which, subsequently, I really think was bedrock. But the first meter was quite easy to drill and then it was very difficult to get the stem any farther. I got about two-and-a-half stems in and, in trying to remove the drill, the chuck had frozen, and I think that's because of the high amount of torque put on the stem themselves and the chuck just biting into the stems and locking up.

71-H-840

We'd never seen this in training nor had we ever seen any material that was compacted or as hard as that material I was trying to drill in at that time. The recommendation from the ground that came up subsequently was to drill slower which was a good idea, and just let the drill do the work. We should have probably discussed that possibility before flight because I hadn't really thought about it. That seemed to help get in a little ways. It did on the second probe. In order to get the drill off the first probe, why, I had to get the vise, the little wrench off the treadle, or off the stand where the drill stems were, and get down on my hands and knees and force it off. I finally ended up physically breaking or bending the top half of that third stem to get the drill off. But it did come off and that was a good call from the ground. I had never practiced that in training. I took the probe, the heat flow probe, and inserted it into the stems, measured it with a rammer, and after that the ground had recommended terminating drilling at that point because of the hardness. I think they called us off it this time as I remember. They said okay that's enough for that side. Or did I go in then and plant two sections? Yes, I guess I went back over to the other side, to the western side, and put two stems in over there, or a stem and a half. Again the drill locked up on the stem. I was having difficulty getting it off and the ground called a halt to the drill at that point on EVA-1. They said they would like to review it and see what would be the best thing to do.

In any mind, I thought at that time we should go ahead and dig a trench and put the heat flow probes in the trench, as we had discussed prior to the flight, if the drill didn't work. It seemed to me that the amount of time being invested in that particular experiment was already becoming excessive. Because of the ground calling, wanting to reevaluate, I terminated drilling at that time and proceeded to deploy the LRRR and take the ALSEP photos. Before the flight we found that during our training sometimes we'd finish at different times, so we planned to use the LRRR and the ALSEP pictures as a buffer, and that was a good plan because I had the procedures in my checklist, and I had only deployed the LRRR once, I think, during our training, and I had never taken the ALSEP pictures. I did have in my cuff checklist all those procedures and they came in very handy because they were straightforward and it took very little time to deploy the LRRR and take the ALSEP pictures. I could do that while you were finishing up. I think, in the end, we ended up at just about the same time. And I got all the ALSEP pictures with the exception of the heat flow, which I didn't take because I could see then, we weren't through with it yet.

IRWIN — I might make one comment, Dave. You know, coming back to the LM in preparation for the ALSEP, I felt that I was thirsty and kind of hungry, and I tried to get some water out of the water bag as we were approaching the LM. Couldn't get any water out of it, but the food stick was there and I gobbled that down. I think that was the thing that pulled me through and gave me the energy to get through the ALSEP deployment. That really perked me up. I felt great after that.

SCOTT — That's a good point. I, too, when we got back to the LM, tried the water and the food stick, and my water worked fine. I got several gulps of water. It was very refreshing and I ate about half of the food stick at that time. That helped quite a bit. I think in looking at it, the problems I had with the water bag were related to tie-down to the neck ring with only Velcro. On the second EVA, that came loose and I could never get to the water bag because it caught under my chin. I think, maybe, if we had snaps in there, or some firmer method of tying it down, it would have helped me. Can you sort out why you couldn't get to the thing?

IRWIN — I could get to it. I just couldn't suck the water out. I just couldn't make the valve operate.

SCOTT — I'll tell you, the water bag is really a valuable asset because one quick swish of water and it really refreshes you. I think, if you really got thirsty, you could stand there and drink the whole thing, if it worked right. There was no problem putting it in the suit, no problem donning the suit with the water bag full, or with the food stick.

That gets us to the end of the ALSEP with a somewhat incomplete drill operation. I took the LRRR a good hundred feet away because of the interest in keeping it clean and the ALSEP was deployed somewhat north of my line of sight in order to get to a level spot. So I took the LRRR farther south than we had planned in order to try and keep it out of the trajectory as we took off, to keep the dust off of it. Another problem along this way was, I didn't have a yo-yo, which complicated things relative to working with the UHT and the drill. It took both hands to drill and it took the UHT to disconnect all the Boyd bolts. I ended up just sticking it in the ground and it didn't seem to hurt it any. Okay, back to the Rover, and driving back to the LM. Did you ride back or walk back?

IRWIN — I think I walked back.

SCOTT — You walked back. Because of all the craters we couldn't drive very fast and it took a fair amount of time to get on the Rover. The closeout went fairly smoothly. We excluded the polarimetrics because of the time problem. As a matter of fact, we did very little other than just gather up the samples and the film, load them in an ETB and ingress. I think this is the point at which I transferred the LEC and SRC with the LEC and got all dirty.

IRWIN — I think we only had the SRC and one rock bag.

SCOTT — One rock bag.

71-H-846

IRWIN — Bag

SCOTT — And the ETB.

IRWIN — Yes.

SCOTT — That gets us in after EVA-1. Think of anything else on that, Jim?

IRWIN — One small comment as far as aligning the central station after it has been erected. It's quite easy to do. I don't know whether it was just the soft soil where we had the central station, or whether it was typical one-sixth g. Even though it's erected, it's easy to shift to line up the shadow device.

SCOTT — Oh, yes. That reminds me of aligning the electrical box on the heat flow. After the initial alignment and all the shuffling around there with the probes and all, at one point I tripped over one of the wires to the probe and I moved the electrical box from its alignment position. I think the ground called up and asked some question relative to the position of wires or Boyd bolts around the electrical box. Maybe they were trying to get data, and the thing wasn't properly aligned. I did realign it after we went out the second time.

10.4 EVA-2 EQUIPMENT

SCOTT — At the beginning of EVA-2, we loaded the Rover as planned. I don't remember any anomalies. I'll go through my events here. The first thing that I did was to change the LCRU battery, and that went very smoothly. It's very easy to do. We attached the geology equipment to the harnesses on the PLSS, and I had no problem there. We proceeded on to EVA-2 traverse. Got any comments on that?

IRWIN — Because of my antenna problem, of course, we did not deploy it. As far as the storage of the LCRU battery the plus Y footpad was in the sunshine, so I changed it there and put that battery in the plus Z footpad, which was a change. I wrapped it in the blanket and put it in the plus Z. Then I configured the bags per the checklist. We had at least one extra bag under my seat, but I don't believe there was any confusion because the bags were clearly labeled.

SCOTT — I think the ground did a good job of keeping track of all the bookkeeping for us. With all that equipment, that was a good thing to have everybody on the ground keeping track, because we didn't have to worry about the way things went and because it could get very confusing. We started and drove and proceeded on the EVA-2. One of the things they were looking for at the end of EVA-2 was Station 8 and a return to the ALSEP site. Since that's related to surface hardware, why don't we step ahead to the return to ALSEP site and the attempt to do a Station 8. You discuss the things you did at the ALSEP site when we go back, and I'll go through the drill again.

IRWIN — Refresh my memory. I don't think we did a comprehensive sample.

SCOTT — Not at the ALSEP.

IRWIN — We didn't do the double core, and we didn't do a pan. The first thing I did was configure for the start on the trench, and we didn't have all the photos. I started digging the trench.

IRWIN — I dug it about 18 inches deep. At that point, I encountered a very hard subsurface layer, and it was of adequate size so it would accept the penetrometer. I collected the contingency sample. The SESC. We filled that. I don't know whether you were there when we filled the sample bags for the geology sample or not. Or did I do that myself?

SCOTT — I think you did that yourself. You did all the penetrometer yourself. That was an interesting departure from our preflight plan, because we had planned to do the Station 8 together. I had all of the procedures in my checklist, since I just walked you through them, as far as you doing them. Apparently you made out all right without having all those detailed procedures.

IRWIN — I had just enough abbreviated details that I could follow it through. I got all the penetrometer tests at Station 8. Then I went down and took pictures of all the activities, the Rover tracks, and the trench. Then, they asked me to take pictures of the heat flow. I did that, and I also took a pan at the ALSEP site. They asked for those additional photos. That concluded my activities.

71-H-832

SCOTT — I went back to the drill and proceeded to implant another stem in the second site. They suggested drilling very slowly so the chuck wouldn't hang up, which I did after I finally got the drill off with a wrench again, and which required about as much force as I could give it. The wrench worked pretty good, though, at that point. I finally got the third stem at the western site. The ground suggested that was plenty and to put the probes in. This I did, and I was surprised that the probe didn't go in any further than it did. I was surprised at the indication on the rammer, and the ground subsequently called up and said that's as far as it should go. I'm still surprised. I thought it should have gone in farther, but I guess they had it all figured out. From there I went and realigned the electrical box with the UHT. I proceeded to take the drill back to get the core. At this point, it was time to take the chuck off the drill. I took the wrench and put it in the opening in the chuck, and I couldn't get the wrench to engage the chuck. It just didn't fit. I finally took the corner of the wrench and bent open the little ears that hold the chuck on the wrench, and I got it off the way by unscrewing it.

IRWIN — Let me make one comment relative to the penetrometer. The ground plate would not stay extended. It seemed like the tension in the cable was too great, and it would always work its way back up about 3 inches from the fully extended position.

SCOTT — The next order of business was drilling the stems for the deep core. As I started out, the soil was very soft, and the drill went very easily and too fast down to the bed rock. The ground gave a call on the rates, which I had forgotten in my haste to finish up the drill. We were supposed to go an inch per second. I got about a stem and a half in before the ground reminded me of the rate, and I slowed it down to an inch per second. I hit bed rock, or the very hard soil, which was a step-jump in hardness as I drilled. From that point on, it was easy to drill on at an inch per second, because that's about as fast as I could get it in anyway. I could feel layering as the drill went in. Some places, it was easier to drill than others as I went through. As a matter of fact, in some places, the drill pulled me down. I could just feel the drill pulling right through the underlying material. I got all the sections in, and I noted in the process that it was more difficult to screw the sections together than it had been in training. I don't know whether it was the thermal problems, or what, but it took quite a bit of motion and patience with them to get the stems all the way to the joint. When I got the drill all the way in, I attempted to pull it out, and not surprisingly, it was very difficult to pull out. We expected that from our training. In certain cases during training, people observing in shirtsleeves couldn't get the drill out of the ground at the Cape. I wasn't at all surprised to find that, after having drilled through bedrock, I couldn't pull the drill out. I got it maybe a foot back out, and at that point, the ground recommended coming back another day to finish. I was somewhat sorry to see that we couldn't get it out any easier than that, because we'd invested so much time in it, and it seemed like a shame to lose that time. On the other hand, there was a question in my mind as to whether we should spend any more time on it at all because of the amount of effort involved.

IRWIN — At that point we tried another grand prix - drove the Rover.

SCOTT — That's right.

IRWIN — Before that, on the way back from the Front, I think we hit one mag that did work. It was mounted on the Rover, pointed straight ahead. I think that mag did drive, and I think we probably got the pictures on that. When we tried the grand prix at the ALSEP on EVA-2, the film would not drive.

SCOTT — The LM closeout was nominal.

IRWIN — We picked up the activities that we missed on EVA-1; namely flag deployment.

SCOTT — That was the only thing we really picked up because I never did get to the polarimetric photography.

IRWIN — I did the pans around the LM. I did the engine bell photography. I think that's when we deployed the Solar Wind.

SCOTT — No, I deployed the Solar Wind at the end of EVA-1.

IRWIN — That's right.

SCOTT — And that was fairly straightforward. I guess we got the engine sample there too.

IRWIN — Yes.

SCOTT — The contamination samples from the engine.

IRWIN — That's all recorded. I don't think we had any problems with any of those things. Once we got to them, they worked fine.

SCOTT — We cleaned off the tool harnesses with no problem. I unloaded the ETB, and I guess we ingressed again without any problem. Did you remember anything off-nominal?

71-H-833

IRWIN — Let's go back to closing the SRCs. The SRC-1 was very difficult to close, to lock the handles. I ended up pounding on both handles to get them locked. Then when I got around to SRC-2, I had about the same difficulty, and you came over to help me.

SCOTT — Yes, I tried to do it too.

IRWIN — We found out subsequently that, apparently, part of the bag was caught in the rear hinge.

SCOTT — Subsequently being now. It looked to me like the lid was closed on the seal.

IRWIN — In the front. Yes, but we never looked at the back of it.

SCOTT — I didn't either. I should have. It looked like the handle was just mismatched completely from its lock. There was no way we were going to get the lock over the handle, because it was too far away. Yet, the front looked like it had been sealed.

IRWIN — We couldn't get the same box stowed. We couldn't get the rod and the pins engaged in the side of the bulkhead on the LM to stow that box. So we eventually lifted off with that box sort of loose, although I put a piece of tape across the thing. But we never could get that box stowed.

SCOTT — That was probably the reason - because the hinge wasn't right.

IRWIN — But it was the upper SRC. You wouldn't think that would interfere with the engagement pins.

SCOTT — Well, it never got stowed in the LM.

IRWIN — It was very warm too. I was surprised how hot the SRC was when we got it in.

SCOTT — That's right. It really was. Let's get your camera on EVA-2. Your camera stopped, right? On EVA-2, because we worked it over that night and then...

IRWIN — Yes, you got it to work that night.

SCOTT — ...and I guess the problem with the camera - we brought it back for the people to look at - I think the problem is definitely dirt in the drive mechanism. I fiddled with it that night and got it going. The next day, it hung up again. After we got into orbit, we worked on it some more, and you could see that the wheel exposed by the Reseau plate was hanging up. If you put your fingernail in there and triggered it, it would get going. I think with the amount of dirt that you have, and the fact that the camera is level with the area in which you work when you roll up the bags, you get dirt, in the camera. I think we ought to put some little Beta booties over the top of the camera to keep it clean, at least over the point there where the film mag goes on. They were getting so dirty that every time we reset our f-stop and lens, I had to brush mine off with my finger. I had to wipe it off, because I couldn't see the settings on the camera, it got so dirty. I'd recommend maybe Velcro tabs and a little piece of Beta right up on top of the camera to keep that mechanism clean.

IRWIN — Dust accumulation also gave a problem as far as removing the film mags from the camera. There were several times where it was very difficult to release it.

SCOTT — I think the camera would be better off if we'd protect it a little bit better. We used the lens brushes on the cameras, and they were very good.

IRWIN — On the TV also.

SCOTT — On the TV also. That lens brush is really a good brush. It cleaned it off very well. The dust brush, to clean off the suits seemed to work pretty good. It got the gross dirt off. It didn't get everything. I guess it also worked quite well on the LRV and the LCRU mirrors - cleaned them off pretty well.

71-H-1102

10.5 EVA-3 EQUIPMENT

SCOTT — With our new plan, we headed for the ALSEP site and the drill again. The object was to extract the core and bring it back. We spent an awful lot of time doing that. Loading the Rover wasn't any problem.

IRWIN — It wasn't nominal because we had your bag on the back of the tool carrier for the drill operation.

SCOTT — Other than the shuffle of the bags, there weren't any problems with the equipment. We finally extracted the core stem. Each of us had a handle of the drill under the crook of our elbow, and we got it up to the point where we could put our shoulders under it. Then with each of us with one handle of the drill on top of our shoulders, we pushed as hard as we could - it must have been at least 400 pounds - and finally got it to move and got it out. Because of the significance of drilling in the bedrock, it was probably the way to go. We could only accept the ground's evaluation, but at the time, it seemed like we were investing an awful lot of energy and time in recovering one small experiment, however important it may have been. But at that stage, I guess we had so much invested in it that we couldn't afford to leave it. It sure was expensive. When we got it out, we put it up on the back of the Rover on the geopallet and attempted to break it down with a wrench and the vise mounted on the geopallet. The vise on the pallet just didn't work. At first, I thought it was on backwards. I knew darn well we'd discussed it before and that it could only go on one way, but I just couldn't believe it was that bad. It just didn't grip at all. Jim got on the other end of the stem and moved it horizontally and vertically, and he put every kind of torque on it I think he could, to try and get it to lock in there. The hand wrench worked fine. It would grip the stems and hold them very well, but the one mounted on the pallet provided no torque at all. I guess we got a couple of stems. I don't know how we got a couple of them loose, but we got enough. We got three stems separated and ended up with three stems joined. We capped them, called out the caps, and took them back that way, but that was a real chore. We fiddled around with the treadle some. That was somewhat of a chore also, but I think that is inherent in the design of the equipment. If the drill works as advertised, it really isn't bad, but in summary, the ground being very hard tightened up the drill stems much harder than we'd seen before, and the vise not working on the back of the Rover complicated the extraction, or the separation, of the stems. Finally, we had the number 4 stem off about half way, and I finally, just in gripping the thing, unscrewed it by hand.

IRWIN — I'd taken my protective covers off my gloves before I even went out on EVA-1 so, of course, they were off for this operation. I was kind of reluctant to grasp that drill very hard, afraid I might rip the gloves.

SCOTT — That's a good point. I had to leave mine on the whole time because of the drill. The protective covers can restrain your hand movements even more than the gloves. I had sort of degraded mobility because of those protective covers, all the way. I finally took them off after we got through with the drill. The LCRU battery change was nominal. We got back to the LM and started the closeout. I don't remember anything that did not work at this point. We unloaded the Rover, and I proceeded to drive it out to the TV site. I don't remember any off-nominal conditions there or any hardware problems, do you?

IRWIN — No.

SCOTT — Ingress. We had a number of bags to carry up at that time two collection bags, the BSLSS, and the ETB.

IRWIN — And the core stem.

SCOTT — I carried all those up by hand. You took a couple of them on the way.

IRWIN — I had two bags plus the core stem up on the porch.

SCOTT — That's right. And all those went in all right. One thing I want to add. I asked you to check me to see if my PLSS was loose, and you couldn't see anything wrong with it. I had the distinct feeling the lower straps were disconnected from the PLSS because it was bouncing on my back when we got to the LM. When I got off, I could feel it bouncing on my back, and I never did figure out what that was. I just went slow, and when I walked back from the Rover, I took very small steps that kept it from bouncing around. During the bouncing steps that we were using, it was really flopping back there.

IRWIN — Well, you hadn't walked much before that, had you?

SCOTT — I would have noticed it before because of getting on and off the Rover. I noticed when I got off the Rover that time it banged on my back, and that's why I asked you to check it. I still don't know why it felt loose.

IRWIN — I retrieved the Solar Wind. It really had so much of a set that it wouldn't roll up properly, and the bottom part of it ripped. It was just like in training.

SCOTT — Really?

IRWIN — Yes. But I was able to manually roll it up with my fingers, trying to avoid touching the foil itself, and put it in the ETB. The bottom of it did rip. I retrieved the penetrometer drum. I took care of all that before you drove off. I don't know, did you take all the tools off the Rover, or were they still on the Rover when you drove off?

71-HC-710

SCOTT — I left them just as you replaced them.

IRWIN — They were all on the Rover.

SCOTT — Yes.

IRWIN — That was one period where we probably could have collected another 50 pounds of rock if we had wanted.

SCOTT — You're right.

IRWIN — We didn't plan that too well.

SCOTT — I don't think we had planned on having that much time when we got back to the LM. I think we really got called back too soon, because once we got back there, we really had more time than we had ever planned on for that closeout. Therefore, I think we wasted a lot of time. I remember when I was out at the Rover, I could see you back at the LM just watching me.

IRWIN — Yes.

SCOTT — You could have been collecting a whole bunch of rocks at that point.

IRWIN — I spent the time transferring as many bags as I could up to the porch, but there was still plenty of time left to collect maybe 50 pounds more of rocks.

SCOTT — They got us back too soon. We wasted considerable time back there. I had the distinct impression people were getting awful itchy about us getting back in time and getting closed out in time. We felt pretty warm about that final closeout because we had run through it so many times, and we knew that we could handle the time line as prescribed. We never did have a problem. Once we got back to the LM, we always had plenty of time to get everything done.

IRWIN — The unknown was your taking the Rover out to the right position and getting all that taken care of.

SCOTT — Rover procedures. I left the NAV system in RESET. That's what it was. It says NAV system in RESET, checklist, and then drive to a heading of 096. I had left the NAV system in RESET, and I got half way out to the Rover site for the TV, and I looked down and it was still zero as it should be if it is in RESET. Then I realized it didn't make any difference, because had I gone to the exact position, I'd have probably been down in a crater because there were so many out there. So I took it to a high point which was like a tenth of a kilometer away, and I think in the final analysis you got pretty good pictures of the lift-off. It was probably better just to select a point somewhere to the east of the Rover that would give a good TV vantage, rather than try and be precise on the distance and the heading, because there are so many craters out there that we might have been in a hole anyway. So that, in the long run, worked out all right. I think all you need to know is which way the Rover needs to be pointed, relative to the LM, so you can get good TV. When I got out there again I had problems aligning the antenna. The ground desired to have us pointing down-Sun for the TV. That meant that when I pointed the antenna, I had to look up-Sun because of the position of the high gain on the Rover. Looking up-Sun, I just couldn't see the Earth. In the pointing device there, it was just too dim, even with the sunshade extended and the filter up and open. About 50 percent of the time, I used the AGC signal strength to get an indication that I was pointing to Earth and just sort of visually eyeballed the thing. It would be a significant improvement if they could open up the light passage through that sighting device. I had the same problem when I got the Rover out to align it the final time and in trying to get the Earth in the exact position in a field-of-view for subsequent TV programs. I finally did find the Earth and got it aligned according to the checklist. The Rover was left in the prescribed position with circuit breakers as planned. Then back to the LM and INGRESS. That's about the sum total of the hardware operations on the surface.

86-11603

71-H-1131

This summarizes the tool operation - mechanically - how they work. They all work just fine. When my palms got dirty, I had a difficult time manipulating the handle squeeze and the opening and closing because of all the dirt in the tongs. And so, about half way through EVA-2, I switched to the other set of tongs which were clean. That helped quite a bit. The problem in not having a yo-yo is that I had to stick them in the ground while we were gathering the samples. The cameras mountings - taking them off the RCU, seemed to work fine. The 500 millimeter worked fine. I used my helmet shield as a base to steady it. I noticed it made some light scratches in the gold material, but it didn't really bother it. How about the extension handle with the scoop? Did it all work okay?

71-H-1389

IRWIN — It worked fine, and digging the trench went much faster than I had expected. I estimate, in 5 minutes I had the trench dug. Then, of course, the TV caught the action as I used it. The connector for the dispenser sample bags came off of my camera once.

SCOTT — Oh, did it?

IRWIN — Yes, I retrieved it. It was on the ground and I picked it up. It was a good thing I used the tongs to retrieve it and put it back on. It stayed on for the rest of the time. The operation of the rake went just like our simulations in the K-bird. It worked good for collecting the rock fragments as well as for transferring the soil. I thought it went real well.

SCOTT — The gnomon worked okay. The gnomon bag worked okay, except for the problem of having the bag disconnected all the time, which we also experienced in training.

IRWIN — A comment on the stowage of the scoop on the extension handle. Rather than tethering it, we mentioned already that the yo-yo had come off. So I just positioned the scoop extension handle on the left-hand side of my seat, kind of under the bracket, at the attachment of the seat to the Rover frame. It did ride fairly securely there.

SCOTT — I took the tongs and stuck them under the left-hand side of my seat between the seat post and the bag, and they rode very securely there, too. Lets see, the core tubes worked fine. It was the first time we'd ever seen the core tube caps and they were a little different. I got the impression they were a little harder to put on, but once they were on, they stayed on better. I would recommend that the future crews get to see the flight hardware some time before the time they arrive on the lunar surface. But they worked all right. We never had to point the low-gain antenna. I mounted it in a not-quite vertical position. As a matter of fact, I mounted it as it was stowed and never had to change that, which was nice. The covers over the LCRU battery worked fine. We could put them in any position required in the checklist, and they stayed. The Rover battery covers didn't close automatically one time. They were relatively easy to operate manually.

IRWIN — I guess we commented about the general dust condition on the Rover. We just took one series of pictures of the dust accumulation on the Rover after the EVA.

SCOTT — Certainly. I got a picture of it there at the end when I parked it, which will show it. The geopallet gate worked fine. It looked to me like there wasn't any problem with that. Hanging the bags on worked all right. We had a little problem with the bag on the inside of the geopallet between the seats and the pallet - getting it locked.

IRWIN — The BSLSS bag. We had trouble getting that locked. I think that's because the bag was larger than the collection bags, and you just couldn't get your finger in there with a suit on to get it locked. Wasn't that the problem?

SCOTT — Yes.

IRWIN — In order to get it unlocked there, I ended up jumping up on my seat and reaching over to get it unlocked. I guess the vise was really in the way, to some extent. It didn't give.

10.6 EVA-1 GEOLOGY

SCOTT — It was only that one bag that we had a problem with. That finishes up the hardware part.

SCOTT — We'll start out on the geology portion of the EVA number 1, and I guess we'll organize it relative to discussing our general impressions on the geology as we went from station to station. At each station, we'll discuss our general and specific impressions and try and go along with the actual traverse that we conducted, rather than related to the planned traverse. I guess I might make a general comment in looking at the map of where we apparently actually went, which is a preliminary event in order for us to have a reference. I think the time we made on the Rover throughout the excursions was as good as we expected, and probably better in between stations. With the problems we had on the surface, we lost some time in between EVAs, but I was very happy with the Rover performance. I think we probably had planned more than we could ever have accomplished, as far as distance was concerned, if we were to spend any time at the stations at all. I thought the stops were, in general, fairly efficient. We didn't have any hangups in the procedures or the equipment, as far as the geology goes. I don't know where the time went. I guess it went on ALSEP. We didn't cover as much as we had expected to prior to the flight, but I feel we covered as much as we could have in the time allocated. I was very happy with the Rover traveling speed. I think the geology tools and the concept and manner of sampling were just fine. Didn't you?

IRWIN — Yes. No problem.

SCOTT — We did everything we could have done in the time allocated, and there weren't any particular hangups. Shall we start out driving down on EVA-1 toward the rille? The ground called us and said to skip Checkpoint 1 and go right on down to Elbow. Our general technique was for me to drive and keep my eye on the road as well as was possible. Jim would do the navigating and commenting on what he saw geologically, and if I had a chance, I'd fill in a comment here or there on the geology. Jim really did most of the navigating and discussion. Most of the stuff is on the tapes, I'm sure. Why don't we head on down south, and why don't you make your comment there on the traverse to Elbow Crater.

IRWIN — We were supposed to be looking for a possible ray, and I saw no evidence on that leg of any ray. I didn't see any lineaments. There were probably frequent fillets around rocks, but I did not comment on them. I did not see any mounds, didn't see any mounds at all in the entire area.

SCOTT — No, I didn't either. Never saw what would be comparable to an Apollo 12 type mound.

IRWIN — We were looking for a raised rille rim or a levee, and I think we commented on that later on. There might have been suggestions of just a very slight end of maybe a levee. I think that was more evident on EVA-3, really, rather than EVA-1. The block distribution, I didn't see any pattern at all, other than distribution related to individual craters. I didn't see any difference as we drove down there.

SCOTT — Occasionally we'd see a block - not a block, but a large fragment. I wouldn't even call it a block, a "foot" kind of fragment, like the one we ran over. That was an occasional kind of thing.

IRWIN — Again, that was associated with a particular crater, I think.

SCOTT — Yes. Probably was. We didn't see the excavation of bedrock by 25-meter craters. We did see some fresh craters that size, but the general surface was very hummocky and had a relatively heavy crater density, but all subdued and rounded with low rims, no raised rims. But they were larger than 25-meter craters which had not excavated bedrock or showed no signs of outcrop of bedrock.

IRWIN — We were about a half a kilometer from Elbow when we saw the rille, and at that point, I think we were heading a little too far southwest. We changed our course a little - swung around to the south. And we did that, a short while later, we could see Elbow very plainly. We then headed toward Station 1.

SCOTT — We might have been pretty close to the first leg of the EVA-1 traverse on the map, which shows us heading more southsouthwest than south, because we did see Elbow Crater from the side of the rille, quite a ways away. And there again, I think the distances were somewhat deceiving and that it looked closer than it really was. When we did see Elbow Crater, I felt like we were almost there. Then there was a fair amount of driving before we got there. Everything looked closer, and as I look at our landing site, relative to Pluton, I would have thought Pluton was just right around the corner from the site. I think the distances, again, as everybody has said in the past, they're really deceiving up there with no other objects to measure and compare. On the lineament thing, you and I both discussed that. I think you can see lineaments if you look for them. That's true as a function of Sun angle and the angle at which you're looking, because you can imagine them in almost any direction. I could almost say there are lineaments anywhere, if I really used my imagination, although some places

they appeared more evident than others. I saw one place where it looked to me as if they were running along the lunar grid, northeast-southwest, northwest-southeast.

IRWIN — I guess it was on EVA-3 where I really thought I saw them. Down on the edge of the rille. I saw them parallel to the rille and perpendicular to the rille.

SCOTT — I think I saw them on EVA-2, driving back. The distribution of soil, grain size, and that sort of thing indicate the ray. I didn't see any significant change of granularity for the soil at all. I think it was deeper up the side - of course, that's getting further down the EVA here - but on EVA-1, I couldn't recognize any change in soil. Could you?

IRWIN — No.

SCOTT — I believe you mentioned that the block distribution or fragment distribution did increase somewhat as we got to the rille rim. There were more fragments.

IRWIN — I'm wondering, was that a function of the rille or a function of craters there? I know that was true when we got to Elbow. There were plenty of rock fragments there.

SCOTT — I think we commented on that on the voice tapes. I guess my general impression of Elbow was that it was much more subtle and subdued than we had expected.

IRWIN — You know, when we first saw Elbow, I think we were kind of downslope, down on the rille side of the levee. We saw it, and we went back up on the top of the slope. It was smoother driving there.

SCOTT — You're right. Matter of fact, I think we commented at the time that it was better driving back up on the ridge line, or the raised point if you don't want to call it a levee which, I guess, I agree wasn't really a very profound levee, if it was at all.

IRWIN — We had the sense we were going up and down into these valleys as we were coming back to the LM, but you didn't have that impression as you were driving south to Checkpoint 1. So maybe this was a levee, but yet you had these undulations - little valleys - east of the rille rim.

SCOTT — Okay, on the visibility of the far-wall rim, I think we had a fairly good look at that and decided that it wasn't exactly as we expected. There was some apparent layering in the upper levels, but it wasn't a clean breakout of three or four levels on our way down. And you commented, I think, when we got up here, on what was called Bridge. Isn't that where you got a good look at that.

IRWIN — I think it was Station 1 where I said the Bridge Crater looked more like a shallow depression on the northwest wall of the rille, but certainly not a place where you could actually drive across the rille.

SCOTT — We stopped at Elbow, and I think we sampled radially, although it was a short radial sample. But I think we did pick up three separate bags of frags.

IRWIN — The rim wasn't very distinct there; it was a very subdued rim.

SCOTT — Sure was.

IRWIN — And I guess the first sample was probably what, 20 feet or so from the rim? Hard to tell, but it wasn't right on the rim.

SCOTT — You couldn't define the rim. It was a fair distance from where the slope ended and the bottom of the crater began. There wasn't a raised rim at all.

IRWIN — And no bedrock exposed in Elbow. There were some rocks in the crater, but not clearly bedrock.

SCOTT — That's right. You couldn't define any big in-place outcrops. I think the three samples we got there we described as we went along, and that was the one place I think we saw what appeared to be olivene in that rock. Remember? It was green - looked like a crystalline rock which had a lot of green in it.

IRWIN — I can recall that that came from Station I.

SCOTT — I sure would like to see it again, to see if we really saw what we saw. Those visors might have fooled us a couple of times there, but it was colorful. Well, that was a short stop. I guess, we had planned to have a short stop. You got a pan here, didn't you?

IRWIN — Yes.

SCOTT — Did you get a pan at every stop?

IRWIN — Yes, I did.

82-11168

SCOTT — Crater wall - stratigraphy; we didn't see any crater wall stratigraphy. I don't think we recognized any significant ground pattern around the crater. Station 2 we selected as a large boulder on the surface. It was very prominent, very unique, and it was the one large boulder that was visible anywhere in the area. I think our idea of going up to the rim of St. George Crater would not have been worth the time, because there apparently wasn't that much on the rim to tell us anything. There was no ejecta blanket, and there was no distribution or increase in fragmental debris anywhere that I can remember.

IRWIN — I guess, the only thing that would have been significant would have been if we could have gotten up to the very fresh light colored crater.

SCOTT — We considered doing that for a while.

IRWIN — Yes, you did.

SCOTT — I was going to drive up there, but I think we ran out of time by sampling the block. That would have been a very good crater to sample, because it's quite visible, as I remember, even from orbit.

IRWIN — Yes, you can see it from orbit. It was a fresh crater, and it was very light albedo. I just wonder if that light material is that very light from the anorthic site. You know, it was underneath the big rock and it was kind of powdery white.

82-11450

SCOTT — It could have been. Too bad we didn't get up there.

IRWIN — Too bad we didn't get a lot of places.

SCOTT — Yes.

IRWIN — Well, on traverse I was supposed to observe Elbow ejecta distribution. I think the ejecta distribution was uniform around Elbow, optimating one crater diameter.

SCOTT — Yes, but there wasn't much of it.

IRWIN — No, not much.

SCOTT — I'm not sure you could call that Elbow ejecta though. I don't think we could distinguish a radial or circumferential ending of the ejection. There was just a lot of debris.

82-11471

IRWIN — I thought we drove out of the ejecta and it got fairly smooth again.

SCOTT — Yes, you could definitely see the slope increasing as we went up toward St. George.

IRWIN — I didn't see any change in rock type.

SCOTT — No, there wasn't any.

IRWIN — The ground texture as far as we went up the slope - well, as we looked upslope, we saw maybe a suggestion of horizontal beds from this downslope movement.

SCOTT — Yes, the slide of the material.

IRWIN — And I didn't see any St George ejecta.

COTT — That's why I think that it would not have been too fruitful to go to the rim of St. George unless we could have gotten to that fresh crater up there. But there really wasn't an ejecta distribution per se.

IRWIN — I can't think of any particular explanation of that large block.

SCOTT — It didn't look to me like it came down from above. I didn't see any tracks. My best guess would be that it came from a secondary cluster, or was the secondary from below. It's hard to relate to these secondaries. It did look like it came from the top. I didn't see any outcrop up there which could have produced it. Did you?

IRWIN — No. I guess we'll have to look at the pictures in context.

SCOTT — Yes.

IRWIN — We didn't sample regularly there.

SCOTT — No. We just sampled that block. I think we filled all the squares for block data. Yes, you even tried to turn it over.

IRWIN — Yes. I tried to turn it over, and we did a comprehensive sample there.

SCOTT — Yes, but it wasn't very fruitful. You raked and raked and raked, and I think we got about a fifth of a bagful, because you were not shaking little rocks out earlier.

IRWIN — Yes. I wonder how it came back, probably all soiled after the transport. We did do a double core there.

SCOTT — And it was an easy double core.

IRWIN — We did mostly soil there.

SCOTT — Well, we got the boulder. We got one fragment from one side where there appeared to be a linear contact within the boulder itself. Whether it was a clase within a much much larger rock, or whether it was actually a contact, I don't really know. But, we sampled on each side of the contact. We also sampled the soil near the boulder. We sampled the fillet and underneath the boulder.

87-11845

IRWIN — You know, the surface soil there was very soft. It was the same textured soil as we saw on the slope down near Spur. And this related to the softer upslope.

SCOTT — Another thing I remember was that the fillet was clearly on the downhill side of that boulder. There was no fillet on the uphill side, which was rather interesting. But, you could almost see underneath on the uphill side. And there was no doubt (*Transcript obscured Ed.*) on the surface. It looked like it had been deposited on the surface and the fillet had somehow accumulated on the downhill side, like the wind was blowing from upslope. I guess that other crater, that light-colored one, was almost halfway up St. George, wasn't it? We would have had a tough time getting it all the way up there. I think the Rover would have made it, but I think it would have taken a great deal of time. We came back then, almost directly to the LM. As we drove away from Station 2, I think one of the things that really impressed me was the very gradual slope into the rille, just to the north of Station 2. It was almost a very subtle V-shaped depression or slope into the rille. I got the impression you could have driven down into the rille there.

IRWIN — Very easily; just a little more slope there.

SCOTT — Right between Elbow and Station 2 was a neat slope. I'd say it was at least 5 degrees less than the rest of the rille. That's why I get the impression that that's a portion of the fracture along there filled by a slide. I felt that, if we'd turned left there, we could have very easily driven down into the rille and back out. The drive back was a fairly easy drive. We were following the NAV system almost all the way.

IRWIN — There wasn't any change in block distribution or rock distribution on the way back. I didn't see any rock flows or any suggestion of rock flows coming out of the front. Again, no pattern.

SCOTT — Had we attempted to go to Station 3, we'd have picked a place somewhere and stopped, because it didn't look like we'd see contact up there. It was all too weathered. Everything seemed very uniform, as far as the surface texture.

IRWIN — The only observation that we really made was the fact that we were going down in one valley and up over a hill.

SCOTT — Right.

IRWIN — It was quite a topographic change between one and the other. You could really have parallel bridge lines there, although very subtle. We could very easily look over the EVA-2 route and see that it was just as smooth or maybe smoother than the EVA-1 route.

SCOTT — That's right. I think we commented on that at the time, that there would be no problem getting down to the EVA-2 route or driving along the Front. The general distribution of the craters was about the same. There was a wide variety of sizes, all very subdued, with an occasional fresh one which had almost 100 percent coverage of fragmental debris within the inner walls, and maybe a quarter of a crater in diameter out over the rim. It didn't go very far.

IRWIN — With the glass portion in the center?

SCOTT — Yes. That was the one we were going to sample sometime along the way, but we never sampled it.

IRWIN — We sampled one along the Front - the first one we stopped at.

SCOTT — That's right. One time I had to stop and fix my seatbelt. We picked up that rounded vesicular basalt fragment that was setting there. Through the windows, prior to leaving the LM, we had seen a large black fragment on my side. And you had seen a black frag on your side, as you looked out the front window. These were unique to the local surroundings. I don't think we'd seen other fragments that black and prominent. I picked this one up; it was probably 60 or 70 meters in front of my window. When we got back we tossed it in the bag. It was about 8 or 10 inches across by 6 inches thick. It almost looked like one big piece of black glass with a rough textured surface.

87-11835

10.7 EVA-2 GEOLOGY

SCOTT — We drove directly to the Front. There was a certain wander factor there as we went by Crescent and Dune.

IRWIN — I saw one crater there I estimated was probably half a kilometer out. I thought I saw a bedrock exposed, probably 10 to 15 feet below the surface. They probably would have wanted us to stop there on the way back, if we could have found it. But, we never saw it on the way back.

SCOTT — They commented on it, and we did, too. We could pick that one up on the way back, but we never saw it. We followed our tracks on the way back. On our way down, the surface relief appeared to me to be generally the same, a variety of crater distribution, all subtle, subdued, and an occasional fresh one with all the debris in the bottom on the glass in the center; less than 1 percent, much much less than 1 percent. When we went by the big craters, Crescent and Dune, they were really subdued. They sure weren't very obvious.

IRWIN — But, it was obvious when we saw Dune. We also remarked that we didn't see that rampart on the southeast side.

SCOTT — I think, we recognized Dune by the notch on the side. Although it was more subdued and there was no rampart, I think it was quite easily recognizable by that notch on the south side. We proceeded to drive on up the slope. I didn't realize we were going quite that steeply up the slope until we got up to Station 6. Driving down there, we discussed the secondary sweep up onto the Front. It's obvious that the only craters within the Front itself appear to be due to the secondaries, because it's a straight line right up to the side of the Front with maybe a dozen craters up there on the slopes, in line with what is expected to be the direction of secondaries. I didn't see any other craters anywhere. There was a large block down in the vicinity of Front Crater, up on the slope.

IRWIN — It must have been a huge one.

SCOTT — It was really big. I got a 500 of it, so maybe we'll get a chance to see what it looks like. We drove up to Station 6.

IRWIN — That was one of those small fresh craters with a glassy center. We sampled the center first.

SCOTT — We can sort out the rocks and easily identify them instead of trying to remember them now. It seemed like a significant stop. There were worthwhile assortments of things to be sampled. We saw Spur Crater. The idea was to stop there and to press on down to the Front.

IRWIN — Somehow, we turned around, though, instead of pressing on down to the Front. What did we do, run out of time?

SCOTT — Well, time was getting short.

IRWIN — It all looked the same.

SCOTT — It all looked the same, except for that very large boulder down there. Other than that, it all looked the same. I didn't think we would gain anything by going in that direction that we couldn't expect to see at Spur. It just didn't seem fruitful to head off to the same type of surface that we'd been seeing all along. We had three things in this area that we could sample which were representative: a young fresh crater which we were on, the boulder which was upslope, and Spur. It appeared to me that to go any further would have really compromised the sampling at the other places.

IRWIN — I was thinking that the boulder was more in line with Spur.

SCOTT — No, we went up to the boulder.

IRWIN — We went up to it?

SCOTT — Yes. We were driving uphill to the boulder. Then we went down the slope from there to Spur.

IRWIN — The large rock was on the upslope side of the crater.

SCOTT — You went up and looked at it and said it was green. I came back down to the Rover and went back up with the tongs. I pried a piece off with the tongs.

IRWIN — There was a layer there.

SCOTT — In the central part of the boulder there was a very loose surface covering, which could be scraped off. That's what you scraped off. You could see beneath it the lighter colored material, which we interpreted as green. I think that was because of the visors.

IRWIN — Yes.

SCOTT — It was really a light gray, similar to the type of material we'd seen at the rim of the fresh crater and which we sampled.

IRWIN — After that, we went down to the downslope side of that crater.

SCOTT — We went down to Spur. The downslope side of the first crater, at Station 6, was where we got the light albedo and where you dug the trench. We were going to trench on the uphill side of the crater. You said "Hey, there isn't any light-colored material here." So we went back down to the downslope. That was the same material that was on that rock, although it was very loosely consolidated in the central portion of that large boulder. It could be scraped up as sort of a crust material on the boulder. I scraped it up with the tongs and put it in a bag. I also pried off a chip of the boulder which appeared to be an Apollo-14-type breccia. The boulder appeared to be sitting on a surface. If I had to call that one, I'd call it the upslope from the secondary. There where we noticed the difference in the Rover tracks and the depth of the bootprints.

IRWIN — I think we photographed that. I guess it was one of these stations where we looked over Hadley and saw all the organization of the beds over there.

SCOTT — Yes, as a matter of fact, that was back at Station 6 because I think we did the 500 there. Didn't we? I think I pulled out the 500 and either did that at Station 6 or at Spur.

IRWIN — I think we did it at 6.

SCOTT — And I got the whole organization there on the Front. I took a couple of horizontals and a vertical strip of the thing. If all that 500 works out, it will be a pretty enterprising operation. We might discuss the 500 since we're on it. We had trained with the trigger and the handle. We both decided not to use the trigger and the handle because it seemed to require enough torque to move the camera when you took the picture.

So, I tried the first EVA without it and it seemed to work better. I felt more stable without the trigger and I never did put the trigger and the handle on it. I just used the straight pushbutton method. It felt fairly stable while I was taking pictures. On down to Spur, Station 7. We did quite a few samples there. There is where we found what we're calling a lot of plage in that rock, an anorthicitic rock if there ever was one. That was the one on the other rock on the sort of pinnacle. It was different. It looked white, dust covered, with white spots on it, which indicated it was different from the general gray fragments around it. There was a nice 3-foot boulder there on the surface, which was another breccia which we were going to work our way up to. We never got to sample that. I did get one piece of it. It looked like it had fallen off on the ground. I think you raked there, didn't you?

IRWIN — Yes. I had a good place to rake. Good comprehensive samples.

SCOTT — Yes, you got a good rake and a good soil.

IRWIN — Yes.

SCOTT — Okay. It seemed like that was a very fruitful place to obtain samples. I wish we could have spent more time there sampling because I'm sure we'd have found more of the anorthicitic type or the plage. But time being what it was, we pressed on back with a thought in mind that we'd stop at Dune Crater to pick up a secondary sample and take care of that requirement. Summarizing the observations of the rock types collected at the Front, we saw breccia and crystalline. That was about it. Did we see any good pieces of basalt?

87-11839

IRWIN — You mean up on the Front.

SCOTT — Yes.

IRWIN — No. We didn't see that until we got down on the Dune.

SCOTT — I think that's right. There wasn't much block distribution. There weren't very many. All along the Front, there were half a dozen blocks that you could see on the whole base of Hadley Delta. There were no mounds. Did that big boulder we sampled up there have a fillet?

IRWIN — I don't remember, Dave; you'll have to look at the pictures.

SCOTT — Any patterned ground that you remember?

IRWIN — No.

SCOTT — No apparent flows or slides. That takes us down toward Dune and we backtracked. We found our tracks and followed them back. It was interesting to us concerning the ground's interest in finding our tracks. Every time we headed back from any point they said, "Find your tracks and follow them." I guess there was some doubt as to the Rover nav system, but I felt very comfortable about where we were. I never felt that we needed to find our tracks. Did you?

IRWIN — No. Particularly from the Front because we could see the LM.

SCOTT — Another factor was the mountains in the background and the horizon. We could pick a point on the mountain and drive towards that point and we knew we were going toward the LM. I never felt disoriented or lost. I think we could have completely lost the Rover nav system and I wouldn't have had any apprehension about finding the LM.

IRWIN — Yes, as you remarked, you could see Pluton all the way back; just head toward Pluton.

SCOTT — We could see Pluton, and we knew the LM was on a slight rise; topographic high, anyway. So, I didn't feel tracks were necessary. As a matter of fact, I think we deviated from the tracks to find better routes or more direct routes.

IRWIN — Yes. We certainly did on EVA-3.

SCOTT — Well, on EVA-2, also. After Station 4, if we had followed our tracks, we would have had to do some weaving in and out of the craters there. Let's see, in approaching Station 4, did you get the feel of any buildup on the downsweep side of the secondaries? Did you get any directional kind of feel for those secondaries? I didn't.

IRWIN — No. But it was just obvious that we were coming into an ejecta pattern there from Dune Crater. Concentration of rocks increased as we approached the rim.

SCOTT — But I didn't notice any grain size difference. When we talk about grain size, I don't believe the grain size change would be obvious to the eye. If there is any difference in grain size, it's probably micro because I never noticed any.

IRWIN — It's hard to see when you're driving.

SCOTT — We got to Dune Crater and there was one obvious boulder with large vesicles right there in the southern side of the notch that we hadn't sampled. That was probably one of the most prominent rocks we saw during the whole time.

IRWIN — I saw another rock with exactly the same size vesicles right at the edge of the rille.

SCOTT — You're right.

IRWIN — That was probably the bedrock.

SCOTT — Yes, I'd say definitely it was bedrock. We sampled that one in the center near the vesicles, and on the edge where there were smaller vesicles on the outside of the rock. They were millimeter-size vesicles. That rock was about 6 feet high and 4 feet across, with rounded 3-inch vesicles, very clean with plagioclase laths in it which were centimeter long and millimeter wide; random orientation. In contact with that was a highly vesicular-like, maybe half-centimeter, uniform spherical vesicular rock, which was a lighter gray and had not been chipped. I took a picture of it. It's too bad we didn't get to sample it. But, it was a different flow entirely. A different rock and they were in contact.

IRWIN — I remember observing that the largest crater in the south cluster was one that ran east-west. I got the impression that it was elongated that way.

SCOTT — You really don't get that from the photos.

IRWIN — Yes. Looking down on it from up here, it was one oriented this way (gesture). I think we'll be able to see it from the pans.

SCOTT — The other frags we picked up at Dune Crater, we just didn't have time to look at. We didn't give them a TV stop there either. When you cut down the time to the point we had it's just too bad. We spent a lot of time there, too. Then we proceeded on back to the ALSEP, and you picked up a couple of rocks back there. You picked up your black one.

IRWIN — What I refer to as a pink.

SCOTT — Pink.

IRWIN — Pink with light plagioclase in it.

SCOTT — Did it really? You got to pick up rocks while I had to drill. You have all the luck.

IRWIN — We both had our thing. I was doing other things. I dug the trench.

SCOTT — Yes. That I guess summarizes EVA-2. I didn't notice anything in particular driving back. You could see albedo changes where we'd been. Any disturbance of the soil was apparent. The Rover tracks were a little different.

IRWIN — A little darker.

SCOTT — Yes.

**10.8 EVA 3
GEOLOGY**

SCOTT — On EVA 3, we started out after exercising the deep core drill again for Station 9 and Scarp Crater at the edge of the rille. We had to take a circuitous driving route, going around the craters, which seemed to be elongated north/south. We were going again up over depressions in topographic highs which trended north-south. You felt like they were pretty much circular.

IRWIN — I felt they were circular. Perhaps the photos will tell us some more mainly because the three circular features were all lined up there.

SCOTT — Look at that this way, though. Why don't you comment on what you saw of the terrain because I was just pretty much trying to drive and to avoid the big holes.

IRWIN — There was just a gradual drop down. We drove through one or two of those depressions. I would not say shallow depressions because actually the bottom was probably 150 feet below the general surface of the plains. It was about a 5-degree angle into the bottom. I remember one in particular that seemed like a fairly fresh crater in the very center of it, with no rock debris, no ejecta on it. It was right in the center of the large, shallow depression. There were three of those as we headed west. I did not see any change in rock distribution as we proceeded to the edge of the rille until we came up to that very fresh one which we incorrectly called Scarp, initially. It probably was Rim Crater.

SCOTT — Probably, because it was too small to be Scarp, and it was fairly fresh.

IRWIN — That was the first place we stopped - on the western side of that very fresh crater.

SCOTT — That was the one that had a very soft rim. Soil was just much softer than we had seen before. The frags that we picked up there were clods. I mean they fell apart - were very fragile.

IRWIN — They all looked the same, sort of angular, but they did have some glass in them. I guess we disagree there, for I say that crater is similar in characteristics to the very small ones that we saw earlier, except there is no concentration of glass in the very center.

SCOTT — I guess I thought there was not that much concentration of fragmental debris. The smaller ones appeared to me to be pretty nearly 100 percent covered with frags; and this, I would say, had maybe 30 percent frags.

S71-42951

IRWIN — Well, I thought it was 100-percent coverage.

SCOTT — Did you really?

IRWIN — It will be interesting to see what that picture shows.

SCOTT — We may be talking about two different craters.

IRWIN — I don't think so. I remember that you went on ahead because I was working on my camera, trying to get the camera to work; and you went on up the rim. You were sinking in, and I came up about 5 minutes later. I was impressed with how soft the rim was because you would sink in almost 6 inches. That was a very unique crater. It was the only one of that size and that type that we saw on any of the EVAs.

SCOTT — It surely was. I hope some of those clods got back intact because they really fell apart easily. I think the photos will describe the rim better.

IRWIN — In fact, one of the photos that we saw this morning was of that crater.

SCOTT — That's right; it surely was. I guess you are right. If that is the case, then that indeed was covered as much as the others; but it just looked different to me. It looked like more of a tan or brown or darker gray.

IRWIN — The color could have been slightly different. The fresh ones were a very light gray. The fresh ones looked like hard, angular, fragmental debris covered on the inside; and this one just didn't look quite so hard. It had a different color. It will be interesting. We sampled both types so we could compare them.

SCOTT — We headed on to 9A, which was on the terrace, and made a rather lengthy stop there. We did the photography. In looking at the rille, I can remember seeing, on the upper layer about 10 percent down, exposure of bedrock with internal layering, quite discontinuous and irregular, but all across the same level and with different characteristics within the layering laterally. I took the 500 vertical/horizontal strips and also other targets of opportunity down within the rille. It is unfortunate your camera was not working there because that really slowed us up. We did a comprehensive sample there which I think was the best one of the whole series. That was where we moved the gnomon to get better coverage with the comprehensive; and we each got a big rock.

IRWIN — That one went in the B-SLSS bag.

SCOTT — We sort of came up a very slight incline to the rille rim. It was not anything I would call a levee. I think we were quite aware of coming to the rille rim when we got there, and it seemed to me that it was a very slight incline. Then it broke to maybe a 3- to 4-degree slope down towards the rille to the edge where it broke on down to another inflection point, down to 25 degrees into the bottom.

IRWIN — I got the impression that that next break point, from which we were looking down to where the big blocks were, was a very steep break, maybe 60 degrees.

SCOTT — Well within that layer of bedrock. You could look back down the rille towards the south and you could see that we were on a layer of bedrock.

IRWIN — Looking to the south and also to the north, you could see the bedrock slightly above us. Maybe we were on a terraced portion that had slumped down because you could see the top, the level surface, the top of the bed both to the north and to the south.

SCOTT — I am confident that the large rock that we sampled there was bedrock.

IRWIN — Yes, the one with the very large vesicles?

SCOTT — Yes. We chipped off of it and got a couple of frags off the side of that nodule.

IRWIN — I was hoping that we would get down lower to where it was obviously bedrock - either down lower toward the rille or to the north or south - but we never had the chance.

SCOTT — The color looked darker black. Those very large, almost rectangular fractured rocks, as you called them, looked a little bit like columnar jointing. Those big black ones down there were darker black than the ones we sampled.

IRWIN — Yes.

SCOTT — Then, I guess we got a couple of cores there. Then we proceeded on up the rille rim to get the stereo, the stereo pan, and the 500 mm photos. Did you think of anything else as we went up there?

IRWIN — No. The distribution of fragments seemed to be uniform along the edge of the rille. By uniform, I mean about 20 percent.

SCOTT — Yes.

IRWIN — On the surface.

SCOTT — Yes. That is about right. And a variety of sizes from the 1- to 2-inch size up to the large 1-foot to 1 1/2-foot size.

IRWIN — The fragment distribution probably was very similar to what we saw at the south side of Dune.

SCOTT — Yes.

IRWIN — Those two might relate very closely.

SCOTT — You are right. Those big blocks that had the jointing or the linear fractures in them did not have the vesicularity that that big block did in the bottom, south side of Dune Crater. I don't remember seeing the 3-inch large well-defined vesicles, do you?

IRWIN — Yes.

SCOTT — In those big blocks?

IRWIN — Yes.

SCOTT — You think so?

IRWIN — I documented one.

SCOTT — Did you?

IRWIN — It had the same size vesicles. The ones off in the distance you mean?

SCOTT — Yes. The ones off in the distance.

IRWIN — It seemed that there were some large vesicles, but we really were not close around any.

SCOTT — It did not look to me like it had the very large ones that we saw at Dune Crater. It was the same color.

IRWIN — You got some 500s that probably took in that field of view.

SCOTT — I think so. It seems to me that we terminated that rather hastily, and it is too bad we could not have gotten a Pluton. I might comment here, in looking at Pluton, that I did notice that the inner walls were covered with some large fragments. These were on the order of probably 2 meters or so. Maybe 5 percent of it was covered somewhat uniformly, and the outer walls did not seem to have any debris at all.

IRWIN — You probably saw that best from the SEVA, though.

SCOTT — Yes. I could see it fairly well. I did take 500s of it, so we will see if it shows anything. Then we drove back to the LM, and the NAV system took us right straight back with no problems. Did we do any rock collecting when we got back there? I guess we got the DPS engine valve back there on the SESC. I guess we filled all three SESCs, didn't we?

IRWIN — No, I think we had one left.

SCOTT — Did we?

IRWIN — Well, I thought we had done that on EVA-2.

SCOTT — That finishes EVA-3 as far as the geology goes that I can remember. Do you think of anything else?

IRWIN — No, we had, of course, a lot of rocks that are probably not documented too well.

SCOTT — Yes, we were running out of time there. Time and camera. Did you get your pans there? You took my camera and got the pans.

10.9 LM LAUNCH PREPARATION

SCOTT — We'll start into launch prep after the jettison of LM equipment. There's really not a lot to say about it, except that it went as per checklist. We skipped a P22 with the command module because we were somewhat behind time there and that didn't seem to be too necessary anyway. With that elimination, we were pretty much right on the time line all the way up to lift-off, and everything went as per schedule. I think we had run this a number of times in the simulator and felt pretty comfortable with it, even though I remember commenting that that was probably the fastest 2 hours we spent in the whole flight. The alignment went well. The stars were good.

IRWIN — It was interesting to me that the star angle difference with the gravity vector was somewhat more than it had been with two stars.

SCOTT — Same; .03 and .04, something like that.

IRWIN — Here it is. Plus 08 here.

SCOTT — Oh, yes; the first one was. That was a gravity vector. I'm talking about the star angle difference down here at NOUN 05. It was somewhat larger than two stars. That one surprised me a little bit.

Lift-off. All the checkouts of the systems went very well. We powered up as advertised.

IRWIN — We were really pressing now, with not much extra time. We had a change on angles here for the rendezvous radar that they voiced up real time.

SCOTT — Which didn't help us any, I guess.

IRWIN — There's a question there. Why'd they come up with them real time?

WORDEN — I think it was probably because of the orbital changes.

SCOTT — Could be. I guess our antenna drifted. We can talk about that when we get to the ascent portion. The checklist looked all right. The switch settings all worked. I remember it being very busy throughout the time line, but we were never behind. We were just about 5 minutes all the way. We had time to get everything stowed properly. Then, we got down to lift-off.

IRWIN — I thought the battery management during the surface went just as planned. I was surprised it worked as smoothly as it did, because before the flight they said they were going to call me on real time and tell us when to switch the batteries.

SCOTT — You did that just as per checklist all the way through.

IRWIN — Yes, but I usually asked them if they were ready for it. They always said yes, do it per checklist. It worked out real well, real smooth. We checked in with Al two or three times there on the surface.

SCOTT — Two times. Once each day. That worked pretty well. It was obvious they were keeping him informed of what we were doing. We knew pretty well what he was doing, so that played pretty well.

IRWIN — Did we mention that we ran out of food there in the LM? We could have used a little more food.

S71-44666

11.0 CSM CIRCUMLUNAR OPERATIONS

WORDEN — My impression of the operations of the spacecraft was one of complete confidence in the equipment on board. Things worked very smoothly, and I didn't have to keep an eye on all the gauges all the time. There was very little noise on board. The only things I recall hearing are the suit compressors. I ran them most of the time with the three sets of suit hoses out and screens on the return, so the suit compressor noise was there. Also, I could hear a pump operating in the service module, which I assumed was the water glycol pump. Those are about the only two continuous noises that I had during the lunar orbit operations.

The rest of the spacecraft ran just beautifully the whole time. The fuel cells ran without a problem. In fact, everything ran just beautifully, and I really had no concern for the operation of the spacecraft during the lunar orbit operations. The only things that were off-nominal, of course, were the burns - the circularization burn and the plane change burn, where we had the problem with the SPS main A pilot valve circuit breaker. I made both of those burns on a single bank and they were nominal, except for that particular circuit breaker being left out.

Navigation was about as it was on translunar coast and up to and after that point of the flight. The guidance system was very tight. I never had any problem getting a star pair. Whether I was doing a slow ORB rate maneuver or whether I was inertial. P52s worked very well.

I still had the problem with the sextant. Even on the back side of the double umbra, the sextant was very difficult to use - to identify constellations and to identify the stars. The attenuation in the sextant was really much more than I had anticipated. I could look out a window and see the star field very clearly. In fact, it was much brighter than I expected it to be. There were so many stars in the field of view out the window that, in a way, it was a little difficult to find a constellation and to find the navigation stars. But through the sextant, only the very brightest stars came through. I was able to identify the stars after a while, after I was used to the star pattern, and I did the alignments just about the same place every time.

Even with the light attenuation through the telescope, the guidance system was so tight that every time I did a P52, I could look through the telescope and grossly identify where I was in the sky. Then when I looked in the sextant, there would be a star right in the middle of the sextant every time. It maintained its orientation beautifully the whole time. The drift rates were very low. The only thing, I guess, that I'd want to comment on concerning navigation, and that in regards the Flight Plan, is that, when we did an Option 1 reorientation, for example, to the plane change attitude, there was no place in the Flight Plan to write the gyro torquing angles for the second P52. Of course, each of these is done with an Option 3 realignment for drift reasons, and those gyro torquing angles are recorded. But then, when you do the Option 1 to go to the new orientation, there's no place in the Flight Plan to record those. I guess there may not be any valid reason to keep those gyro torquing angles. Possibly the ground doesn't need them, but I was in the habit of writing down the gyro torquing angles, and when I got to the Option 1, I did just this. I recorded them in a blank place in the Flight Plan. I feel that we might consider putting those in the Flight Plan, because they are some indication as to how the coarse align works.

This reminds me that, on each of the reorientations, I used a coarse-align option in P52, and in each case, the coarse align was good enough to put the star in the sextant, except for one instance on the way back home when we went to entry orientation. The star was just outside the field of view of the sextant, and I had to look for it a little bit. However, the coarse align worked very well. In almost every case, it put the star within half a degree of the center of the sextant.

Next item is LM acquisition. After the P24, after the circularization maneuver, the next pass over the landing site was a LM acquisition pass. It was made on REV 15, and that all went very well. The pad was sent up, I went to the attitude, and there were no problems with any of that. Everything went nominally. As I came over the landing site, I saw the LM shadow very clearly, and once I had identified the shadow, then I could also see the LM in the sextant. I watched the LM until I was near nadir, until I was almost to TCA, and then I took out the visual map, the 1 to 25 000 scale, in the CSM Lunar Landmark Map Book, and marked the spot where I saw the LM. That was BR .5 and 75.5, in the Lunar Landmark Book. One more comment on the LM acquisition, and, again, it's a comment that's been made before on landmark tracking. Once the LM was spotted, there was no problem at all tracking with the optics. Of course, at this time, I was in a 60-mile circular in ORB rate, and the rates were very low. But even at the low altitudes, there was no trouble tracking any landmark that you selected with the optics, in either ORB rate or inertial hold. The optics were very smooth in tracking and very positive. As long as the trunnion angle is great enough so that you don't go through or close to zero trunnion angle, any landmark you pick is fairly easy to track.

The next item is update pad and alignments. I've already covered the alignments. I used the update pads in the Flight Plan almost exclusively. As a matter of fact, the whole lunar operation was oriented toward using just the Flight Plan for all updates and for all information that went back and forth from myself to the ground, and I found that worked very conveniently. I checked things off in the Flight Plan as we went and wrote all of the corrections and changes in the Flight Plan. This meant that I only had one book to go to all the time, and it did work very conveniently for me.

Next item is mass spectrometer deployment. At the first part of the lunar orbit activities, the mass spectrometer was deployed and retracted almost as I had anticipated, knowing the approximate times the boom should take to deploy and retract. Those times came out very close. Only along towards the end of the lunar-orbit activities did I start to see those times varying. In fact, at one point, the mass spectrometer failed to retract. I never did get a gray indication. I turned it off, turned the retract mechanism off, and extended and retracted the mass spectrometer in short bursts, cycling it until I got a gray indication. This meant that the mass spectrometer was very close to being fully retracted, but yet something was holding it from the final retraction. Looking in the Flight Plan, I noticed that the first time I saw a problem with the mass spectrometer boom was at approximately 119 hours and 20 minutes in the flight, when I got no retract on the mass spec boom. At that time, I had retracted the boom and waited approximately 2½ minutes and then started watching the talk back, expecting it to go gray so that I could turn the switch off. Instead of going gray, it went to a half barber pole; the gray shutter in the talkback dropped about halfway, and it stayed there. I cycled it to extend three or four times, maybe bursts of 5 or 6 seconds, and then to retract. And after about the third cycle, the talkback went gray, indicating that it had fully retracted.

I ought to clarify the operation of the talkback. On all the extensions, the talkback was full barber pole until the boom was extended, at which time it went gray. On the retraction, it was full barber pole until the nominal time for full retraction had elapsed, at which time the talkback went to half barber pole. That was the only time, or that last bit of the retraction, when there was anything unusual about the operation of the talkback.

HENIZE — Did you ever notice a half barber pole in later retractions? We never heard anything more about it.

WORDEN —Yes. On each succeeding retraction, after that first one, the mass spec boom operated exactly the same way. I always got the half barber pole. After 4 or 5 cycles, for a considerable amount of time at least, could always get the gray indication. Along towards the end, it finally got to the point where I never could get full retraction on the mass spec boom. In fact, during the EVA, we had cycled the boom to extend and retract on the short cycles several times. I never could get a gray, and when I looked at it during the EVA, the cover was tilted about 30 degrees on the hinge, and the guide pins in the mass spec were just barely coming through the guide slots. It was on the guide rails, but the pins weren't fully extended through the guide slots.

The next item is bistatic radar test. That was all nominal. There were no problems with that. It was a P20 type maneuver, which was conducted during one complete frontside pass; I think two times. That was done all as per Flight Plan with no problems.

Solar corona photos: they were done as per Flight Plan. There were no problems with any of the solar corona passes. Everything worked very well, except that - I should make a comment at this point - that the solar corona photography was done using a countdown clock on the DSKY, which I called up by using P30 and loading the T-start time into the computer, in P30, and then letting the computer keep track of the time for me. At this time, there was no lighting in the LEB for the mission event timer, and solar corona photography required that the lights in the spacecraft be turned low. Because of the light problem in the LEB, the rheostats that adjust the integral and numeric slidings were taped in the position that they were in when we had the problem with the AC. This meant that the DSKY in the LEB was at a higher intensity than I would have liked for the solor corona photography. There was considerable light inside the spacecraft as a result of the lighting in the DSKY and the LEB. I turned all of the other lights out and monitored the DSKY in the LEB to do the solar corona photography and all of the other low-light-level photography.

Another comment on the use of P30 for the timing of some of these things in flight, and that is that, after I had used P30 for a while to time the events, I was called by the ground and told not to use P30 so extensively, because I interrupted the integration of the state vector in P20, which meant that the orbital-rate attitude was varying and was actually drifting outside the limits that we required for flight.

WORDEN — My recommendation is that we somehow devise a way of monitoring time on the DSKY, since it's a very convenient way of doing that particular thing. The digital event timer on the main panel is too far away and it's unusable for that type of activity. It means that the DSKY is really the simple solution, if we can somehow load the computer to count down to a time and then to count minus time to zero and then count plus time so that these activities can be monitored.

Next item is the UV photos. I think all of the UV photos went as per Flight Plan and went on schedule. There were no problems with the UV photography. Most of the UV photography was done when all three of us were on board. Jim handled all of that. I read the checklist, and it worked very well.

Window number 5 was covered with a Lexan shield, which acted as an ultraviolet filter for those portions of the flight when we weren't taking ultraviolet pictures out that window. Because of the distortion and the poor optical quality of the Lexan, pictures would have been greatly degraded if they had been taken through the Lexan shield. There were some portions in the Flight Plan where it called for the Lexan shield to be removed for visual or for orbital-science photography, which was not ultraviolet photography. At some portions in the Flight Plan where some of that photography was being done, the Lexan shield was left off the window for periods greater than the time prescribed in the Flight Plan. I observed no effects from any ultraviolet radiation. I don't believe there's anything that was observed after flight either.

The lunar libration photography was performed using the 35-mm camera with the very high speed black and white film. The camera was mounted in window number 4, the right-hand rendezvous window, through a shield that was placed in front of the window to screen any interior lighting from the lens of the camera. I always had to take considerable time and patience to put the lens in the slot in the opening in that filter to make sure that I got a good field of view in the camera. The filter or the shield really didn't seem to fit as well as I thought it should. So it took me a little bit more time to make sure that the shield was around the lens and the lens was in the window properly. Once that was done, I found the 35-mm camera very easy to use, and all of the low-light-level photography was done as per the Flight Plan. We had just the right amount of the film. There was some earthshine also taken with that film, on MAG T, and the other low-light-level photography was done as per Flight Plan with no problem. Once again, I used the DSKY in the lower equipment bay as a clock and turned the other lights in the spacecraft out.

The orbital science photography, for the most part, went as per the Flight Plan. There were a few instances where some other activities were scheduled real time which interfered with orbital photography, and in those places, the photography was not accomplished. In the Flight Plan, the orbital photography was almost invariably strip photography, with the camera being held in the window and pictures taken at some prescribed interval, such as 15 seconds or 20 seconds. Those at 20 seconds were done with the intervalometer, and those at the other times were done just by counting on the clock.

In almost every case of orbital photography, the ground site had been analyzed preflight, so that I knew what the targets were. In flight, rather than just take pictures looking straight out the window, I concentrated on taking pictures of the sites that we are interested in. That worked in almost every case. There were several strips of photography taken from Crisium to Serenitatis. I think there were five strips scheduled to cover some of the Lunar Orbiter photos that were of very poor quality. We got all of those, except one strip, which was replaced real time by some other activity. I don't see it in the flight right now, but, as I recall, there was one strip that we didn't get. The rest of the orbital photography all went pretty much as planned. I had no difficulties. The targets of opportunity, I found fairly easy to handle, as far as the camera settings were concerned. The camera settings were both on the Fullerton wheel and on the lunar orbit monitor charts. I found that all very - fairly straightforward and easy to accomplish.

S71-44672

Monitoring lunar activity: There was ample opportunity for me to observe the lunar surface from the spacecraft. There were some periods specifically set aside to do nothing but that. During periods of SIM bay operation, I had ample opportunity to look at the surface. I found no problems with that.

Next is SIM bay daily operations. I want to talk first about the Flight Plan. I found that the Flight Plan for the solo portion of the operation in lunar orbit worked quite well. I think that the only reason that it worked well is because there were very few updates to the Flight Plan during that period of time. The SIM bay operation is a monitoring operation as much as anything else. It got to be rather difficult at times to keep track of the times and to do things at the times prescribed in the Flight Plan. To do that meant that full attention had to be devoted to just keeping track of the time and switching the instruments on and off at the proper times. One thing that was used in flight was that the ground would give me a 30-second or a 1-minute warning on when to do some particular switching, and that seemed to work quite well, because, as I was off doing something else, doing a visual sighting or operating something else in the spacecraft, the 30-second warning gave me ample time to get to the SIM bay station and accomplish the things that had to be done. I found that a great deal of my time was spent in monitoring the SIM bay operation, in getting all of the experiments running, in deploying and retracting booms and the mapping camera.

Essentially, the basic instruments in the SIM bay were started in the morning after first getting up. They operated rather independently all day long, except for some changes in the gamma ray and some changes in the mass spectrometer, in the gains and that sort of thing. Most of the time in the SIM bay was devoted to operating the cameras, and that's where the clock watching was most important. The SIM bay operation is a very complicated operation. We attempted to simplify it with the Flight Plan that we used. The plan was to use the checklist as an operational guide and to use the Flight Plan only as an event guide. This meant that, when the event was about to occur, you go to the systems checklist and perform that particular function, such as extending the boom and operating the experiment. I found it rather unwieldy to do. It left you with a feeling that you weren't really aware of what that instrument was doing with respect to the rest of the SIM bay.

Each operation was an individual operation, and a lot of the SIM bay activity had to be done roughly at the same time.

For my own use, an integrated flight plan or an integrated switch list, such as we had in the Flight Plan, was most efficient for me to use when I was by myself. We did have some real-time updating during flight when there were three of us inside the spacecraft. We did use the checklist for some of those portions of the operation. Those operations worked about as well as the operations where we used nothing but the Flight Plan. It's a personal preference on my part that everything appear in the Flight Plan. It does pose problems if you have a lot of real-time changes, because it takes a great deal of time to write the changes down in the Flight Plan.

I found that my biggest single problem with the operation of the SIM bay was in not being continuously aware of the state of the various experiments in the SIM bay. The only indicators on board are the talkbacks associated with each of the instruments. In both the stowed position and the operate position, those talkbacks are always gray. You really have no way onboard of identifying the mode of operation of each of the instruments without going back and referring to the Flight Plan and knowing that you've performed the functions on the Flight Plan as prescribed. This caused some confusion at times when we had real-time updating, because then, the SIM bay got in a nonstandard configuration with respect to the Flight Plan. It was very difficult, without a lot of discussion from the ground, to determine the mode of each of the instruments and what had to be done at the next step. It would be a great help if there were some indication on board of the mode that each instrument was in at the time.

A comment about the solo portion of the Flight Plan, with respect to the amount of activity involved. That portion of the Flight Plan was not too crowded. Outside of monitoring the spacecraft and doing the visual sitings, my main function was to monitor the SIM bay. I found that that all worked out okay and that there was no undue amount of work associated with it. It was just time consuming, and so much of the operation involved sequential switching and monitoring of the clock. As far as the time was concerned - the work load was concerned I found it not to be excessive.

A comment on the Flight Plan, in general. Something that should be factored in carefully into the Flight Plan, particularly during the solo portion, is the fact that it takes longer to do things in flight than you'd anticipate. For instance, to eat a meal seems to take me considerably longer than I had thought would be required before flight. Invariably, there were things to do right in the middle of an eat period which took your attention away from that part of the flight and extended the eat period considerably from, that shown in the Flight Plan. The same thing was true of the exercise periods. I tried to get the exercise during those periods when it was called out in the Flight Plan, but again, there were things that had to be done almost inevitably during the exercise periods. Every effort should be made in keeping those periods free from any other activity, and sufficient time should be allowed for those things in flight, so that you can get them done, get them out of the way, and get the spacecraft cleaned up again before you get back into the working part of the Flight Plan. That would

88-120021

pay great dividends in the orderliness with which the SIM bay operation is conducted, not having to intersperse that operation with the general housekeeping operations that have to be done on board.

WORDEN — Let me now talk about the individual experiments. Most of the experiments operated just as we had planned preflight, and there's nothing to say about them. The mapping camera works just as I had expected it to work. I kept track of the extend and retract times, and everything was nominal until the very end of the flight, when the mapping camera failed to retract. It just stopped, and it looked to me during the EVA that it was fully extended. I looked around to see if there was anything that had jammed it or anything that could have interfered with the mapping camera to cause it to stay extended, such as the covers being jammed against the side of the mapping camera. I could find no evidence of any of that happening, any jamming at all from an external source causing the mapping camera to fail to retract.

The X-ray, laser altimeter, gamma ray, and alpha particle experiments all worked as per the Flight Plan. The laser altimeter apparently was failing somewhat in flight, but it didn't affect the Flight Plan and didn't change the Flight Plan, except for a few real-time changes, as concerned the laser altimeter itself.

The mass spectrometer was the most troublesome, in a way. The experiment itself apparently worked well, but the boom failed to retract properly. That started almost from the very beginning of the lunar orbit operations, in that the mass spectrometer would not retract properly. I think I've already covered that previously.

The gegenschein calibration photos went as planned, and as I commented before, the lights inside the spacecraft were primarily from the lower equipment bay. There was some afterglow in the floodlights, and I taped the floodlight above the righthand rendezvous window to reduce the light from that source.

The zodiacal light and the gegenschein were about the same, in that the experiments went as we programmed them preflight. They went as scheduled with no problems.

General photography within the spacecraft: I had inserted in the Flight Plan at the beginning of each day's activity those magazines that would be required for that day's activities. That worked very well in helping me organize the photography for the day. I used one of the fabric containers just to the left of the side hatch as a storage bin for the magazines and for the cameras when I wasn't actually using them. That worked very well, because, with the center couch out, I was standing in a position which was very accessible to that particular compartment and to window 5 and window 4 and to the side hatch for taking the pictures. It was very convenient and worked very well.

The plane change was a nominal burn, except for the single-bank portion of the burn, this being an off-nominal condition. The plane change went as planned. Realignment was very simple and worked very well. There were no problems associated with it. Residuals on that burn were 2/10 ft/sec, which was the trim lower limit. So that burn was not trimmed.

A comment on the plane change, and it applies to other maneuvers that have to be done, particularly solo. Sufficient time should be allowed before those maneuvers to get the SIM bay operation cleaned up and ready for an SPS maneuver. I didn't notice any particular bind in the timing of getting the SIM bay powered down for the plane change. Of course, the SIM bay wasn't in operation for the circularization burn. For the plane change, I powered down the SIM bay approximately 15 minutes before the burn. If at all possible, that time probably should be extended a little longer before the burn just to allow time for any anomalies that might arise before the plane-change burn. Although, in this case, I had no problems. I do recall thinking at the time that it would have been nice to have had a little more time there.

Communications during lunar orbit operations were very good. I don't recall having any problem getting the high gain locked up at the times prescribed and that whole operation went very smoothly. I manually switched the DSE after LOS and I don't recall any case where the tape recorder wasn't already operating the way it should be; the way we had expected it on ground command. So, that was merely a manual backup to a switching action which had already been performed.

The rendezvous portion is next; maneuvering support lift-off. We did the vhf check. That went well. Prior to the rendezvous, I was asked to do a P24 on the LM for the rendezvous targeting. This was the result of an insufficient time line in the LM to allow them to do a P20 or a P22. I did the LM visual at about 170 hours and was never able to identify the LM on the surface. Two things I think caused that particular result. One was that the Sun angle was very high and that there was no discernable shadow from the LM, which helped me on the first LM visual to recognize and locate the LM. The second was because of the Sun angle, or at least I assume it was because of the Sun angle, the landmark line-of-sight part of the optics cast a very red or bright pinkish to red image in the sextant, which was very difficult to see through to actually look at the terrain. The landmark image kept sweeping through the sextant as I was looking at the landing site in the sextant. It was so bright at times that I couldn't see the actual image of the terrain. That also added somewhat to the confusion. Maybe that can be explained in terms of the geometry of the optics and the particular Sun angle at the time. I wasn't able to pick up the LM and I don't feel that that was a very successful landmark tracking pass.

The rendezvous was as nominal as any rendezvous we ran in simulations. We did have some communications problems on the rev prior to rendezvous so that the rendezvous pads were read to me by the ground at AOS on the rendezvous rev. That all went very smoothly with no problems. We did the vhf check and that worked okay. I did notice after lift-off that it took several attempts to get the vhf ranging reset so that it would stay locked up. Once I got it locked up, it was about 136 miles. It stayed locked up from then on. I think it broke lock only once and I got the tracker light. From the LEB, all I could see was a caution and warning light, a PGNS light. When I checked the DSKY on the main panel I determined that the tracker light was on and then I knew, of course, that the vhf had broken lock. So, I reset it and everything was fine.

I got the state vector of the LM and started looking for the LM, but I couldn't find it. The LM was not in the sextant when I started looking. I went to the telescope and could not see the LM. I called in to make sure that the rendezvous light was on; that the beacon light was on and it was. About that time I picked up a very faint flash in the telescope, about 10 degrees away from the center of the telescope, and I slewed the sextant over to that point and picked up the LM. The tracking from there on went very nominally and, in fact, ended up prior to TPI with 19 vhf and 18 sextant marks. The solutions were very close on the recycle; the X and Y solutions were very close, and in the Z solution I had about a 4- or 5-ft/sec difference from the LM. On the final comp, the Z solution was within 1.6 ft/sec and X and Y were within a few tenths of a foot-per-second difference. Rendezvous was very nominal and, as I said before, was one of the most nominal that we've ever conducted.

I backed up the TPI burn by maneuvering to attitude and following the cue card for backup burns. I got all the systems on line, except that I did not turn on the EMS and I did not turn on the DELTA-V thrust normal switches. I let it count down to zero at which time the LM did the burn and nulled out the residuals. I went on into P76 and on into P35. The MINKEY program worked without a flaw during the whole rendezvous. I never had a problem with it. It sequenced automatically and everything worked just as planned. During the first midcourse correction, I had 11 vhf marks and 18 sextant marks. and the solutions from the CMC and the LM had maybe 1½ to 2 ft/sec difference. The LM executed the maneuver on their solution. The TPF phase of the rendezvous was nominal. The LM came to within about a hundred feet of the command module and started stationkeeping. I then did a VERB 49 maneuver so that some SIM bay photography could be accomplished and went back to the docking attitude. After that the docking was nominally completed. There were no problems with any of the docking. The predocking checklist was carried out as listed and I thought that the whole thing went very smoothly.

12.0 LIFT-OFF, RENDEZVOUS, AND DOCKING

SCOTT — The LM lift-off preps down to T-0 were nominal. The ignition occurred automatically. The pitchover of the LM was very smooth. The spacecraft seem to be more stable than we'd seen in the simulator. The oscillations due to the PGNS fuel saving program were somewhat less than I had expected. Everything went smoothly and very slow. We had a great view of the rille as we went across. I thought the ascent was pretty nominal all the way up. Do you remember the numbers?

IRWIN — They were very close to chart values.

SCOTT — I don't remember anything that led us to suspect a problem during the ascent. When we attempted to get lockon with the radar - we had previously been given different numbers for setting the antenna - I pushed the circuit breaker in about 4½ minutes and didn't get a lock. I waited until about 5½ minutes and still had no indication of signal strength on the AGC. I slowed it up, down, left, and right in high for about 5 seconds in each direction. I received no response on the AGC. I don't have an explanation for that.

WORDEN — I confirm that. On board the command module there was no indication of the systems test meter that you'd locked on.

SCOTT — We either had an antenna drift or the numbers they gave us were incorrect. I would suspect the antenna drifted like it did on Apollo 14, which it wasn't supposed to do. Just prior to insertion we received a call to trim the AGS rather than the PGNS. This was somewhat of a surprise, as we had no indication up to that point that there was anything wrong with the PGNS. You closed the interconnects at 500 and that worked as advertised. Everything was nominal with 200 ft/sec to go. After automatic shutdown, we attempted to trim the AGS. I couldn't get the X-axis less than 2 ft/sec, because it kept building. That was not unlike what we'd seen in the simulator on previous rendezvous. We did ask at some point what caused that and I don't ever remember getting an answer. It seem like the AGS continued to build like it was still calculating and still projecting the orbit to the insertion parameter. We terminated trimming AGS at about 2 ft/sec, trimmed Y and Z, and informed the ground. Shortly thereafter they came up with a no-tweak call. It was pretty quick. We

S71-41512

confirmed comm with the command module before lift-off. Ground had their handover and that was something like 2 minutes before lift-off. This was a little later than we'd been used to in the simulations, and I think that was because the mountains blocked the VHF.

WORDEN — Yes. I could hear you sporadically until just prior to liftoff, then you came in loud and clear. I could hear portions of conversations up to that point, but it would keep breaking up.

SCOTT — I think that was due to the mountains. We had PIPA bias to load before lift off, and I had some additional PIPA bias coming into TPI. I guess we must have had a bad PIPA all the way. Let's go ahead through the rendezvous navigation. We did the automatic P20 to the track and the attitude and I got a visual on the command module with the COAS. The radar needles were aligned, the PGNS needles were a little off, and the AGS needles were fairly well aligned. The AGS really had a better vector than the PGNS at that time. So with confirmation of good angular data, we began to update the PGNS and the AGS automatically. The PGNS had one NOUN 49 on the first mark. It was a small one and I incorporated it. Then, we proceeded to take marks automatically right up to the first recycle point. We cleaned up the cockpit as per checklist, and everything seemed nominal at that point. Can you remember anything up to the first recycle, Jim?

IRWIN — Just that the AGS warning light came on and we've already talked about that.

WORDEN — I waited until I got a state vector from the ground, and the first thing I tried to do was get the VHF locked up, but it wouldn't lock up. I guess I reset the VHF range four times before it finally locked up. The first good solid range I received was at 136 miles. That was a closer range than what we normally saw in the simulators. I guess it was because of the orbit that I was in at the time.

SCOTT — They told us - I'm not sure they told you - that we were going to have an off-nominal trajectory. Did you know that?

WORDEN — I recall them saying that the rendezvous was going to be off nominal because of the orbit that I was in. I was expecting something a little bit different. When you called and confirmed the 136 miles, I believed the VHF and started taking marks.

SCOTT — You locked up before we got insertion. On the way up, in spite of the fact that we had no radar, you gave us the VHF range, we checked the PGNS and the AGS, and they all agreed.

WORDEN — That's right.

SCOTT — They all agreed, so we got that confirmation before insertion.

WORDEN — That was just before insertion. It was a little later than we'd seen in simulations.

SCOTT — I kept trying to get the radar locked up. Did you proceed on your tracking schedule as planned, Al?

WORDEN — I did. I received the state vector and went into MIN Key, called up P34, and loaded the TPI time. Then, I let the CSM do an automatic maneuver back to the tracking attitude. I looked into the sextant and I didn't see anything. I looked into the telescope and I didn't see anything, but there was still some sunlight shafting into the telescope. It was still pretty bright out there, so I couldn't see anything. When we finally got into darkness, I'd estimate 12 to 15 degrees from the center of the telescope, I picked up a flash out of the corner of my eye. I manually drove the telescope over to that point and picked you up in the sextant. You came in loud and clear in the sextant. The light was really bright. I asked you about your tracking light about that time and you said that you had it turned on and you could see it flashing. You could tell that it was on.

SCOTT — That is one of the things you can see from the left window. You can see the light flashing on the hand rests. Jim couldn't see it on the right side because there was nothing for it to reflect on., but you can definitely tell the light is working on the left. That means you didn't see us until we got into darkness.

WORDEN — That's right. I didn't see you until you got into darkness. I had two large NOUN 49s and after that everything was right down the line. I had 18 optics marks and 19 VHF marks until TPI.

SCOTT — Let's discuss the recycle first. How many marks did you have when you went through the recycle?

WORDEN — I had seven marks when we did the recycle.

SCOTT — Seven of each?

WORDEN — I had nine VHFs and seven optics. In simulations, I always did the recycle when I got seven optics marks and accept whatever VHF marks I have at the time.

SCOTT — We did a recycle when we had 15 marks. There had been some discussion prior to the flight on the comparative values at the recycle point. We have the ground solution, the PGNS, the AGS, and your solution for TPI. Jim, why don't you give them for reference. I think we expected to see a fair disparity in the Z-axis and we did. This confirmed the preflight data.

IRWIN — Do you want me to read them into the tape?

SCOTT — Yes.

IRWIN — The PGNS final was plus 70.3 plus 5.9 and minus 17.7.

WORDEN — Jim, give the recycle first.

IRWIN — Okay. PGNS on the recycle was plus 70.6, plus 5.9, and minus 16.9. CMC on the recycle was minus 69.4, minus 6.2, and plus 12.0. The AGS solution that I have written down was the solution right at insertion; it was plus 67.5, minus 6.4, and minus 30.4.

SCOTT — Did you have one at that recycle point?

IRWIN — I didn't write it down for that point.

SCOTT — How about that ground solution? What was the first one that they gave us?

IRWIN — The ground solution was plus 66.3, plus 7.8, and minus 31.2.

SCOTT — The reason I wanted to put that in there is we're the only ones that have some of these solutions. So at least the trends from all of these were the same. We knew at preflight that we were going to be within three sigma. It was comforting, because there were some pretty big deltas. The ground had told us that our TPI would be somewhat different from the nominal, even to the extent that we wouldn't have to pitch all the way around. We would not have to do the YAW-ROLL maneuver but we would break radar lock. I think their first cut on their solution was a little off, because their subsequent TPI and ours led us into an almost nominal TPI. We could see this trend coming as we were doing everything nominal, just as planned.

IRWIN — The recycle gave me a warm feeling. We knew beforehand that we would see some differences in the Z-axis, but I think it gave me a good feeling that the X and Y components were almost nominal. I felt we had good solutions going then.

SCOTT — After the recycle, we continued to take automatic updates. Could you see us all the way in from there?

WORDEN — Yes. Once I picked you up and had you in the sextant, I never lost you again.

SCOTT — As soon as you went into darkness, I lost you visually. I had a very small reflected image on the order of a second magnitude star before we went into darkness. As soon as you went into darkness, I lost you. I know your light was on because I saw it again at about 18 miles. All the way into TPI, I had no visual. I was glad we had confirmed the radar and the PGNS before we got into darkness because your light did not give us much. Did you add any manual updates to the AGS, Jim?

IRWIN — Not until the end of TPI.

SCOTT — It was all automatic into TPI. We got down to the final count point, and you gave us a call, Al, on your final comm point; and we proceeded after 26 marks, which was about 9 minutes. If we had one more mark, we would have passed 8 minutes. We came up with everybody's final solution, and that is one I know the ground does not have. Why don't you read those, Jim?

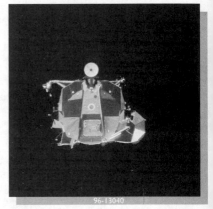

IRWIN — Final solution for TPI. PGNS was plus 70.3, plus 5.9, and minus 17.7. The AGS was plus 70.4, plus 5.9, and minus 19.1. The CMC was minus 69.1, minus 6.1, plus 16.1. Dave, these are negative values for Z, and nominal was plus; so we knew the Z values were still quite a way off from nominal.

SCOTT — Did the ground give us a second solution?

IRWIN — Yes, they changed the Z to a minus 19.0, which is what the AGS came up with.

SCOTT — This pulled everything together, and I think our only difference was in Z. That was much, much less than the acceptable deltas in the burn rules, so we accepted the PGNS solution and passed it to Al. We then proceeded into the nominal procedures for the burn. We made a 3-second automatic burn using the APS. We had the prescribed 10-second ullage, and everything went nominally. We had an overburn of 4.3 ft/sec. We had somewhat of an overburn in Z, too. We did have to trim out X for about 4 ft/sec and a little bit out of Z. You had plus 0.2, plus 0.2, and a minus 0.4 when we finished trimming. At that point, we proceeded to go into, P35 and passed the data down. We started the burn and began taking marks again. Did you have any trouble picking us up after the burn after you had loaded your P76?

WORDEN — No, I did not have any trouble picking you up. We were close enough then so that there was no problem seeing the light in the telescope. You were out of the sextant field of view on the first marks that I took, and so I had to go to the telescope. There was no problem at that point. The range was close enough so that I could see the light without any problem.

SCOTT — Here are the pretrim residuals on TPI, minus 4.6, some small number out of plane which I did not write down, and a minus 4.2. That is pretrim and that put us in a 64.2 by 38.2 orbit. We proceeded into the solution for midcourse 2. We got a few marks to get the tracking going, and I rolled back around to a heads up. Did you go automatic all the way or did you add some manual marks in there?

IRWIN — That was a combination. I went automatic for range, and manual for range rate to make sure I had enough. And the PGNS, of course, went automatic. I guess you got your prescribed number of marks.

WORDEN — The program MCC-1 had nine VHF and 10 optics marks.

SCOTT — I think we had eight in the PGNS, or something like that.

IRWIN — In AGS, we had eight and seven.

SCOTT — Okay, that would add up. At 12 minutes after TPI, we all proceeded and I gave Allen the call. He asked for a call in the PRO and the final comm so we would be synched. We were 3 seconds difference in time break which I think is exactly right because we got a 3-second burn and we came up with the first midcourse solution. Do you want me to read those off?

IRWIN — Yes.

SCOTT — PGNS minus 1.1, 0, and minus 1.1; CSM plus 1.5, minus .2, plus 1.9; and AGS minus 1.5, 0, and minus 3.0. At that point, it looked like the CSM and the AGS were both trending towards a higher midcourse than the PGNS; but since we were on the PGNS and it was apparently running all right, we accepted the PGNS solution and burned it. Now, did we get a PIPA bias update prior to TPI?

IRWIN — Apparently, we did. I wrote it down right here. 14/52 and a 14/56 which looked like we had some sort of PIPA problem still with the PGNS. It was not a problem, but with the bias changing like that, there was something different.

SCOTT — We burned the PGNS solution and pressed on into midcourse 2. I might add that the first time I saw the command module was prior to midcourse 1 at about 18 miles, and I could see the CSM light very dimly. I guess we all proceeded for final comp at the same time for midcourse 2. Do you want to read those numbers, Jim? They were a little bit larger, indicating that we might have had a better solution had we burned the CSM or AGS at midcourse 1. This sort of indicates the PGNS was a little behind on the solution.

IRWIN — Solutions midcourse 2: PGNS minus 0.8, plus 0.6, minus 2.6; CSM plus 2.8, minus 0.3, plus 6.2; AGS minus 1.4, plus 0.3, and minus 4.1.

SCOTT — Al, did you get a previous set of marks between MCC-1 and 2?

WORDEN — Yes, in fact the best set of marks I had probably was between midcourse 1 and midcourse 2. I had 11 VHF and 18 optics marks.

SCOTT — Wow! That's good. I have 10, I think, on the PGNS and Jim had -

IRWIN — Coming in to midcourse 2, I had seven and seven.

SCOTT — Did you do manual there, too? That sort of says that PGNS is in the ball park, and they are all generally in the same direction and have the same trends; but the PGNS is not giving quite as heavy a solution from midcourse 2 as the other two. Nevertheless, we were within the bounds and we burned the PGNS and proceeded on into TPF. I guess you could see us all right after we popped into daylight.

WORDEN — Yes, after MCC 2 and we got back into daylight, I tracked you in the sextant visually the whole way.

SCOTT — Then you must have seen us in daylight prior to midcourse 2?

WORDEN — Yes.

SCOTT — We saw you in the daylight all the way in. The ground told us that we would probably approach somewhat off nominal and that we would be almost horizontal during TPF, which we were. As we approached the braking, we came through the first gate at about 25 ft/sec, and our final solution in PGNS had given us a TPF of 25 ft/sec; so that ought to match very well. There was no braking at the 6000-foot mark. At the 3000-foot mark, I braked down to 20 ft/sec. At 1500 feet, as I was coming back to 10 ft/sec, I noticed I had a visual line of sight rate up and left. The radar needles were not giving me any indication; so I checked to make sure we were on low mode on the needles, and we were. If anything, the vertical needle was displaced just a little bit to the right. I could see by our attitude in the ball that we were coming in out of plane. We had some out-of-plane correction at the beginning of the TPI. What was surprising was that I had to start making corrections up and left in a tight deadband attitude hold to keep the COAS on the command module. The radar needles were not giving me any indication of out-of-plane rates - line-of-sight rates. I guess I don't understand that one right now. I do not know why we are getting that, but to maintain the CM fixed inertially, I gave a fair amount of up and left thrusting as we came into the braking attitude. I came into the final stationkeeping position. When we got to stationkeeping, did you have anything on the braking? I guess that was sort of a nominal thing.

WORDEN — Well, that was all pretty nominal. The only thing I recall was that, starting at about TPF, I went to attitude hold to watch the line-of-sight rates. As far as I was concerned, you had a rate slightly up, with respect to what I was seeing.

SCOTT — Yes, you could see some line-of-sight rates.

WORDEN — In fact, you ended up a little high on me when we finally got into docking, and I had to do a pitch maneuver. I do not recall now how far it was, but it was maybe 5 degrees to get the COAS back on the line of sight. You were a little bit high with respect to my attitude hold at the time.

SCOTT — Your attitude hold and our attitude hold ought to be showing our line-of-sight rates if they are holding right. We ended up out of plane in attitude. We were about 20 degrees off the proper axis attitude when we got on station. You maneuvered to the SIM bay attitude, and the ground called us to take a look at the V/H sensor on the pan camera and to take some pictures. I thought your maneuver worked out very well. We put ourselves in a tight deadband attitude hold and just watched you maneuver around and we ended up looking right at the SIM bay. We took a look, and I could not see anything wrong with your V/H sensor, although I have to admit neither Jim nor I knew exactly what to look for. It was there and wasn't obscured. The next little funny occurred when you maneuvered back to your original attitude, which I assume you did.

WORDEN — No, that is where the confusion existed in that attitude. When we were at the first stationkeeping attitude, I did not check that against the Flight Plan to make sure the gimbal angles were all the same. I did the pitch-around maneuver for SIM bay photography per Flight Plan and just put the numbers in from the Flight Plan. Then when I did the maneuver back, I went back into the Flight Plan attitude and that is where the difference in that position was.

SCOTT — I could not figure that one out because we ended up pointing at you eyeball to eyeball and did a maneuver over and a maneuver back and we were not pointing at you any more.

WORDEN — That is right.

SCOTT — That is because you went back to a different attitude. You started from TPF attitude and you went back to the Flight Plan attitude.

WORDEN — Right.

SCOTT — That was no problem because we just maneuvered around there facing you and then we set ourselves up for the docking by the checklist procedure. We pitched down and yawed left, and it looked to me like that put you in a good position for the docking. I could see out the overhead window, and it all seemed to line up. Why don't you go through the docking part?

WORDEN — Docking, except for one thing, I guess, was completely nominal all the way down the line. I went right through the predocking checklist, and you got lined up. I got lined up on target and closed on you. I think maybe the closing rate was a little bit low. I think it would have been better if I had had a little faster closing rate. I guess maybe I was about 0.1 ft/sec when I came in - maybe even a little less than that. There was no problem with the control. I felt it was pretty smooth all the way in; but when we made first contact, the probe did not slide right in the drogue as I had sort of expected it would, so I thrusted a little bit more after contact before we finally got it all the way. That is a note to make. While docking with a light ascent stage like that, the closing rate really should be a little higher; and it probably would work better if it were a little higher.

SCOTT — I could see out of the top window, and it looked to me like you were coming very slowly. You got the barber poles and pulled us in, and there was no question when we got hard dock. I turned off the mode control switches. You took over the attitude holding and we proceeded to do the power down and transfer stuff. We plotted ourselves on the relative motion plot several times. During the post TPI period, we were somewhat low and forward. We might have expected that because of their call we would end up roughly in a horizontal plane at TPI.

13.0 LUNAR MODULE JETTISON THROUGH TEI

IRWIN — Yes, we did get a call that we could not use the chart solution for TPI.

SCOTT — That is because we were that far off nominal, so we did not even check it. Oh yes, and another thing. When we went to TPI attitude (even though we were still well within radar coverage when we maneuvered to the attitude) at the completion of the maneuver when the spacecraft stopped and went into the attitude hold, the radar broke off just from the impulse of the stop in the attitude hold. That surprised me. It stayed off until we reselected P35 after the TPI maneuver. I was surprised to see it break off just with that little jar.

WORDEN — The post docking sequence went very smoothly. We had no problems getting the stuff transferred back and forth. Right after docking when we were trying to transfer some of this stuff, we were faced with a SIM experiment prep cue card and a lot of SIM bay activity. It really confused things because I was trying to do the SIM bay operation and you were trying to talk to me through the tunnel. Our coordination, I thought, was hampered quite a bit by the fact that the SIM bay was being fired up at the time. That is at least one point in the Flight Plan where maybe we should not be fooling with the SIM bay. It is the same as before PDI when we finally eliminated the SIM bay activity because there were too many other things, going on.

SCOTT — Actually, that is an absolute requirement because when we got docked with you, we were depending on you to take care of all that stuff. Everytime I looked in the tunnel, you were down in the LEB or somewhere doing SIM bay stuff. I kept having to say, "Hey, Al, how about a hand?" I think that really compromised the operations. Even though we had an extra rev to get transferred, we had more gear to transfer and one less man to really help us do it. In the future, you ought to take that period of time and just terminate everything to get all that stuff transferred, because with all those rocks and everything, that is a pretty good job.

WORDEN — I don't think we had any extra time, even though we had more time in the Flight Plan before LM jettison. I don't think we had any extra time.

IRWIN — No, I think it would have been much better if we had had about 3 revs. We could have done it comfortably and checked everything out.

SCOTT — Or if we had had Al free. You would have helped. You were loading in the LM, but I was going up into the tunnel; and everytime I went into the command module, he was down fiddling with the SIM bay.

WORDEN — It was confusing in the command module because you have a probe and a drogue all floating around in the command module, you are trying to transfer equipment back and forth, and you're trying to do a SIM bay operation at the same time, It is just too darn much.

SCOTT — Yes, and we had to get everything configured for the burn, too. The next interesting point was when we got ready to jettison the LM. This is where we ran into a little confusion, so I wrote down that night what we had seen. I would like to do a quick summary of what I wrote down. We ran the hatch check as per the checklist. I was doing some stowage in the command module, and the DELTA-P when we finished with the hatch was about 3.5. We left it for a while and went over to configure for the pressure integrity check. About that time, the ground wanted to know the DELTA-P. I checked the LM/CM DELTA-P, and it was 2. I called and asked them if that wasn't a little low. They said they thought it was a little low and that they wanted more than that. Somehow, we got some oxygen in the tunnel. The first thought Jim had was that the LM dump valve was open and we were dumping oxygen into the tunnel. We checked the ground, and they confirmed that the LM pressure - I really don't know how we got that extra pressure - did not indicate any leak in the tunnel. We went back and checked the seals on both hatches, which we should have done earlier. I think we waited too long to do this because it was a simple thing to do. With two of us in the tunnel, it was easy. We pulled both hatches out, and I ran my hands around both seals. I felt nothing, but had there been something in one of the seals, it could have blown out or drifted out when the hatch was opened. Al, you looked at the command module hatch seal.

WORDEN — We pulled the hatch down into the center couch, and Jim and I both went over the seals on that hatch as carefully as we could. We found only one very, very minute nick in the rubberized seal portion of it.

IRWIN — I think that was a manufacturing bubble.

WORDEN — It wasn't even a nick in the seal, and we could not find anything on the hatch at all.

SCOTT — We put the hatch back in. I might add at this point that we were trying to go very slowly and very carefully, because we knew that everybody on the ground was tired. We were tired, and we wanted to make sure everything was done exactly right. We did not want to blow it at this time. In the process of getting the pressure integrity check on the suits the first time around, we could not get the suit pressure above 6. We had a leak somewhere, and I guess Jim called it. The first idea he had was the LCG connector. Before that, we all checked our helmets and gloves, and everything looked good. The first thought was to put one of those LCG plugs in the suit - the interior, inside plug. Jim undid my suit, reached in, disconnected the LCG, and put the little plug in. You did verify that the LCG was locked.

IRWIN — Yes, it was locked.

SCOTT — You put the plug in and locked it. During this process, everybody took their helmets and gloves off. We figured it was going to take us a while. I guess we thought we were going to have to unsuit. We all suited up again to try the integrity check, and it worked fine. So whether it was somebody's glove or helmet or the LCG I don't really know. We were a little surprised that it might have been the LCG. That was the only thing we could think of at the time that was not firmly attached. Everybody checked their connectors and plugs. I might add that my restraints were pretty dingy at that time. We all had dirt on those things. They were getting a little tough to work. That was prior to the hatch operation. After the hatch operation, we ran another pressure Integrity check on our suits. We had good flow for about 5 seconds - less than I psi. Then it came down to about 6 or 7. That took about 5 seconds. The ground called and said, "You have a good pressure integrity check. Press on." Then the pressure went back up to I psi which violated our onboard 15-seconds requirement. We decided that was not a good pressure integrity check, rechecked all the helmets and gloves, and found one glove unlocked. We locked the glove and ran a fourth pressure integrity check. That one worked just fine. We finally reached the point where everybody was satisfied with the hatch seals and the suit integrity. The jettison went with a bang and worked as advertised. You could see the LM drifting out your window. Did you ever run the SEP burn in the simulator?

WORDEN — Yes, I did the SEP burn in the simulator per the Flight Plan. We did that the last day. If we had been at the right attitudes, it would have worked the same in flight. By the time we did the jettison, we were in a different place, and there was some confusion about which direction to make the SEP burn. We needed to get some words on that, because I was confused then as to which direction to go.

SCOTT — Yes, the thing that was somewhat confusing was that the LM was right straight out the front window, and part of the burn was directly toward the LM. There again, we thought, let's be careful and not blow it here at the last minute. It did not look too good. I guess that is why we got into the confusion factor. Our concern was to be sure that we made a good SEP maneuver.

WORDEN — I recall getting a call from the ground, saying burn the numbers as the SEP pad called. When we were at that attitude and we called up P41 to do that, in body axis, that turns into a burn which was directly at and above the LM. That's when we decided that we had best get this straightened out.

SCOTT — I guess we could have made that one, and you could have made sure that we did not hit by deftly maneuvering around the LM. It just didn't look right. Subsequently the ground figured it all out, and we got a 2 foot per second retro, which was a nice burn. It gave us a warm feeling, and then we all went to bed.

WORDEN — I thought the second call was a little bit confusing, too. The second call we received from the ground said, "We want you to burn retrograde behind the LM. Get behind the LM and burn retrograde trailing." We were way out in front of the LM at the time, and it would really have been a major maneuver to get around behind the LM at that point. So, there was some confusion by the ground as to what our positions were in the orbit at that time.

SCOTT — You could see the LM for quite a while afterwards.

WORDEN — I watched the LM until we got busy doing other things.

SCOTT — The vacuum cleaner worked pretty good I thought. We brought the vacuum cleaner over to the LM and just turned it on and let it run. It did a pretty good job of clearing the dust out. We were pretty dirty.

WORDEN — The vacuum cleaner is a big bulky piece of gear, we were all surprised at how effective it was in flight. It really worked out well.

SCOTT — I thought it did, too. We stowed the CDR and LMP suits in the L-shaped bag, to get the dirt out of the cabin. We left Al's suit out because of the bulk. Al's suit was still clean. We put the filter on the cabin fans and turned the cabin fans on. We already talked about the foreign object in the cabin fan which we heard periodically. When the cabin fan was running with that filter, I thought it did an excellent job of cleaning the cabin. You could sure see the particulate matter floating around there after we finished with the transfer.

WORDEN — When we got up the next morning, the cabin was as clean as it was before the initial separation.

SCOTT — The high gain antenna was working well. We might talk about the next couple of days. I started with the SIM bay operations. We might comment that with the updates we were getting to the Flight Plan, the SIM bay operations kept us very busy for those last few days. There was really no time to sit around and gaze out the window at the scenery. Somebody was always on the SIM bay. I think we were late on a number of items on the SIM bay. That was primarily because we were trying to get other things done. We did not realize what concentrated attention was required to the Flight Plan in order to keep up with SIM bay. We were trying to give Al a break, because he had been hustling for so long with the SIM bay. We finally decided that the best thing to do was to let Al do the SIM bay.

WORDEN — That's right. I'd like to make another comment about the SIM bay. One of the comments I made was the fact that you never know what the configuration of the SIM bay is. I think that's particularly true with the three of us in the CM and all of us operating the SIM bay. We never knew if the booms were in or out, whether the experiments were on or off, and just what was going on in the SIM bay. I think that added to some of the confusion.

SCOTT — That's right. And in retrospect, it seems to me the best plan, with three people in the CM running the SIM bay, is to assign one man to do nothing but SIM bay operations. Let him concentrate 100 percent on SIM bay, and the other two people can do the stowage, cleanup, and fix the meals. With three trying to run the SIM bay, I'm sure we all weren't very well coordinated.

WORDEN — I had some Flight Plan photography to do, and you had decided at the time that you and Jim would run the SIM bay and let me take the pictures. You had to switch that around a little bit to get the rest of the things done. I think that the SIM bay requires one man's complete attention.

SCOTT — When you get to eat and cleanup periods, you need to turn the SIM bay off and forget it. It was forever making us inefficient in our eating and housekeeping. You would get half way through preparation of a bag of food and you would have to do something on the SIM bay, which means you didn't do either thing well. You need to optimize that SIM bay operation such that when you get to an eat period, or an exercise period, or a presleep period, everything is off and you can concentrate on the housekeeping task. You took the lunar photography, we took some lunar photography, and everybody enjoyed the view. You had the general science part well in hand. I thought the view was spectacular. Every time we came around the corner and had another chance to look at the surface, I saw something entirely new and different.

WORDEN — It was interesting too, from my standpoint. I'd been there for quite a while just looking at the surface go by while you were on the surface. I did the plane change at 6 hours before rendezvous, and I never had a chance to look at the ground track from the time I did the plane change until after we all got together in the command module. It was completely new terrain to me, too. We were all sitting there looking at something very new.

SCOTT — The terminator is the most interesting part, by far. You can see so much. It is just spectacular. I saw something at Hadley as we went over that was surprising. It's a continuation of the rille into the mountains. As you looked out, it was quite obvious that Hadley Rille was much longer than we had thought before the flight, from the Orbiter photos and the maps. It goes right into the mountains.

IRWIN — There is a parallel rille there too.

SCOTT — Yes. I think you took some 250s of that.

IRWIN — Yes, the pictures ought to show some of that.

SCOTT — The last two days were not at all slow days. They were pretty fast, with all we had to do.

WORDEN — Everything that I said yesterday applies to when the three of us were in the CM. I think that the Commander and the LMP should get involved a little bit more in the SIM bay operation before flight.

SCOTT — Yes, I think that's probably a good idea. Once things settle down, I think people can put more time on the SIM bay, since it is a very useful operation.

IRWIN — I had one session in the simulator, at the car lot, that was very good. I agree. I could probably have had more time. You probably commented on the temperature increase on the front-side. It seems that it got hotter and hotter. It got up to about 83 degrees.

WORDEN — I didn't make any specific comment on that, Jim, because I had been briefed preflight to expect this. That is something that has been pretty general on all the flights - that the temperatures vary like that. As a matter of fact, on the radiator outlet temperatures, the upper limit was raised to take that into account, so that we wouldn't have to bring the evaporators on. That has been a pretty standard operation in the last few flights, so I did not make a specific comment.

SCOTT — Yes, I think I remember hearing that, too. It was just that Jim and I were sort of surprised at the cabin temperature being that warm.

IRWIN — We were kind of surprised, because there we were sitting on the lunar surface and thermal conditions were warmer there, and it was comfortable on the surface.

SCOTT — The cabin temperature on the surface was very comfortable. Just perfect, I think. Then we got up in the command module and it was a little warm.

WORDEN — There is a difference in the cooling, though, in the heat transfer.

SCOTT — There should not be that much difference.

WORDEN — Because we were using all the radiators, and you were using the evaporator.

SCOTT — TEI. The pads and the updates and everything were timely for TEI. We reviewed all the procedures for the SPS relative to the short in the switch, and the ground gave us one new update, I think, procedurally, as far as the shutdown technique on TEI. We did a single bank burn, the Bank B burn for the shaping burn, and that worked very well. The subsatellite came out as advertised. It appeared to be working very well. I think the movies will pretty well show them how it works. It looked quite stable. The TEI preps went very well, just like the rest of the SPS maneuvers. Attitude was right. I really do not have any comments on TEI, other than the burn status report which we read down to the ground. Looked like a good, straight forward, smooth burn, to me.

WORDEN — Yes, sir. It was very smooth.

SCOTT — The residuals were a little larger than LOI, but it turned out that it was probably a pretty good burn, since we didn't get any midcourses until we got down to midcourse 7. How about the PUGS, Jim?

IRWIN — I had to operate it in MIN to keep it in the green band. I just put it in MIN DECREASE once, and left it there. It stayed stable.

14.0 TRANSEARTH COAST

SCOTT — That got us out of lunar orbit. Then we turned around to take a look at the Moon, and that was one of the nicest views we had the whole trip - knowing that we were on the way home, and getting to see all the terminators from the Moon. We made a number of comments, that we recorded, on what we saw. It was quite obvious that we were going straight up. You could see the results of the burn immediately. There was no question that we had a significant change in, our velocity. That gets us on the return leg.

SCOTT — Systems-wise we had no significant problems on the way back. We got a little leak out of our chlorination port one time. We tightened it up again and it was fine. Everything else worked very well. NAVIGATION - you might comment on your P23s.

WORDEN — We tried to follow the no-comm schedule on the P23s, and there were some periods where we couldn't follow that because of some other things going on. But, I felt that the on - the - job training on the way out was very valuable, because when we started those P23s on the way home, I had a pretty good feeling for what had to be done and how to handle that whole program. Even after the first set of P23s, we had a pretty good feeling about the computation of the onboard state vector because the ground called up and said that they weren't going to update our state vector because our vector was almost as good as theirs. I haven't seen the numbers on the P23s yet, but I think the reason the P23s worked out as well as they did was the fact that I'd done considerable work at MIT on their simulator practicing P23s. That made a great deal of difference to me. I had a much better understanding of which horizon to look for and mark on and of how to maneuver the spacecraft with minimum impulse, which can be kind of tricky.

91-12343

SCOTT — Especially with a light spacecraft.

WORDEN — Especially with a light spacecraft. It is really responsive to even minimum impulse. The system of doing the P23s, the maneuvering that we did, and the procedures for going through the P23s worked even smoother in flight than it ever had in the simulator.

SCOTT — The overall concept of how the state vectors were updated and continued on board worked very well. It was obvious that we kept our onboard state vector comparable to the ground state vector all the time. There was no question that we could have completed the navigation on board and made a very acceptable, if not precise, reentry with an onboard vector all the way.

WORDEN — Yes. I definitely had that feeling.

SCOTT — The PTC worked as advertised, the same as it did on the way out. There wasn't any problem there.

WORDEN — One more comment about the P23s before we leave that area. The Earth was a very thin crescent when we did the P23s on the way back home. We had some discussion preflight about taking marks toward the limb of the crescent on the Earth. You don't want to get out too far on the limb. All the stars that were picked were pretty much in the center of the crescent. I never had any problem locating the horizon working on that part of the crescent or taking those marks which kind of surprised me. It's a lot easier than I thought it would be.

On boom retraction and deployment, we could see both of the booms from window 5. As far as the problem we had with the mass spec retraction, I talked about that.

SCOTT — Consumables seemed to be going along very well. We were well up on everything. We had no midcourse corrections until MCC-7 and that was in RCS. I think the DAP loads worked out very well. Your alignments seemed to work very well. You had no problem with the PTC at any time during alignment.

WORDEN — I was really quite surprised, with all those alignments. There were several periods where we went a considerable amount of time between alignments. When I did a P52 and an alignment, looking into the sextant, it was still very hard to verify the stars in the telescope on the way back home. We'd look into the sextant, and the star would be right in the center of the sextant. That really surprised me that it maintained its alignment that well for that long a period.

SCOTT — We ought to discuss the CSM EVA with some detail. First, on the night after TEI, Al started configuring the cabin, stowage-wise, so that we'd be set up for the EVA. I think he put at least 2 hours into configuring the cabin the night, before.

WORDEN — Yes. There was a lot of detail stuff, like putting things in to the EVA bag, getting the purge valve out, and getting a lot of the little stuff out of the stowage containers. We tied the rock bags up to the sides of the spacecraft, rather than tying them down on top of the lockers. That way we could get in and out. Rearranging the stowage was kind of the detail part of some of the EVA prep that we did the night before.

SCOTT — The point is that when we got into the EVA day - when we got up that morning for the EVA - the cabin was already in good shape. Everything was set up so that we could proceed into the EVA prep according to the checklist with a minimum amount of shuffling. At the outset, I'd like to say that the check list was excellent. The procedures ran very smoothly. I don't think anything was out of order. We had a very complete set of procedures overall. Everything got done according to the book, and it was very good. The only problem was time. We got up that morning and we had a few SIM bay things to do. Al had some P23s to do, and as soon as he finished his P23s, we started into the checklist and the portion called "Cabin Prep for EVA." We started that at 237:30 g.e.t. . It's really called in the Flight Plan to start at 239:30, so we started 2 hours early in the cabin prep for EVA. We went through every step line-by-line to make sure it all got done. It flowed very smoothly with no hitches, and it just took a little time to get everything done. We ended up just about on time for the pressure integrity checks. That means that it took us almost 2 hours longer than preflight planning. We were very happy that we had started early. We were glad that we had Al configure the cabin the night before to take care of the little details. I think it will pay off if you get started early on the EVA, because it really takes a lot of time making sure that you get everything done.

WORDEN — Most things you do on board take a little longer than you would expect them to preflight. That's because you take a little bit more care with what you're doing in flight. You do it much more methodically than you do preflight. That was particularly true of the EVA prep. We went through the checklist very carefully, very methodically, and we never rushed at any time. It flowed very smoothly but a little slower than we anticipated.

SCOTT — Which I think was good in that case, because it was the first time through for that EVA. It was nice to have a comfortable time pad all the way through. We knew we had a good time pad all the way through, so perhaps we were not operating at maximum efficiency relative to time. We were taking our time because we knew we had the pad. Hatch opening occurred about 5 minutes after the planned hatch opening. The integrity checks went very well. The procedures played just exactly as they were laid out in the checklist and just like we've seen them in the chamber. I'm glad we ran those chamber runs because that helped us, Jim and I, to understand what you were doing with your equipment.

WORDEN — I think so, too. We were all well prepared for the EVA. There was a lot of discussion about cracking the side hatch valve to maintain the cabin pressure during the EVA prep. That's particularly true when I was flowing through the umbilical. I thought that operation worked very well. I didn't see any problem at all with opening the side hatch valve just a little bit to relieve the cabin pressure.

SCOTT — Except that it was easier for me to do it than it was for you.

WORDEN — That's right.

IRWIN — My tether was too short. I would have had real difficulty if I had really tried to change the MAG on the sequence camera. As it turned out, I thought I pushed the button. I did push the button on the sequence camera, but apparently it did not drive.

SCOTT — We depressed the cabin, got all the integrity checks, and everything worked fine.

WORDEN — Once we got the side hatch open, from that point to the time we closed the hatch, the whole operation went almost exactly as it had in preflight training, both in the zero-g airplane and the Water Immersion Facility. I don't recall anything during the EVA that I thought was off-nominal. As a matter of fact, it was so much like preflight, that I really had no anxieties about the EVA at all. The whole thing went just as smooth as it could. The mapping camera was in the FULL EXTEND position. That was expected at the time, since we had some trouble retracting the mapping camera up to that point. But we practiced all that, and it was no problem. It was just as we practiced. That's the key to the whole thing - good solid practice before flight. Be well prepared for what's going to be out there and for the kind of body motions that are required to get back into the SIM bay and into the foot restraints. I opened the hatch. After getting the hatch open, the first thing I did was take the TV and the DAC and mount them on the bracket in the hatch. The hatch didn't get fully opened the first time. When I got part way out, I guess you opened the hatch the rest of the way so that the camera was pointing down along the SIM bay. I just went outside the hatch, grabbed the first handrail, and positioned myself just outside the hatch until Jim got in the hatch to observe and to watch the umbilical. I went hand-over-hand down the SIM bay and to the left around the mapping camera. I just floated myself over the mapping camera instead of going around it down into the SIM bay. I put my feet in the foot restraints and just stood there for a minute, resting and looking at the SIM bay, and waiting for Jim to get himself positioned in the hatch.

As far as the cassette operation, the pan camera went just as I had anticipated it would go. I pulled the metallic cover off the pan camera and released it. Then, I pulled the fabric cover off. The force that it took to pull both of those covers off was just as I had expected and remembered from preflight. It was the same operation. I pulled the pin on the pan camera cassette, tethered myself to it, and pulled the release handle. It came out even easier than I had expected. The mass of the pan camera cassette was a little bit more than I had expected, but it was no problem handling it. I just very carefully drifted it back towards the hatch, keeping my hand on the handle and maneuvering myself back. I did release it at one time, because I had to use both hands to maneuver myself over the mapping camera. But I didn't release it clear to the end of the tether. I just let go for a minute, repositioned myself, and then grabbed it with the handle again. I thought that went very smoothly.

The transfer back through the hatch went just as we'd done before, too. I handed the pan camera cassette back in through the hatch. You tethered it and then released my tether. That was pretty much as we'd done before; no problems there.

SCOTT — I put it down in the LEB and it stayed. I left it on the tether and it never got in your way. No problem.

WORDEN — After that, I turned around and went back out in the foot restraints. I don't recall now whether I looked at the mass spectrometer between camera film cassettes or whether I did that before. I think it was between the pan camera cassette and the pulling of the mapping camera cassette when I leaned in and looked at the mass spectrometer to see if I could determine why it had not fully retracted. The Inconel cover on the mass spectrometer was cocked about 30 degrees from the closed position. I reached over and grabbed the cover and moved it a little bit. It's a fairly flimsy cover, but I wanted to see if it was jammed against anything. One corner seemed to be hung up. I released it, but the cover stayed where it was. I really couldn't close it. Then, I looked down inside the mass spec itself and noticed that the guidepins were through the guide slots in the experiment itself, indicating that it had at least positioned itself on the base of the boom itself. I wasn't sure at the time. That's something I hadn't looked at preflight. I wasn't sure just how far those guidepins should come through the slots to indicate that the mass spectrometer was fully retracted. So I called down to the ground and said that the tip of the guidepins were just through the guideslot. They called back and said that it wasn't fully retracted then, because the guidepins should be through the slots far enough so that the cylindrical part of the guidepin could be seen. So, that indicated to me that the mass spectrometer wasn't fully retracted. That was all I could see on it. I couldn't see around the mass spectrometer. I couldn't see down into the SIM bay at that point because the cover was obscuring the view.

I left the mass spec and went back to the mapping camera. Then I pulled the cover off the mapping camera, I noticed that that particular cover was a little more difficult to release than I had anticipated. That particular cover is set under a flange on either side. It's held down by some pins at the release end of the cover. I had to twist it a little bit and pull it a lot harder than I had anticipated to release it from the flanges on the side. But, it did come off all right; there was no problem. The fabric cover underneath got hung up on a corner. The fabric has a rubber slot that it fits into around the edges, and it's almost an airtight seal. That rubber-slotted flange hung up in one corner, and I had to pull it three or four times before I got it released. After that, everything was just as I had anticipated.

I tethered the mapping camera cassette, released it, and it was a very easy operation after that. I brought it back into the hatch, as we had practiced preflight. When I got the mapping camera back to Jim in the hatch and he took it, I asked the ground if there was anything else that needed to be looked at in the SIM bay. There ensued some discussion about looking at the mapping camera to see if we could determine what caused the mapping camera to stay in the EXTEND position. I think the concern at the time was that the laser-altimeter mapping-camera contamination cover was binding against the side - forcing it to stay in the EXTEND position. I went back out and looked, and there was about 3/4ths of an inch to maybe 1 inch clearance between the cover and the mapping camera itself. From that, I concluded that the cover hadn't anything to do with it. I looked underneath the mapping camera, and I looked around all of the edges to see if there was something binding, maybe something that had lodged alongside the mapping camera. Everything looked clean to me. There was nothing that was impinging on the mapping camera at all. The stellar shield was still out, but of course, it would be with the camera extended. At that point, it was maybe 12 to 15 inches away from the SIM bay mold line. So, there was nothing I could tell from there that would shed any light on why the mapping camera did not retract.

After looking at that, I went back in the hatch, pulled the quick release on the TV camera bracket, which we had decided to do preflight. Rather than releasing the handle itself, we pulled the Marmon clamp, releasing the pole. I sent the pole in the hatch, backed into the hatch myself, and pulled the hatch closed. I thought that went very easy. It took hardly any force at all to close the hatch. It operated very smoothly and very freely. I pulled it right down to the point where it was closed. A couple of pumps on the handle, and the latches were over and off. It was a very simple operation.

SCOTT — Your PCV flow did not in any way hinder the hatch closing.

WORDEN — That's right.

SCOTT — No buildup of pressure inside.

WORDEN — I didn't notice anything as far as the hatch was concerned. It was a simple operation.

SCOTT — The pressure equalization valve was open.

WORDEN — That's right. The equalization valve was open.

SCOTT — The repress went nominally, with no problem.

WORDEN — In true zero-g it was really much easier than it had been even in the zero-g airplane. I think there's some rotation that you get in the zero-g airplane that does effect your motions a little bit. True zero-g is just much easier. If you can do it in a zero-g airplane and in the Water Immersion Facility, in flight it is easy.

SCOTT — The next order of business was to make sure the contaminated gloves and articles were cleaned and stowed in a separate bag. We ended up with two sets of EV gloves, one set of IV gloves, a purge valve, the washcloths, and the tethers in the contaminated bag, which we so marked. We doffed the suits and put all three suits in the L-shaped bag by taking out the center couch, stuffing them in there, and then putting the center couch back down on it. It seemed to work pretty well. We got all three of them in.

WORDEN — A comment on the cleanup. One of the things that had to be done, after we got the suits off and we got straightened around, was to stow the cassettes. Before we stowed them, there was some taping that had to be done on the cassettes - taping up the slits and taping up the opening in the mapping camera cassette where the two halves of the shelf joined. The only surprise I got was that the tape wouldn't stick at all to the mapping camera cassette, and I finally had to wrap it almost like a Christmas present to keep the tape on that slot. There's a rubberized coating on the outside of the mapping camera cassette that just wouldn't adhere to the tape.

SCOTT — We must have used at least 100 yards of tape. That gets us down to the rest of the TEI activities. I think the SIM bay operations were pretty much standard by that time, with no unusual things that I can remember. Can you think of anything in the SIM bay that was really unusual on the way back?

WORDEN — No, except that we really seemed to have a lot of SIM bay activity on the way back. That was because of some of the X-ray experiments that were added to the Flight Plan. I think on the SIM bay operation on the way home - with us rotating the operation in the SIM bay - we were still in the position where we didn't always know what was going on with the SIM bay. We didn't always know what the configuration was. At that point, it really wasn't any problem, and the ground was very good about giving us reminders on those things.

SCOTT — I guess that kept us busy. There wasn't time on the way back to sit around and look out the window and reflect about it all, because we were doing something almost all the time. I think the housekeeping kept us up against the wall because every night when we finished an eat period, we had on the order of 20 minutes to start the rest period. During that 20 minutes, we had to get into PTC, clean up the SIM bay, chlorinate the water, sometimes dump, get our hammocks out, give a crew status report, record consumables, cycle the H_2 fans, and often, change the canister. That takes a lot of time. I think 20 minutes isn't enough to do all that. We sort of bit into our rest period every night, and I think there's just no way around it. I don't think it hurt us any. Cleaning the screens was another thing we were doing quite often. Especially post-lunar-orbit activities, when we had all the dirt. We had to get around and clean those screens at least daily. We'd lost, somehow, two of them. One or two?

IRWIN — One.

SCOTT — We took netting off the LCG and put it over the return hose on the suit circuit and wrapped it with tape - made our own little screen. This seemed to work pretty good.

WORDEN — The cabin fan filter is on the output side of the cabin fans. We noticed, on the way home, that the inlet to the cabin fans seemed to be the thing that was collecting all the dust and the dirt. That's like a register in a home; it's just a metal grill. There seemed to be a lot of dust and particles collecting around that, and we could see it on some of the hardware inside that metal grill.

IRWIN — It was an inner grill. There was an inner grill there that seemed to be collecting a lot of dust and debris, and we couldn't get to that. I don't recall seeing anything in the cabin fan filter. It didn't seem to be collecting anything, because everything seemed to be collecting on the return line.

SCOTT — Yes. We looked in the cabin fan filter, and we took it off on the inside.

WORDEN — That's right, and it looked pretty clean to me. It just surprised me that we had a cabin fan filter there.

SCOTT — We did another light-flash experiment on the way home, and I don't think we talked about the one on the way out. That was noticeably different, in that we saw fewer flashes. All of us did, on the way back, when we got closer to the earth. We commented on tape during the flight on what we saw. They should have that data. Jim and I commented that we did see flashes on the surface. I thought the response was a surprised one at that. We could both see them every night.

IRWIN — Yes. I did one solo light-flash experiment during lunar orbit.

SCOTT — We did the lunar eclipse photography and the TV. The ground had a fairly good picture of that for a while, until it got too bright.

IRWIN — I thought the TV interrupted that at an inopportune time. It came right in the middle of lunar eclipse photography, although I think we probably got all the pictures required.

SCOTT — That's right. We had the nighttime requirements, and 250 requirements, and then the TV requirements on top of that. It was a pretty big shuffle there at the end of the press conference to make sure we got it. The priority was to get the photographs and then, if we could, get the TV. But, it ended up I was doing the 250 and Al was doing the night, so I did the TV out my window for a while and then passed it to you, out your window, so we could work it in between the photos.

WORDEN — I don't recall any plans prior to that time to look at the lunar eclipse with the TV. I think that was something that got added at the time.

SCOTT — I think we got the photos all right.

WORDEN — At the time, we really thought it would be nice if we could keep the TV going and take the pictures too. So we re-juggled a little bit, and I think we found that we could get it all done without any trouble.

SCOTT — EATING, RESTING, EXERCISE, and COMFORT. We continued eating and started running out of food, except for bacon squares, which we never managed to run out of. Everybody did a little exercise on the way home. As we got closer to the Earth, we found that the cabin got cooler and cooler. I don't know what that's associated with, but the last night was quite cool.

WORDEN — As a matter of fact, I remember the comment that the cabin was down to 62 degrees on the last night.

SCOTT — Yes. It was pretty chilly.

WORDEN — It was pretty cool.

SCOTT — But, the sleeping bags and coveralls and CWGs were adequate to keep us warm enough. FLIGHT PLAN UPDATES. There were a lot of Flight Plan updates. Then we get ready for ENTRY. The procedures for entry all went very smoothly. We did the final midcourse maneuver number 7. It was an RCS burn, 5 feet per second. It didn't complicate the time line in any manner. We had all the entry stowage done, almost all of it done, the day before entry. On entry day, we found ourselves with an awful lot of time on our hands, which I think was a good idea. Everything was cinched down tightly, and I thought the ground pretty much agreed with where we put everything. I thought they had a fairly good handle on our stowage locations, and we stowed just exactly like we had it on the stowage map. Our comment going into the entry was, "Gee, we've never had this much time in a simulation before." We sat there and coasted along.

WORDEN — All the events flowed very smoothly, and there was plenty of time between events to get set up for the next one.

SCOTT — Yes, that was very comfortable. We got ready to do the next-to-the-last GDC align - GDC to align and ROLL. I remembered that, in lunar orbit when I last aligned the GDC, I had to jiggle the PUSH button to get the thing to align and roll a little bit. I mentioned that to you, and you played with the button for a while and finally got it to realign and roll.

WORDEN — That's right.

SCOTT — I think you've got a bad GDC align pushbutton in the spacecraft. They ought to take a look at that.

WORDEN — I'm glad you brought that up, because there's another thing about the UDC ALIGN, too, that surprised me a little bit, particularly the first time I did it. The CDC align pushbutton has two détentes, and, you can push it to the first détente and nothing happens. You have to push hard on it. In fact, I ended up using my thumb because it took so much pressure to push the button in - to get it clear to the second détente and to get the GDC ALIGN to begin with. Then in addition to that, we had the problem with roll.

SCOTT — The alignments went very well prior to re-entry. Star check was good. All the attitudes worked out as planned. We went through all the systems checks, and I can't remember any anomalies in there. Can you, Jim?

IRWIN — No.

SCOTT — You checked over all the systems per the checklist. You configured the camera. We discussed the fact that the checklist said to take pictures of the chutes, and the Flight Plan said to take pictures of the fireball and the chutes. There was some question in our mind concerning the setting of the camera. I guess you got that squared away. No pre-heat on the RCS. We went into the SEP checklist, and that all worked well.

IRWIN — The CM RCS check, prior to separation.

SCOTT — We went through the CM RCS check, and when Al did the minimum impulse on the hand controller, none of us heard anything. We heard the propellant run through the lines as expected. As I had remembered from Apollo 9, we could hear very positive minimum impulses in the CM RCS. I was very surprised that when Al ran around the stick we didn't hear anything. I thought we might have heard solenoids in some cases. I was very surprised that it wasn't a positive squirt out the CM RCS. We quizzed the ground, and they confirmed they could see the solenoids. Then I realized that they had second thoughts about it and decided that they couldn't really verify we had RCS. We were thinking along the same lines, too. "We ought to do something else here before we go SEPARATE to confirm we've got CM RCS." So, we had transferred SM, transferred back to CM RCS, and the ground had suggested we try ACCELERATION COMMAND - we were about to reach the same conclusion - to get some good rates. You put in an ACCEL COMMAND and checked both rings and got some good solid rates. We could see the flashes then. Then after that, particularly after we separated, we could hear the minimum impulse, and there was no question. We had good, solid burst on minimum impulse.

WORDEN — That's right. I think we were all surprised at the noise level when we first checked those. It was quite different from what we heard in the simulator.

SCOTT — Yes. The simulator is much too loud. Except after we got separated and you were pulsing around the entry attitude, then you could hear them.

WORDEN — That's right.

SCOTT — They were very positive and very sharp.

IRWIN — Again, on that first check, it seemed that we could hear ring 2, but not ring 1.

SCOTT — But only the solenoids, you couldn't hear the firing line through it, later on.

IRWIN — But there was still a difference between the two rings.

SCOTT — Yes. That's one that somebody ought to think about. I'm almost sure that on Apollo 9 we could hear the minimum impulse right away. But, it all worked. We got the separation and that went very clearly, another big bang. All the transfers were automatic. You maneuvered us around to B entry attitude. We just waited for the time of the entry interface, followed the procedures as per the checklist, and everything ran very smoothly. The timing worked out. The earliest check I got on the G&N was about 11 minutes prior to DEI, and it was tracking the 29 seconds we had on the pad for RRT. So, all the times agreed, and all the guidance systems looked like they were in good shape. The next event was 0.05 g.

WORDEN — Before we get to 0.05 g, I have one thing to comment on. We had a dark horizon. One of the things that I was curious about, one of the things I checked very carefully, was to see if there was a horizon out there; and there was, in fact, a horizon. It was very clear. We turned the lights in the spacecraft down a little bit to help. The horizon was very clear. When I first saw it, it was about 7 or 8 minutes out from entry interface. It was about 5 degrees above the 31.5-degree line in the window. As we got closer to entry interface, it was obvious to me that it was progressing right on down to the proper point at entry interface. It was very easy to track, and it was a good indication of attitude.

15.0 ENTRY

SCOTT — The reentry went as planned. The g levels agreed all the way into the 6gs and back out. The G&N was given control after we confirmed that we had a good g time. I compared the G&N and the EMS range to go and they looked close all the way, within about 20 miles. You were watching the scribe and it looked very smooth and nominal all the way in.

WORDEN — Because of the AC problem we had in the circuit breaker, we didn't have any backlighting in the EMS. There was some concern that we might not be able to see the scribe; but, it was very clear the whole time through entry. One other comment about the EMS was that you called .05g and just as you called it, the .05g light came on on the EMS. It was very clear. There was no problem in seeing that as soon as it came on.

SCOTT — All automatic.

WORDEN — All automatic.

SCOTT — That's a good point. I gave a call right after blackout. We ended blackout at 3:37 and I called about 3:45 or so with our delta between the EMS and G&N and everything was in good shape. I got no response. I never heard from the ground on any of the calls. I made probably four calls on the way down. I gave the NOUN 67 values when they first came up and then later on I called them. We heard the Recovery forces radio, but I don't think we ever had two-way contact with anybody until we got in the water. I think they heard us in MCC, but we couldn't hear them. We could hear the Recovery forces and apparently they couldn't hear us.

S71-41999

WORDEN — That was my impression, too. I don't recall any conversation with anyone. There was a terrific amount of radio chatter going on. I don't recall having two-way communications with anyone.

SCOTT — We saw the ionization prior to .05g. You could see it out your window, and you started your camera.

IRWIN — That was about 5 seconds after RRT had started.

SCOTT — Very clear. We mentioned our control modes with G&N automatic all the way in, very smooth and very positive. I didn't feel any oscillations on the way down. The drogues came out automatically at 24K. Right?

WORDEN — Exactly at 24K.

SCOTT — Then, the mains came out exactly at 10K automatically, and we did not push any buttons on the panel; just let them close because it was all automatic. Why don't you talk about what you saw on the main chutes out the window?

S71-43543

WORDEN — The main chutes came out at 10 000; the drogues released just a few seconds before that. The main chutes came out at 10 000 and at about 8000 we got the cabin configured and started the fuel dump.

SCOTT — When you saw the mains come out, you saw three chutes. Is that right?

WORDEN — That's right. I saw the three chutes, come out, reef, then disreefed, and I had three full chutes in view. When we started the fuel dump, my view of the main chutes was obscured by a cloud of fuel that was going by the window.

SCOTT — A big red cloud.

WORDEN — A big red cloud going by the window. When we finished the fuel dump and the view cleared, I could see that one of the chutes was not fully inflated anymore.

S71-43542

SCOTT — I think that was just about the time we got a call from Recovery that we had a streamer. Right?

WORDEN — That was just about the same time.

SCOTT — Yes.

WORDEN — I think they had seen it before and gave us a call. It was just about that time when I picked up the chutes again in view.

SCOTT — We might add that that was a very good call; for them to inform us at that time. That was an indicator that we had less time to get everything done on the way down than we had normally planned; especially after you confirmed it. The dump and the purge went according to plan. Jim was going through the checklist and I had the feeling you were having to really hustle to get things done.

IRWIN — No, not really; but, I had to yell.

WORDEN — I think that was noise; that was the problem. Jim and I were yelling at each other, back and forth across the cabin, because there was so much chatter on the radio.

SCOTT — Yes. That's true. There seemed to be an excessive amount of chatter on the radio. I don't really know why. Maybe we're getting too many recovery items nearby with airplanes and ships and helicopters. It might be getting too crowded out there because there was just a constant chatter going on. They were calling us. They were calling each other, coordinating their efforts.

71-H-1235

WORDEN — I remember one voice in particular that was giving a running commentary of everything that was happening on that final descent. Other people were trying to call us over that voice. I thought that that running commentary was really hurting our operation more than it was helping because we didn't need it. What we really needed to do was talk to the other people who were trying to get hold of us.

16.0 LANDING AND RECOVERY

SCOTT — Yes. What we really need is to have one chopper out there, or one point of contact. One vehicle in the air, everybody else maintaining radio silence so that he can call us and we can call him and can coordinate that way.

WORDEN — I would think they could do it on a different channel.

71-H-1259

SCOTT — Something should be worked out because that was sure a busy time voicewise. I'm not sure the reason they didn't hear us, not because somebody was blocking us all the time, but because we were calling.

WORDEN — We came pretty close to that chopper, which could have been disastrous.

SCOTT — We came down expecting to have a rather solid impact, which we had. I had the feeling we hit pretty flat. There was no apparent roll to the spacecraft at all. I could see water up over the windows after we hit. You all got the main release and the circuit breakers, and we ended up in a very stable I condition with no rocking or anything.

WORDEN — That was surprising that we went straight down and straight back up, and there wasn't any motion at all, hardly, except for the sea swell.

71-H-1355

SCOTT — Went through the postlanding checklist, and stood by for the collar and the swimmers. I thought the cabin atmosphere was just fine. Nobody had a tendency, I think, to get seasick.

WORDEN — One thing we commented on is that, when I first turned the postlanding vent on, Dave got a face full of water.

SCOTT — Yes, but it didn't get me any wetter than I already was because when we started into reentry, all the water in the tunnel came down on me. So I got bathed from top to bottom.

WORDEN — When I turned the postlanding vent on, you got a face full.

SCOTT — That's something you might think about in the future, making sure you mop the tunnel up. There's an item on the checklist that says "check the tunnel for water," but it doesn't say what to do about it. We had a little moisture up in the tunnel, but I was very surprised that so much water came up in that tunnel on the way in. Just like a bucketfull. Wasn't any problem. We gave the swimmer thumbs up, which he relayed. I guess by then they had heard us and that everybody was in good shape. We cleaned up the cabin as per checklist. We powered down, egressed, got picked up, and I thought that all went very smoothly. Just exactly as we had trained in the Gulf. The same Scuba team leader was there, and he did the same thing. The only anomaly there was that the swimmer couldn't get the hatch closed all the way. That left me with a rather uneasy feeling leaving the spacecraft - even though the seas were calm - with an open hatch. That just didn't make me too warm. I don't know why he couldn't get it closed. It looked like the dogs were all the way backed off, and I saw him vent the counterbalance. It was open about 3 inches.

WORDEN — It was open more than that. It was open a good 6 inches at the open end.

SCOTT — I don't know why it didn't get closed. That's something that we ought to make sure that the swimmers are maybe briefed on - malfunction procedures with that hatch. It would be a shame to have it sink. The pickup, I thought, went very well. The chopper operation, in my estimation, was smoother than the one we had in the Gulf. I had a smoother ride up and into the chopper than we did out at the Gulf.

WORDEN — I thought that whole operation went very smoothly.

SCOTT — It did. Everybody felt good. When we got on the chopper, it looked to me like nobody had any trouble changing clothes up there and putting on a new flying suit. I think we all recovered from zero g to one g within 5 or 10 minutes.

WORDEN — I felt better on the chopper than I did when I got back aboard ship.

17.0 TRAINING

SCOTT — Yes. It was more stable. Those ships without their fins do a little bit of rolling. In general, the CMS is an excellent trainer. The people are well qualified; it is high fidelity. Some of the oral cues might be a little off, such as the emission of CM/RCS. The launch is sometimes a little loud in the CMS, because we didn't have any problem with the noise during launch. I think you might comment on your optics in the CMS, compared to the real thing.

WORDEN — Let me go back to crew station first. I think we saw the best stowage in the simulator that I've ever seen, on this particular flight. That must be a result of accumulating the stowage equipment for the EVA training. It really created a hardship on everybody, switching that stowage back and forth from one simulator to the other, to try and do the EVA training at the same time we were doing training in the other simulator. This is particularly appropriate for things like cameras, and some of the pieces of stowed equipment that were needed in the normal training and were also needed in EVA training. I think that some effort should be placed on getting stowage for both of those trainers down there, because they're both used quite a bit. The visual systems in the simulator are okay for pro- cedures, but very inadequate for technique. You can run through the programs, and you can look through the optics, but they really don't react anything like in flight. The optics in flight were so smooth, as compared to the optics in the simulator. Get a star in the sextant, for instance, and you can move it around very slowly and very smoothly in flight. In the simulator, no matter how much it was worked on, the star would be jumpy.

On landmark tracking in flight, the optics tracked very smoothly. Once you get on the target, even if you're at low altitude, you can track it very smoothly through the nadir with very little problem. In the simulator, there was a light to be tracked which simulated the landmark, and that was subject to electrical fluctuation. The position of that light would change with respect to the background film, and I found it much more difficult in the simulator than I did in the flight. However, procedurally, it's okay. The P23s were the same way. In the simulator, the horizon was not like the horizon I saw in flight, although it was very close to it - a lot closer than I expected it to be. I found that P23s in the simulator took much longer to do and required much quicker attention to things like small rays and optics drift than they did in flight. I found it much easier and much quicker to do in flight. Procedurally they're okay.

WORDEN — I thought the software in the CMS was fine and it worked just like in flight.

SCOTT — The LMS crew station fidelity was fine. Within the simulator, the L&A was great, but the model was overly enhanced, or the topographic relief on the model was far more than we experienced in the real situation. That's a function of the enhancement of the photography which brought out more shadows than really existed. In the final analysis, that gave us a problem during the landing, and during the EVA traverses in locating ourselves relative to the features on the surface. There's not much you can do about that when you accept 20-meter resolution photography. But, it was quite different. The projection within the simulator was great. The L&A is a very useful thing, and I might add that at Jim's suggestion, they built a Rover simulator - which isn't included here - utilizing the L&A and the television display, for driving on the lunar surface. We both thought that was a very useful simulation.

71-H-1159

IRWIN — We probably spent just about the right amount of time on it.

SCOTT — A couple of times around each traverse was fine.

IRWIN — It really made us feel at home once we got to the moon.

SCOTT — It made us familiar with the sequence of craters we'd encounter and their names and relative positions. Software worked fine. Have you got anything else on the LMS, Jim?

IRWIN — I'm wondering if we couldn't get a little more usage out of the film strips. We finally got the one film strip leading into the landing, going in the right direction, I think, one time. Concerning the CMS, I would like to have a filmstrip projection out the right windows to use, just to get some practice identifying features. Window 4 or 5.

71-H-1160

SCOTT — The integrated simulations we ran worked fine. We had a few problems here in Houston with dynamics between the two simulations. I don't remember ever having had any particular problem at the Cape, other than occasional computers that would go down. That's to be expected. Simulated Network simulations - with Houston. Those are invaluable. The more we have, the better it is. I think we had the minimum number. I'm sure Mission Control Center, and as far as that goes the crew, could both use more. Those are the peak of the training curve, working with the guys in the Control Center, where we really iron out the problems. Once you get yourself trained in the simulators, so that you can handle nominal and off-nominal situations procedurally, then you can step into the operations with the whole Network and smooth that out. I think we had some rough edges as we went along in both vehicles. I'd say the last SIM in each phase was where we got the kinks ironed out. I think it would be nice if you could have a couple of SIMs before launch for everybody together. I know for instance, on our last descent SIM, you and I had a pretty good day, but when the backup crew got in, they had a lot of trouble that day. They had a number of problems. You sure make a lot of money with those things. The ground needs to do more math-model runs, or everybody ought to get together for

71-H-1163

a few more SIM NET SIMs. I think ours were adequate. I think we had enough DCPS - that's a very good simulator.

WORDEN — Yes.

SCOTT — We had plenty of time on it. It was very useful. A lot of launches in a short period of time with a great variety of malfunctions. I felt very comfortable during the launch. I think you did too.

WORDEN — Yes. The CMPS is good for initial training on the programs. They have more capability now than when we were using the CMPS. In fact, the only thing I trained on the CMPS really was the MINKEY operation. I found that that was good program training and good procedures training. The operation of the CMPS is somewhat different than either the CMS or in flight, but it's good procedural training.

SCOTT — Before we leave the electronics simulators, I'd like to say that the support was great. We had outstanding cooperation with everybody on the simulators. I was very pleased with the training in those things, particularly at the Cape.

WORDEN — I couldn't agree more.

SCOTT — Egress training - we ran the standard pat egress training at the Cape. That was fine. Gulf egress training went very well. We'd done a tank exercise on Apollo 12, so we didn't do that. I don't think it was necessary. Spacecraft fire training we really didn't do, but we did as Backup on Apollo 12. We went to the planetarium once, which I thought was a useful trip - not really required, but it was useful. Simulator training plans. With the overall system of spacecraft and checkout, I think it's unnecessary to go to the depth of malfunctions that we tend to go to in the simulator because we just don't see that many malfunctions, or that type of malfunctions in flight. The people in the training program are getting smarter, the people in the Control Center are getting smarter, and they tend to dig a little deeper in malfunctions and go down to double and triple failures in some cases. I think, for the crew to be able to handle all phases of all malfunctions to that detail is really unnecessary at this stage of development of the hardware. I think, the system has matured enough so that the crews can now concentrate on accomplishing the mission objectives and spend their time on learning how to do that, rather than spend a great amount of time, like we have in the past, on malfunctions. We have to be aware of how to handle malfunctions, primarily the dynamic situations which occur making a major maneuver, or the landings, or situations in which you have to make an immediate correction to stay out of trouble. But as far as going through every malfunction procedure in the book, and experiencing all the various paths of each malfunction procedure, I really think that's unnecessary at this stage of the game. If you understand the Systems Book and the Malfunction Procedures Book, and how to use it, you can work through any malfunction with the logic diagrams presented. I think those (logic diagrams) are very well presented and very easy to understand. In constructing the requirements for training, again, I had the feeling that everybody wanted a little bit more out of the crew in each area. The simulator people are a lot smarter than they used to be. They could see more things that we ought to learn relative to handling spacecraft malfunctions. But if we do that, we compromised our learning process and the mission objectives. My feeling is that we should now concentrate on learning how to accomplish the objectives and assume that malfunctions are going to be about as rare as they are in aircraft. Know how to handle the emergencies, know how to understand malfunctions, discuss them with the ground, and rely on the ground as a monitoring system and to solve the non-time-critical problems for you. Only in that way are you going to be able to spend the time to learn how to accomplish the mission objectives. Maybe some things, like for instance, LOT aborts, I think we've put too much time in on them. They're useful, they're necessary, there's a certain degree of proficiency you have to reach; but if we put in our time on learning all the possible combinations of LOT aborts, and exactly how to do them precisely with finesse, then we just don't have time to learn how to do the orbital geology, or the surface geology, or some of the things which I now think we need to concentrate on, knowing that the spacecraft systems are working fine.

WORDEN — When you get down to the final stages of training and you look ahead and formulate a simulator training plan, I think all of the previous training that the crew has undergone should be taken into account. We ran into some problems where only a part of the past training that we've had was against future training plans. We tried to accomplish some things which we had done here in Houston, which didn't have to be done at the Cape. When plans are formulated, I think that the total training the crew has undergone should be taken into account and not just that training which is done at the Cape.

SCOTT — That's a good point, because a lot of the far-out abort modes you learn one time, you know what the checklist says, you can follow the checklist, and that's all you need to do. If you can do that here at Houston, I see no requirement to do it again at the Cape.

IRWIN — I agree.

SCOTT — Systems briefings. We went through the required systems briefings early in the game and I think they were good. They got us to the point where we could understand the fundamentals of the systems and then utilize it on the simulators.

IRWIN — We're referring to the contractor systems briefings.

SCOTT — Orbital geology training.

WORDEN — I thought that the training that I received on orbital geology was better than I had anticipated. I was very well prepared when we got there. The only comment I'd have is that most of that detailed training we had came very late in the game. It had to be sandwiched in with other things at the Cape and some meetings through the isolation booths on the final stages of the training. It would be helpful if we got into the detailed part of that a little bit earlier in the training cycle.

SCOTT — On the part that we participated in, Jim and I, I thought it was excellent, well presented, and very interesting. Didn't you Jim?

IRWIN — Yes.

SCOTT — Landmark and identification training - landmark tracking.

WORDEN — We got involved in the landmark and the site selection at the beginning, so that was kind of a continuous process all the way through. It certainly was a worthwhile thing to do, to get involved with it that early in the game, because then you are quite familiar with it when you get in the final stages.

SCOTT — SIM bay training. Al, you did most of that. I can only say that the CDR and LMP ought to have more SIM bay training. We just didn't have time for it, but it would have been very useful to have more, particularly on the controls and displays.

WORDEN — SIM bay training was kind of a mutual process between myself and the simulator instructors. We were all learning the SIM bay at the same time. It was a boot strap operation. In the future that portion of it should be cleaned up considerably, since the simulator people are quite well up on it. I found the training that I had was perfectly adequate for what we saw on the flight.

SCOTT — Lunar surface. One-sixth-g and KC-135 - I thought all those sessions were very good and very useful. The level that we had and the detail we had were fine.

IRWIN — The cabin work with the PLSS on might be well to work on.

SCOTT — That might be a good thing to add. You could add that without any trouble - get in the cockpit all suited up one time.

SCOTT — One-g walkthroughs. The rock pile. We've discussed this, particularly on the way back. I don't think we would have traded any one minute of that, particularly the suited operations. That really prepared us for the surface work. There were some suggestions toward the end that we run shirtsleeve. We both decided to run suited up to the end, and I'm glad we did. I think every exercise we had out there in suits was well worthwhile.

IRWIN — The work on the lunar surface was not much different from what we experienced on the rock pile. We didn't sweat as much, but it seemed like the work was about the same.

SCOTT — If we could get LCGs in the training suits, and the training backpacks, we'd have an excellent simulation of the lunar surface, in spite of the fact that you'd have the heavy back packs. That was excellent training. I agree with Jim. The surface operations were not too much different from what we'd experienced on the rock pile. You gain an awful lot by going out there and working on the rock pile back of the simulator building. The addition of the geology stops there at the Cape is good. We didn't have the opportunity to exercise all those rocks they'd put out there for us, but I think the following crews will find it very useful to drive the Rover and go through the procedures of getting off the Rover and doing the geology, the sequence of events with the high gain antenna, the LCRU, and everything. It was very good training. The field trips were excellent. We had one a month. We never had a bad one. We got rained out one time. We got dusted out one time. We were very fortunate with the weather. The people who conducted our field trips were excellent instructors. I felt they were very useful; I wouldn't want any less. I wish we would have had one more good exercise out in the field in the last couple of months.

IRWIN — We got cheated out of one.

SCOTT — We lost one. The Rover. I thought the one-g training in the Rover was good. We added the one-sixth-g deployment operations at the Cape, and I'm glad we did that. I'm glad we got the qual unit down to take a look at it, because it was much higher fidelity and had little pieces and systems not in the one-g trainer. We got to learn a few things from it. The program that was finally evaluated with the Rover was excellent.

IRWIN — We had the right level of training.

SCOTT — Running in the centrifuge at one-sixth g was good. Except for the fact that we were heavier and didn't tend to float off the Rover, I thought that was a fairly good simulation.

IRWIN — They just didn't have the crater densities that we had on the Moon.

SCOTT — SESL. We didn't use the SESL. We used OSS's 11-foot chamber. We did about right on that. We had two runs early in the game and one late in the game. I think that was probably a reasonable approach to the situation. The runs that they planned for us were fine. I wouldn't cut any out, and I wouldn't add any more.

Briefings on the lunar surface. We had the briefings associated with the exercises. The people working out the procedures for the surface did an excellent job. It's a very complex operation and very difficult to put it all together for the first time around. We had a lot of loss of fidelity in training equipment, because it was always behind. We never did get our training gear up to equal the flight gear. I hope that's rectified in the next go around for the next flight. We saw a few new things on the lunar surface - I'll have you know that vice was on backwards. They studied the pictures, by golly, and the vice was on backwards.

IRWIN — The vice was on backwards? I believe they can only go on one way.

SCOTT — That's right, there's only one way they can go on, and it was loaded backwards.

IRWIN — Oh, you mean it was assembled backwards?

SCOTT — Sure was.

IRWIN — You put it on the right way, and it was backwards.

SCOTT — If we could have had that equipment a little earlier, we could probably have learned a lot of those things. Contingency EVA training. KC-135, WIF, and one-g walkthroughs. Our contingency training we did in the WIF. We did it shirtsleeve at the Cape. I think it was adequate. Once you run through it, you find out that it works well, that the procedures are well developed, and you need to do it one time.

WORDEN — I thought the training program for the EVA was just right. There wasn't too much. I thought there was an adequate amount of training. The sessions in the WIF could have been reduced somewhat because the sensation of neutral buoyancy is sufficiently removed from zero-g that with too much training in the WIF, it almost turns out to be negative training. The operation is so much more difficult in the WIF than it is in flight or in the zero-g airplane that fewer sessions in the WIF would have been in order. Maybe one or two sessions in the WIF, instead of a large number of them, would be perfectly adequate. The zero-g airplane was invaluable. The one-g trainer going through the prep and the post was completely adequate. I thought that particular training program was outstanding.

SCOTT — EMU familiarization and chamber training. We could have honestly done with one more EMU session before we went, particularly on the PLSS.

IRWIN — They wanted to brief us on it, we just didn't have time to fit it in.

SCOTT — During the last couple of months, you should have one more run through on that. We have no simulator for the PLSS at all. All we have is the chamber run, so we never really go through the malfunction procedures in any simulator. I would recommend, during the last couple of months, to sit down and go through all the PLSS, particularly the malfunctions, to make sure that the malfunctions included in the cuff checklist are thoroughly understood. The chamber training was good. Mockups and stowage training equipment.

WORDEN — I thought, that was the best stowage I've seen in a simulator at the Cape. My only complaint about it is that it had to be shuffled from one simulator to the other.

SCOTT — On the LM side - the mock-up at the Cape was a little slow coming up to speed. When we left here and went to the Cape, we stepped down as far as high fidelity in the mock-up. I didn't think the one at the Cape was quite up to speed when we got there, but it came around. Other than the lack of flight configuration on the mock-up on the training equipment down there, it went pretty good. Photography and camera training equipment. That was available, and we utilized it. We had a rather slow response on the film, I thought, particularly when we were trying to work out the surface 16-millimeter procedures. With new procedures and trying to get the Rover and everything - I thought the response on getting film developed was very slow, until the last couple of weeks. Then it was very fast, but we could have sure used a little earlier evaluation of some of that.

SLAYTON — Did you have any problem getting the film exposed at the Cape?

SCOTT — No. We just didn't get it developed.

SLAYTON — You didn't have any camera problems at the Cape like you had in flight?

SCOTT — No, as a matter of fact, the camera at the Cape worked fine. We got some great film that we took down at the Cape. The film worked fine at the Cape, and it didn't work in flight at all. I don't know what the answer to that is. Except we had one mag, I guess, that worked on the surface pretty well. The Hasselblads that we used on our field trips in the training at the Cape worked well for the most part. We had some failures during the training, but it sort of prepared us for the failures during flight. Lunar surface experiment training. Our ALSEP training at the Cape went well. We had trouble with the pits in which we drilled holes, but I think they finally got that straightened out. I thought the ALSEP here itself was pretty good. It was good training.

IRWIN — It sure was.

SCOTT — There again I think the suited exercises were the valuable ones.

Lunar landing - LLTV. All I can say is that that's the absolute answer to learning how to land on the Moon. I was very comfortable during the final approach phase. We did it just like we planned to do it - went manual at 400 feet. I felt a little more positive than in the LLTV. The LM was a little more responsive. I put it just the way I wanted to put it, no problem. The reason I was comfortable was because I was comfortable in the LLTV. That machine is excellent. The support out there is superb, and it is an absolute requirement.

The LLTVS is a good simulator for procedures, and I think everybody understands how to use that. The LMS has its role also. I think that the L&A is excellent for giving you visual cues, except for our problem of having too much relief. The system itself is good.

Within the LMS you have a delay in the manual throttle, which at times, I feel, is negative training. They're looking at it. I think that they are going to try and see if they couldn't put some sort of circuit in there to take the delay out.

Apparently, the ACA has some sort of circuit in it to eliminate the delay between the command and response. If they could do that with the manual throttle, I think that would make the LMS much better. The manual throttle practice I got in the EMS was negative training for the manual throttle I had in the LLTV. It's too bad Jim didn't get a chance to fly the LLTV.

IRWIN — I didn't need it.

SCOTT — Planning of training and training program. I thought it went along pretty well. Mike, I think you did a good job. I think you kept up with us.

WORDEN — I think Mike did a superhuman job.

SCOTT — I think the schedule went along pretty well. I don't think we had too many changes. Once Mike got a weekly schedule out, it worked very well. I don't remember having any big glitches. Occasionally we had some little changes, but I thought the training plan went along pretty good.

IRWIN — It worked very good.

18.0 CSM SYSTEMS OPERATIONS

IRWIN — We sure get a lot of confidence in the SM RCS and the DAP as you go along, because it all works so smoothly. The EPS worked very good.

WORDEN — Jim, comment on that gimbal motor transient. When the gimbal motors are turned on, the voltage drops only a half a volt whereas in the simulator, it frequently drops 2 to 3 volts. That was the only difference that I noted.

IRWIN — Okay, ECS. I think we've talked about our problems there relative to the water supply system.

WORDEN — The only thing we didn't cover in that area was the differences in the quantity reading of the potable water as we got closer to entry. We had a blockage in the potable water supply. The potable water tank seemed to be going down at the same time the waste tank was going up, which certainly doesn't correspond to the way the system operates.

IRWIN — I think they called us and told us we had a blockage in the potable water. But they felt like we had enough water to finish the mission, which was like 80 percent with a few hours to go. They did call us and tell us that. So, I guess they've got a handle on whatever the problem was.

WORDEN — Either that, or the sensor wasn't working right.

SCOTT — Waste management - The urine management was a chore, but it worked okay. Having to go into bags and then dump overboard took a little bit of time, but it wasn't any real problem. Those bags seemed to work okay, and the big filter worked okay. I think that's an acceptable mode. It just adds one more housekeeping chore during the day. The canisters worked all right. Telecommunications - I see nothing off-nominal there. Everything seemed to work as per the Flight Plan. Mechanical, tunnel, struts, probe, hatches - all worked very well.

19.0 LUNAR MODULE SYSTEMS OPERATIONS

19.1 PGNS

SCOTT — Lunar module systems operations. We can go right straight down Section 19 here and talk about the PGNS first. I thought the PGNS was fine inertially. The only question I have is that I don't understand why we got the PIPA BIAS updates prior to descent. Optically, the AOT worked very well. No problems. Rendezvous radar worked very well, with no problems except for not getting our lockon during ascent. We did not expect to have the dish drift. We thought that was a unique problem of 14's. Our understanding prior to going on the flight was that it would not drift and we would get a good lockon during ascent. We got new numbers to load prior to the ascent for positioning. I never got a lockon. I attempted to manually slew it 4 seconds up, 4 seconds down, 1+ right, and 4 left and I got no indication of signal strength at all. I don't understand that. Landing radar worked very well. We got altitude velocity lights out right away after we yawed around on the descent and, as far as we know, we got good data all the way. The computer worked very well - no unexplained alarms. We had an uplink and downlink too fast alarm when we initially powered up. I think that was inconsequential. Controls and displays were as advertised. Procedural data was as advertised.

SCOTT — I think all the procedures were excellent. Great guidance system.

19.2 AGS

IRWIN — As for modes of the operation, I guess you checked it out the attitude hold function.

SCOTT — Yes, that was very good.

IRWIN — I did pulse also; it works better in flight than the simulator, I think. We have heard that from previous crews; it was more positive. Initialization went just like it was supposed to. CALS worked great and were within limits every time. Rendezvous radar navigation: we used the automatic updating during the rendezvous, you checked out the needles, and they agreed with the PGNS.

SCOTT — I might add, as far as the overall system there is concerned, that the rendezvous with the new programs in both computers just worked. I couldn't ask for any more. We got into orbit. I had a visual on the command module and pointed a COAS setting. The radar needles were nulled, the PGNS needles were a little off, and the AGS needles, as I remember, were almost null. That had everything lined up, and it was going to work. So we pressed on.

IRWIN — Engine commands, I suppose, were there. Ground can confirm that. The electronics: I suppose there must have been a glitch to give us the AGS warning light. It occurred about at insertion time. For some reason, they told us to trim AGS residuals at insertion rather than PGNS. You commented that we could not get it below 2 ft/sec.

SCOTT — I think we have seen that in the simulator, too. It wasn't a surprise to me because we had seen it in the simulator. I don't think we ever had a good explanation. I remember asking after some of the SIMs, and they said they would check it. We can get into that in the rendezvous part.

IRWIN — Burn programs worked well. Controls and displays were good. As for that AGS warning light, when it came on I immediately reset it. I checked the AGS self test. It looked correct. I immediately got a call from the ground that the AGS looked good to them, so we continued as normal. AGS solutions were very, very close to the PGNS, command module, during the rendezvous.

SCOTT — Controls and displays: I might comment that during descent and ascent both checked - the FDAI on the left side relative to PGNS and the AGS - and they always had good agreement. There was only a slight jump in the attitude. I think the alignments were very good and they held very well.

19.3 PROPULSION SYSTEM

SCOTT — I have no comment on propulsion system descent and ascent. It worked exactly like it was supposed to work. Smooth burns. One question we had before descent was whether we would feel the ullage. I felt it. Didn't you?

IRWIN — Yes.

SCOTT — There is no question that we had ullage when we started descent. The ascent propulsion system was very smooth. It felt as if we came off on a spring when we left the descent stage. It is just a real smooth, quiet ride.

19.4 REACTION CONTROL SYSTEM

SCOTT — Very positive attitude control. Somehow or another, we drifted off attitude during the SIM bay inspection on the command module, or else the command module attitude was different. I am not sure exactly which. We stayed in a tight deadband attitude hold, and the command module did a maneuver to the SIM bay inspection attitude. Then he maneuvered back to his original attitude. We were no longer looking at him. We had about a 90-degree maneuver to do and I am not sure I understand that. I don't understand that one, but as far as RCS is concerned, it worked just fine. No problems.

SCOTT — Translational control was as advertised. Normal. How about electrical, Jim?

IRWIN — On the RCS, you commented we saw some pulses there on ascent feeding.

SCOTT — Yes, the oxygen, the oxidizer manifold readout, was pulsing during ascent feed. I don't know why. Didn't we end up at the terminal phase there of braking with something like 80 percent a side left on RCS?

19.5 ELECTRICAL POWER SYSTEM

IRWIN — As far as the electrical system is concerned, I did not see anything that was off nominal.

**19.6
ENVIRONMENTAL
CONTROL SYSTEM**

SCOTT — Lighting was fine. We talked about the one anomaly we saw in the ECS prior to a separation.

IRWIN — We just used LCG cooling for a short period there before we went on the PLSS.

SCOTT — And it was cool.

IRWIN — After EVA 3, when we knew we had a lot of consumables left, we used it. Water supply: we were short on that because of the 25-pound leak. Let me drop back to the lighting for a moment. We never really needed the utility lights. In fact, we disconnected them and took them out, when on the surface, to get them out of the way.

SCOTT — I could have used them during the rendezvous. I missed them. I missed having mine during the rendezvous.

IRWIN — You should have mentioned it. I would have put it on.

SCOTT — It was not really necessary. We were going along pretty well. I could have gone back and gotten it. I still think they are useful. I got used to using mine.

IRWIN — The water glycol was nominal. That's all on ECS.

**19.7
TELECOMMUNICATIONS**

IRWIN — We both noted the noise associated with the yaw drive on the antenna. At one paint, we thought the command module was firing jets.

SCOTT — Yes we surely did. It sounded just like RCS thruster activity.

IRWIN — Really noisy. It seemed as if it almost smoothed out toward the later part of the flight. We didn't hear it. I don't know whether it actually changed or we got used to it. VHF worked well. EVA antenna was okay.

SCOTT — A comment on the VHF with the command module: we had very broken communications until the command module came up over the mountains. When the CM normal line of sight was obscured by the mountains, we did not hear him until he came over the mountains. It was no real problem, but something to be aware of. That gets us through the LM systems. When everything works that well, there is not much to talk about.

**20.0 LRV
OPERATIONS**

SCOTT — I think the manual deployment system is excellent. All the cues are good and I had a good understanding of how it works. It looks very reliable. The procedures are good. The Rover setup worked all right. Mounting and dismounting was an interesting operation. I found the best way for me to get into the Rover was to sort of back into it, get myself positioned relative to the seat, and then give a little hop while holding on to the low gain antenna mounting staff, and sort of pull myself as I went up into the air and back into the Rover. It didn't take a very big hop, just enough to get off the ground. Then I'd swing my feet over the footrest, grab the seatbelt, pull it across, and attempt to get it attached. I had a fair amount of trouble getting the seatbelt hooked onto the rail, not because it wouldn't hook, but because it was hard to find the rail in a relative position. As a recommendation, I think we ought to have a bar-type affair like in the carnivals - a little kiddie bar - which we suggested prior to the flight, but I guess there was a weight problem. In retrospect, the weight penalty would far exceed the problems associated with the seatbelt. I think a bar which configures to the suit, which can be moved forward against the console when you want to get out and which just folds back and is locked into détente once you get in the Rover, would save considerable time and effort.

IRWIN — I used the same technique that you just mentioned. I grabbed the staff of the sequence camera, gave a little hop, and tried to slide back. Maybe it's because my legs aren't quite as long, but it seemed as if I never did get back far enough in the seat when I used that technique.

SCOTT — Yes, I noticed that. When I was standing to the side of you, you seemed to lean back too far. Maybe that's because your legs were as far as they could be, and that footrest needed to be pulled back further so you could get your back straight up. When you contacted the seat, you were leaning back too far, and the PLSS was at an angle to the backrest on the seat.

IRWIN — Yes. Initially, we had the tool installed on the right side, and I was hanging up on that, both getting on and getting off. I took that off and it seemed to work a little better, I never did get really comfortable at all in getting on and off. Of course, my seatbelt was let out all the way, and I still could not collapse the suit sufficiently to lock myself in. Dave had to come over every time and get my seatbelt. In getting off, I could strain against the seatbelt and collapse the suit sufficiently to release myself, but I just could not lock myself in, which added several minutes to each stop. It was unfortunate. I heartily endorse the suggestion Dave has for a bar there rather than a belt.

SCOTT — As a matter of fact, that was your idea about 6 months ago.

IRWIN — Yes.

SCOTT — Okay. Vehicle characteristics. Unfortunately we didn't get any 16-millimeter of the vehicle, but I'll go through a little discussion of the driving. In general, the hand controller works very well. During the first EVA, we had front steering which was inoperative, and the front wheels were apparently locked in the center position. This resulted in some difficulty in steering in that a sharp turn would cause the front wheels to dig and the rear wheels to break out. It was difficult maneuvering the vehicle and we did lose some driving speed, because I couldn't turn it sharp without having the rear wheels break out. I might discuss what we did there to correct that problem. At the beginning of the second EVA, the ground requested that we cycle the steering switch again, which I had done during the first EVA a number of times. Upon cycling it, the steering worked fine. I have no explanation for the difference between EVA numbers 1 and 2, other than the fact that Boeing must have sent somebody in there during the night and fixed it. The double Ackerman steering was too sensitive for the higher speeds - 10 to 12 kilometers per hour. I think that the max speed we got was like 13. The double Ackerman was too sensitive. Even with the seatbelts fastened, there's a lot of feedback into the hand controller. I think with a more secure attachment to the vehicle, such as a bar that keeps the man firmly attached to the seat, there would be less feedback into the hand controller. Then the double Ackerman would be not quite so overly sensitive. I did attempt to turn off the rear steering with the switch, because I felt that the front steering alone would probably be optimum. I still do. Unfortunately, when I turned off the steering, the rear wheels wouldn't center. They would drift to one side or the other and we'd be in a crab. We could have disengaged the rear steering, but I decided not to fool with it and try something new. I'd leave all the steering as it was and accept double Ackerman, because that was adequate.

The driving was quite easy when we were on flat terrain which didn't have too much in the way of obstacles. When we ran across the crater fields which had a higher density of small craters, anywhere from a meter to 2 meters across. (All the craters had very low rims so they weren't too prominent until you got right up on them.) But these smaller craters presented some pretty tricky obstacles at higher rates. So, we had to slow down. I didn't feel comfortable driving through some of the craters greater than 6 to 7 kilometers an hour. I'm sure the Rover would have handled it. We bottomed out on the suspension maybe three or four times during the whole trip, and it seemed to hold very well. But I just didn't feel comfortable bottoming out the suspension, and I tried to avoid all the craters which would produce that kind of response. In doing that, it took a fair amount of time and quite an attention to the traverse in the direction which we were going. I think the wire wheels worked very well relative to traction. The only wheel slippage that we noted occurred in hard turns, at high rates where the momentum of the vehicle would keep it going straight until the speed slowed enough for the wheels to catch. One time we had the wheels spinning in the soil; they were digging in in opposition.

IRWIN — When we got to the ALSEP site.

SCOTT — Yes. As I remember, we picked it up and moved it to another spot and it worked fine. Did we just pull it out?

IRWIN — We just went in reverse.

SCOTT — That's right. I think the wire wheels are excellent. The bearing of the wheels on the surface must be very light, because the vehicle took us up the slope of Hadley Delta at 10 kilometers an hour. I would guess we were 20 degrees on that slope, wouldn't you?

IRWIN — Yes. I think it was probably 20 degrees.

SCOTT — It went right up there without any trouble at all. When we got off the vehicle, we noted our boots sank in the soft soil a half an inch or so, maybe more. The Rover tracks dust made a very slight surface disturbance.

IRWIN — I'd estimate we sank (boots) in maybe 3 inches.

SCOTT — It was really deep there, wasn't it? The wire wheels are excellent. They picked up very little dust. We did have an accumulation coming up under the fenders. I think the fenders are well designed and quite adequate. It seems to keep the dust off pretty well. You had a chance to see if there was a rooster tail behind the Rover when I drove. Did you see much?

IRWIN — One time I did comment on the rooster tail. I guess it was on the Grand Prix.

SCOTT — How much was it?

IRWIN — It kicked up, I'd estimate, 15 feet in the air. We had one over your head and it impacted in front of you.

SCOTT — Did it really?

IRWIN — Yes.

SCOTT — I didn't notice it looking forward.

IRWIN — It was really impressive. It's too bad that sequence camera didn't operate.

SCOTT — I didn't notice, when we were driving at the higher rates, any dust or dirt coming forward into our view.

IRWIN — I think at that particular time, you were just doing a max acceleration, and that's when it kicked up the rooster tail.

SCOTT — Auto max acceleration. I don't remember at any time feeling a particular wheel slippage. I think the vehicle accelerates very well, probably as we expected and very similar to the centrifuge runs we had under one-sixth g simulation. The breaking is more responsive than I expected. When I did that little Grand Prix exercise and put the brakes full on, it came to a stop I would say comparable to the one-g trainer. I expected to slide more. The braking was excellent. You have to be quite careful at high velocities that you don't turn too quickly, because the rear end will just break out immediately and you'll go sideways. However, in our testing of side-vehicle motion, there was no tendency for the vehicle to tip over. I thought it was very stable in side slippage, didn't you?

IRWIN — I guess there was only one time when I had some reservations; I thought we might flip over. Seemed like we went up to about 30 degrees on a roll.

SCOTT — I don't think it was that much, but maybe so.

IRWIN — It felt like it.

SCOTT — You were on the downhill side.

IRWIN — That's true.

SCOTT — My feeling was that it's a very stable vehicle; the CG is very low. No tendency to turn over. I think you have to slow down when you need to make a sharp turn. If you just pay attention to the surface in front of you, you can control things and make quite good time. I haven't seen the average Rover speed yet, but I was quite pleased with the velocity we could make traveling across the surface. It was more than we expected. I think our last numbers before flight on the LRV trainer were 9.3 kilometers an hour, and as we mentioned, we got up to 13. We could maintain, over most of the terrain, a steady pace of 10 or 11. The 13 was with the throttle full throttle, the 10 or 11 was backing off somewhat. It was easy to position the throttle in one position, or to position it at some throttle setting to keep a constant speed, and steer merely by putting small inputs left or right until you got the turn you needed, and then releasing it for recentering the steering. I had no problem with fatigue in my arm. I had no problem seeing forward because of the suit. It handled quite well, as far as controllability in the suit. It was far more difficult to come down the slope than it was to go up. We approached the down grade very cautiously so it wouldn't get out of hand. On one slope, once we got off the Rover, it had a tendency to slide down the hill sideways. So, we took turns holding it, so it wouldn't depart. It rests so lightly on the ground. Once we got on it though, it was quite stable.

IRWIN — It really seemed to require both of us on it for stability. You wouldn't let me walk down. You wanted me to get on.

SCOTT — Yes. I felt much more secure if we were both on there to keep it firmly on the ground. It's just very, very light when it's by itself.

IRWIN — One time we did a 180; the back wheels just broke loose and they slid around.

SCOTT — Coming down the hill?

IRWIN — Yes.

SCOTT — I think that was just because of the slope. We probably had most of the weight on the front wheels, and I had to make a turn to avoid a crater. There just wasn't much traction on the rear wheels. It was just a matter of going slow when you had obstacles, and catching up on your rate when you had a smooth field in front of you. I couldn't ask any more in controllability of the vehicle. It's just superb. I have no recommendations on any changes In the control system. The reverse switch works fine. The techniques we were taught in how to go into reverse and how to go into forward worked fine. We did use reverse several times. Aside from the fact that you can't see behind you - when you can tell me which way was clear behind me - I was very comfortable backing up. Like the time at the ALSEP site - there was no problem.

IRWIN — We eliminated the Velcro on the seat. I wonder, in retrospect, whether it might have been a good idea to have that there. Maybe give you a little more support, particularly if you went to a bar arrangement rather than a belt. You might want to reconsider the Velcro on that.

SCOTT — Yes. I think it would be a function of how hard it would be to disengage the Velcro getting off. Also, when you get back on, we sort of shuffled around there. We wiggled into position. With that Velcro back there, you couldn't do it. I'd say you're better off without it. You should get a more secure bar-type arrangement to keep you in the seat. I'm afraid with Velcro you'd get stuck to the back in a position which was uncomfortable, or difficult to utilize the throttle. Many times I had to shuffle my position so I could reach the throttle and to be comfortable. I'd get in, and I'd feel like I was in the right spot, I'd reach over and I really didn't feel comfortable on the hand control. I'd vote for no Velcro myself.

IRWIN — As far as coming downslope, I felt more comfortable just holding on rather than being secured by the safety belt. In case there was any chance of going over, I think it's better that one man be unstrapped so he could help turn it over.

SCOTT — I don't think it would turn over. I think it would have more tendency to turn over if you fell off. If it started to go and you came off of it, I think it would have more tendency to go over because the CG would be all on one side. If both people stay on and if you drive it reasonably, I don't think there's any tendency at all for the thing to turn over.

IRWIN — If it did flip over with both guys strapped in, I don't know whether they could release themselves.

SCOTT — With the seatbelts, yes. But if you had those bars on there, you could release yourself. Maybe there's an honest difference of opinion here. I didn't ever feel like we would turn over. And if we had turned over, I didn't feel like we were going to get pinned.

IRWIN — I think you probably have a better feel for it since you were driving it. You could feel, it's like flying an airplane. You know what it's doing all the time.

SCOTT — Yes. That's right. That's probably the same kind of thing.

IRWIN — I had excellent visibility. I could turn my head and I could look back from 8 o'clock around to the 4 o'clock position. One time you commented that we were getting some reflection from the mirror on to the TV.

SCOTT — Yes. We agreed before the flight to point the TV aft and down during the driving. At one point there it picked up the Sun. The mirror on the TV picked up the Sun and put it right in my eyes. That was pretty bright, but no real problem. During EVA-2, you put your visor partially down, the hard, opaque, outer visor. It helped you, and you suggested I do that. I put mine down, and it really helped, particularly driving up-Sun. You can drive right straight into the Sun with that visor down. You probably have better ground visibility than when you're going cross-Sun. But with the visor up, it's pretty tough going driving into the Sun.

IRWIN — After EVA-1, I had a headache because of the glare. On the second EVA, I pulled the glare shield down to protect my eyes and I felt good from then on.

SCOTT — Yes. That was a good suggestion. Once I got the suit attached to the Rover I felt pretty secure. Suspension was excellent. I'd be interested to know more about bottoming out of the suspension. Maybe that's no problem. I felt uncomfortable when we did bottom the suspension. We went over one rock, must have been a 1-foot boulder, one wheel went right smack over it. I was trying to avoid a crater. I missed the crater and picked off the rock, and it bottomed out as we went over it, but I didn't feel anything else.

IRWIN — It was an angular, very angular rock as I recall. Yes. I thought sure we'd tear up the chevrons.

SCOTT — Did not hurt a thing apparently.

IRWIN — I never got a chance to check the chevrons for any damage. Did you?

SCOTT — No. But, even though the wheel bottomed out, I think the chassis stayed pretty level. It was very similar to the one sixth g operation that we ran here in the centrifuge, as far as the bed or the chassis itself. It seemed to remain fairly stable, whereas the independent suspension on each wheel was doing all the work. Lower damping than we had here. We were bouncing more because we were lighter (compared to the centrifuge). Systems operations. I thought the nav system worked extremely well. I don't think we had to make an update the whole time. Yes, we did one time. Was it at the rille?

IRWIN — Yes. There was one on EVA-1 when we were down at St. George. I thought there was one, also, when we were up on Hadley Delta on EVA-2.

SCOTT — Maybe you're right. Well, that's in the data.

IRWIN — Yes.

SCOTT — But, it was no problem. The technique is simple, straight forward. There was one, the ground planned it ahead. They said, "Hey park it down-Sun, give us your reading. When you get back on, we'll give you the update." That was good thinking on their part. The only comment I have about the nav system is: On both trips to the front, I felt that it was pointing us too far to the right of the LM. The bearing was such that it would have taken us to the east of the LM. I commented on that, I think, on EVA-3. As we were going back to the LM, I was going to bias myself a little bit to the left of the LM. But, it turned out that the bearing on EVA-3 was right straight to home plate. When we came up over that rise, by golly, it was pointing exactly to the LM. I think that nav system is just excellent. It gave us a good reference as to where we were. We used that quite a bit, our bearing and distance from the LM. That was a great help in positioning ourselves. I didn't feel like we were ever in any question about our location relative to the LM. Did you?

IRWIN — No. It was a good aid for the ground also to track us, where we were, and assist us in some cases to expect certain craters coming up.

SCOTT — Yes. The batteries. I guess the ground has better data than we do on that. I noticed the amperage readout was always lower than expected. It was never working as hard as people thought it would.

IRWIN — Yes. And the amp-hours sure stayed high too, all the way through. I had some difficulty reading the number 2 readout on the battery. I had to really strain to see that from my position because I was riding so high.

SCOTT — I never tried reading the gauges on your side because you always did that. TV and TCU. We had to help the TV camera there several times when it got hung up pointing up and pointing down. But other than that, I thought it worked quite well. The ground thinks it works extremely well. No problem, if we could square away the antenna pointing device. There's no problem selecting the modes and turning the TV on.

IRWIN — You might want to comment on the cable you had to secure with the tape. Maybe that should be a design fix?

SCOTT — Yes. I think they could put a clip somewhere. The clip could be right down there on the high gain antenna where we put the tape. We put the tape in a position which we had discussed prior to the flight, anyway. We talked about putting the high gain antenna cable around the little loop down there where the shaft comes out when you unstow it. That's where we taped it. Once we got that taped, the problems associated with hanging up on the cables cleared up. The LCRU battery was fine. Electrical-mechanical connections all worked well. Dust generated by the wheels - we'd have to say that the dust was minimum. We did have to dust off the mirrors quite a bit, but it was far less than I expected to see.

IRWIN — Yes. I don't know whether all that dust was created by the wheels. It could have been the dust created by us just getting on and off because we kicked a lot of dust, you know.

SCOTT — Yes, that's right. I really didn't see much dust going forward from the wheels. I could see it hitting the fenders, and it seemed like the fenders did very well. I really didn't see anything going forward.

IRWIN — No. That's why I had difficulty accounting for the dust that was on the mirrors.

SCOTT — Yes. Except it could have been very fine over a long period of time that we couldn't see. The dust accumulation was minimum. It was fine dust. The little decal with the procedures, just forward of the hand controller, was almost completely covered most of the time. If I used it, I had to brush it off.

IRWIN — There was one mirror that was broken on the TV camera. Cracked. One of the small squares was cracked. I don't know when that occurred.

SCOTT — It wasn't there when we started. I don't remember seeing that when I put the camera on. Okay. Payload stowage, I think we have plenty. We never did get everything filled up. I think if you want to go out and make a survey of large rocks and put them under the seat pan, that would work just fine. Because of dust accumulation, I'd recommend those seat bags have a cover on them. Beta bags underneath the seat pan, some firmer cover, because my seat bag got full of dust. I'm glad we had the flaps that stowed over the film mags and the 500-mm, otherwise it would have just been thoroughly dust covered. I'm afraid we would have run into the same trouble as you did with your camera, with all that dust in there. Because almost every time I got under the seat pan, there was almost a solid layer of dust over it.

IRWIN — Yes. And there was plenty of room in my seat pan to stow a lot of rocks. On EVA-3 we picked up a lot of rocks that we just left on the floor pan. We did not put them under the seat. We drove back with them in that position.

SCOTT — Did we lose any?

IRWIN — No, we did not lose any.

SCOTT — That's interesting.

IRWIN — Your seatbelt was hanging up on that screw.

SCOTT — That's right. It was a Cannon plug down beneath the console on my side, that we had never seen before. I guess the one-g trainer just doesn't have it. My seatbelt kept hanging up on that. I'd get on and reach for the seatbelt, it'd hang up there and I'd have to get all the way back off and disconnect it and get back on again. There again, let's eliminate the seatbelt and put a bar in. That will solve that problem.

IRWIN — On the initial deployment of the seats on the Rover, I was surprised that it was so hard to disengage the Velcro.

SCOTT — Yes, it sure was.

IRWIN — It took a couple of extra minutes to pull that loose.

SCOTT — Yes. And the Velcro on the seat bag under my side, there's far too much Velcro for those flaps. It was there in order to hold the seatbelt in during launch. If we eliminate the seatbelt, we can eliminate all that Velcro. In general, the Rover provided us the capability to go places we never would have been able to go on foot. It was an excellent device, with the exception of the recommendations on the seatbelts. I can't think of anything that could be improved. Do you?

IRWIN — Yes. On the map case. That wasn't really an optimum position, because I was sitting up so high that I really had to strain to get to the maps. For that reason, I always just stuck with one map - the 1-to-25 000 scale. So I just had one map to use, always just held it in one hand. I think some improvements could be made in that area.

SCOTT — That's a good point.

21.0 EMU SYSTEMS   SCOTT — PGA FIT AND OPERATIONS. I ended up with a compromise solution on my arm length and my gloves. I had requested, just prior to the flight, for the people to shorten the arms so I could have mobility close to my chest, where I had to do most of the work. If the arms were too long and the fingers were extended at that point, I got hand cramps trying to work the gloves. If the arms were shortened, when my arm was outstretched my fingers were pushing against the inside tips of the gloves. My feeling before the flight was that I'd rather have the tight arms than the cramps in the hands. It resulted in too much pressure on my fingertips, but I'll accept that compromise because it enabled me to continue working without any hand cramps. I never got any hand cramps at all throughout the whole operation. I felt like I had good mobility in cinching up the geology sample bags and in doing all that ALSEP operation. Driving the Rover was also quite comfortable, except for my fingertips. Other than that, I thought the PGA was excellent.

IRWIN — I think I had the same fit that you did on EVA-1, certainly. At the end of the EVA, my fingers were really sore - the fingernails and the end of the fingers. After that, I cut my fingernails back to the quick, just as far as I possibly could with the scissors; and then on EVA-2, my fingers didn't bother me at all. That solved it for me. I didn't have any cramps either.

SCOTT — We had both experienced cramps in training. As a matter of fact, I think when we first started at the Cape, 3 or 4 hours after we'd gone out on the rockpile, we were sitting in debriefing at the CMS and both of us cramped up. You become immobile when your hands get cramped, there's nothing you can do.

IRWIN — They say that is due to a loss of potassium from profuse sweating. I don't think I was sweating at all, except that during EVA-3, I was a little warm. But the fact that we were not sweating, I think, probably helped us.

SCOTT — That's a good thought. We've discussed the suiting and unsuiting. BIOMED INSTRUMENTATION. We've discussed that. I felt LCG was excellent. I think the MIN, INTERMEDIATE, and MAX COOLING positions were just right. I used all three periodically. I used MAX COOLING, particularly after we had sublimator startup on the AUX tank. Because usually, when we got to the AUX tank and went to MIN COOLING, by the time we got the sublimator going, I'd gotten fairly warm. I went to MAX COOLING and that cooled me right away. I think that LCG is going to be one of the significant milestones in the program. That is just really great.

IRWIN — Yes. It's great. I never did use MAX COOLING.

SCOTT — Never did?

IRWIN — Never did. INTERMEDIATE was the most I ever used.

SCOTT — You didn't have to drill.

IRWIN — That's right.

SCOTT — Let's see, HELMET. No comment. Just fine. The visors are all good. They were all useful and it worked. LEVA operation was good. GLOVES, we discussed those. I surely think that a better glove could be made which fits tighter. I think the gloves, in my case, are still too bulky, and there is too much easement inside the glove. I think for an EVA operation you need to have a glove which has a smaller easement than for an IV operation when you don't plan to pressurize. When you plan to run pressurized all the time, as you do an EVA glove, I think they should be designed and built for that operation alone, and not try to compromise by having it comfortable in an IV situation.

IRWIN — I think the wrist ring got to me more than the fingertips. It cut into my right arm across here. Maybe that was a function of my operations. It might have aggravated that. It probably would have paid off in this situation to use a wrist glove, at least on the right arm. Particularly after the first EVA, I noticed it was starting to go raw.

SCOTT — UCTA OPERATION. No comment. Jim, would you like to comment on that?

IRWIN — Well, I used mine quite frequently. Unfortunately, on the first EVA I must not have had a good connection, because it all leaked out in the suit. The LM system took care of any urine smell. I never smelled it once we got back in the LM.

SCOTT — I didn't either. And I guess your suit was dry the next day, wasn't it?

IRWIN — Yes.

SCOTT — Yes. I think that suit drying operation is a good one. The EMU MAINTENANCE KIT. The new antifog application works great. It was very simple and straightforward. Never had any fogging at all. I guess we wouldn't expect any without it, because the flow is pretty good.

IRWIN — We used the lubricant out of the bag. We never used the replacement seals or the rings, never had to.

SCOTT — DRINK BAG. We talked about that. The ANTIFOG. PLSS PGA OPERATIONS. Everything connected and disconnected all right, except when we got the dust and dirt. Then, sometimes, it would stick, but in general, I thought it worked great.

IRWIN — Well, your comment, you know. Yours was riding kind of loose at the end of EVA-3. I thought mine was riding higher than it should. Perhaps because of the adjustment of the straps. Also, the connection of the ECU to the PLSS, particularly on yours, was galled on the surface. It was galled, which made it rather difficult to secure.

SCOTT — The electrical connections? That's right. You mentioned that right away. The first time you put it on you said it was all galled.

IRWIN — I'm surprised that it would go through inspection.

SCOTT — Pressurization and ventilation were excellent. Liquid cooling and circulation was excellent. Communications were superb, except for your antenna. There was another one that I can't believe ever got launched that way. That was about as gross a mess as I think I've ever seen. Connectors and controls, I thought, all worked very well. I think it was a good idea that they put that plastic plate over the flags in the RCU, because that sure got dirty.

IRWIN — I had some difficulty seeing my flags with the visor down.

SCOTT — I did too. I found that it was the dust accumulation.

IRWIN — I had to actually strain against putting my nose against the visor to look down and see the flags. I guess, also, I felt that when I was getting out of the LM when it was in the shade, I preferred to have the visor up so I could see better. Then I put the visor down after I got out.

SCOTT — I did the same thing. As a matter of fact, with the visor down in the shade, you couldn't see at all. RCU. I think the new RCU attachment to the PLSS straps bracket is a good secure one. The RCU won't come off, but it's sort of hard to get on. I don't know what you can do about it.

IRWIN — I thought we got to the point where we were getting on pretty well.

SCOTT — I always had to struggle. We got it on pretty well, relatively. There's not much you can do about that. It's certainly better the way it is now than it was before, because it won't come off once you get it on. I thought the OPS worked fine.

22.0 FLIGHT EQUIPMENT

WORDEN — We ended up without a mission event timer in the LEB. The digital event timer on the main panel - the SECONDS window was obscured. It gradually got worse right from the very start of the flight. It was okay when we first started, and then about half way through the flight the units on the SECONDS window just couldn't be seen. All you could see was the 10-second pulse; you could see that number clicking. But the SECONDS window was completely gone.

SCOTT — Crew compartment configuration - the big stowage boxes are too big. They ought to be partitioned because everything comes out every time you open the door.

WORDEN — I think those mirrors are terrible. They are hard to manipulate. You can never get them in the right location. I had the distinct impression I was going to break them several times. I used those mirrors for some of the photography in window number 4, such as the solar corona, and that required getting different settings, particularly the Hasselblad. I used the mirror to look at the settings because they are on the outside. Those particular settings are on the outboard side of the camera. I had to use that mirror to look at those settings. That really irritated me - trying to get those mirrors turned around.

SCOTT — IV clothing and related equipment - We were wearing the CWGs quite a bit in lunar orbit because of the temperature in the cabin. We suggested maybe putting pockets on the CWGs so you can keep track of your pencils and scissors.

WORDEN — We had a total of five sets of coveralls onboard - the three that we started out with, and then there was a clean set for Dave and Jim.

SCOTT — That was in the LM.

WORDEN — Right. To put on while you were docking. There is nothing quite as refreshing as putting on a clean set of clothes. We ought to consider putting another coverall onboard for the CMP. He wears the coverall more than anybody else.

IRWIN — We had a problem with the wide strut lock.

WORDEN — Yes. That was an internal problem. I wasn't expecting that first détente to be as heavy as it was. After Dave pushed it into lock, it was pretty obvious to me that it was my own fault and my own problem.

23.0 FLIGHT DATA FILE

SCOTT — The checklists Launch and Entry, the Updates, and the Cue Cards were very good. I was pretty happy with the overall Flight Data File. I think it was very timely and very accurate. I don't think we found any mistakes in it during the flight. Towards the end of the flight, especially when we got into EVA, we had a lot of confidence that we didn't before we went. There were a minimum number of changes at the last minute, only a very few pen and ink changes. I thought the guys did a great job. The LM Flight Data File worked very well. It was well organized. Some of us like to use polar star charts, some of us like mercators. I like polar, and I took the CSM backup polar star chart in the LM. That's noise level. It was a good idea to bring the LM tape back to the command module, because we sure used a lot of tape.

WORDEN — Orbit charts. There's a difference between the two of them. I don't think we looked at the Orbit Monitor Chart but once, very briefly, at the very beginning. We never looked at it again because we pretty well knew where we were. We found that the orbit charts weren't good enough for what we wanted off the charts, as far as orienting ourselves.

SCOTT — The Sun Compass. I tried it during the SEVA and it worked great. That was the only chance we had to try it. We carried it the whole way. I felt like we could use it any time. It was a very simple operation and I think a handy backup tool. The Landing Site Monitor Chart. We used that in the simulator quite a bit to get a handle on where we were. We had it out as we went over the landing sites during the activation.

IRWIN — Yes.

SCOTT — It's not very good photography, but it gave us a pretty good handle on where we were over the surface. Horizon Return Chart. That's the one on the surface that I think is a useful chart. The navigation system on the Rover worked so well that we had a tendency not to use all those charts. It would be pretty much crew-preference. To me, the most useful in training, and in driving the Rover, was the Horizon Return Chart which gave me a big picture and a horizon depiction. In retrospect, the optimum for overall location of your position would be, maybe, a 1-to-50 000 with a polar grid on it, with the reciprocal bearings, and 1-kilometer radius circles. Then, any time you wanted to find your position on the surface, you could look at the Rover NAV system and use range and bearing to locate your position.

IRWIN — The only one I used for navigation was the 1-to-25 000, because the surface just didn't look like the enhanced photography.

SCOTT — They were enhanced to the point where they gave much stronger relief than we actually saw on the surface, which somewhat degraded their usefulness.

WORDEN — The Lunar Landmark Maps in the command module were very clear. It was no problem to use them. We had a series of simulated obliques leading up to the landing area. I looked at them maybe once before doing the first set of P24s. I never looked at them again, because the landing area was just so obvious when we got there.

SCOTT — Sure was. Contingency Chart. We had no occasion to use it, but it looks like it's useful. Picking up on general flight planning, the two places in here that will require comment will be from Al on the solo phase.

WORDEN — I was quite pleased with most of the general aspects of the Flight Plan during the solo phase. The checklist was integrated into the Flight Plan, and I thought that worked much better from my standpoint than it would have been if I had used checklists all the time. The only comments I have about the Flight Plan are that we really need to concentrate on making eat periods and exercise periods work-free. That's really important. I found that I was constantly being interrupted in the middle of something to throw a switch somewhere, or to do some other Flight Plan activity. I know there was a concentrated effort to delete all that preflight, but it didn't work out. Continued effort has to be put into that area to ensure that those periods are absolutely free of sequenced work in the Flight Plan. At least one period each day in the Flight Plan should be devoted to a free period so that you can take care of the housekeeping. You could take care of the other functions that have to be done on board, and you wouldn't have to sandwich them in between other periods of activity.

IRWIN — I have a different view of the SIM bay than Al does. It seems like it would be better if the ground called all the actions on the SIM bay rather than having to look at the Flight Plan for them.

SCOTT — Then you're tied to the ground all the time.

IRWIN — That's true.

SCOTT — I'd disagree with that.

WORDEN — That's right. I disagree with that. I did find that I got into a mode of operation where the ground would give me 30-second warnings on something which was in the Flight Plan. That meant that I knew the sequence of things coming in the Flight Plan, but I got a reminder from the ground. If there was a 10-minute period before the next item had to be done, then I could completely forget about the sequencing in the Flight Plan. The ground would give me a 30-second warning, and that would be a cue to me to go back to the Flight Plan and do that function. I found that very useful. In relying on the ground completely, you'd be constantly listening to the communications and waiting for the ground to say something. That would not work too well.

SCOTT — We had good preflight support on all the Data File. The change proposal system had a little breakdown in comm there on a couple of items. We weren't aware of the changes going into the change proposal boards. The boss didn't know whether we supported or we didn't support. I'd recommend in the future that any changes that are being considered by the board first go past the crew so they can pass judgement as to whether they agree or disagree before it gets to the board.

WORDEN — I think you need to split that into two parts, Dave. I think the procedural things should go by the crew, but there are a lot of systems changes that are just mechanical and that you wouldn't want to be bothered with. You need to be aware of them, but I'm not sure that you need to go through all that paper work before the changes are made.

SCOTT — A systems change involves CPCB, which is procedures. And, that's where we had to break down the comm.

WORDEN — Yes.

SCOTT — Real-time procedures changes worked okay. We had an awful lot of Flight Plan updates, but I guess we asked for those. We told them that during solo operations we didn't want any Flight Plan updates. When the three of us were together, we'd accept whatever they thought was necessary, and even though we were busy, I think it worked okay.

24.0
VISUAL SIGHTINGS

SCOTT — We forgot to mention, during the launch phase we went right into the sun. At one point during launch, I put my hand up to shield my eyes so I could see the ball. I was surprised. It's no problem, but it would be a nice thing to be aware of on an early morning launch like that.

IRWIN — It wasn't in my window. Must have been in your window.

SCOTT — Boy, the light was smack in my eyes.

WORDEN — I don't remember seeing anything that we didn't expect to see.

IRWIN — We thought we saw a satellite, but we decided it was probably a planet. Remember?

SCOTT — Yes, that's right. We sure did.

WORDEN — Yes. As a matter of fact, didn't we think that was a satellite because it always showed up at the same place?

SCOTT — Yes it showed up at both sunset and sunrise. You can't do that if it's not a satellite. I mean if it's the same star, I don't think we would see it at both sunrise and sunset.

WORDEN — It always appeared about the same place in the window, too.

IRWIN — Yes.

SCOTT — It's written down in the Launch Checklist at the point at which we saw it.

25.0 PREMISSION PLANNING

SCOTT — Premission planning. Mission plan. The mission plan referred to the requirements document and that was continuously updated. We stopped looking at it about 3 months before the flight because we couldn't keep up with it, and we couldn't prepare ourselves for this much activity with the constant changes that they had. I'm not sure that the Mission Requirements Document was fully completed during the flight because we just couldn't possibly keep up with it. It should be frozen much earlier than it was in our case. For instance, the scientific community came up with a definition of a comprehensive sample on the surface, which included some six or seven events, long after we had completed our training on those type activities. We just couldn't accept a change at that late date. In fact, the MRD has to be frozen at least 3 months before the flight. If there is something that needs to be added, put it on the next flight. I thought the Flight Plan came along fairly well.

WORDEN — I think that for the mass of detail that went into this Flight Plan, and for the originality of the Flight Plan, I thought they did a great job.

IRWIN — Yes, it was well done.

SCOTT — Spacecraft changes. We had a lot of little changes at the end, but none of them really affected the mission.

IRWIN — We mentioned a couple that we weren't aware of, such as the H_2 flow.

SCOTT — Yes, they changed our tape meter and then it broke, and I wonder if that was why. Procedure changes. I think that was fairly minimized toward the end. We had quite a few early in the game, but the last month I didn't see too many.

WORDEN — As a matter of fact, I thought the pen and ink changes that we saw were pretty minimal.

SCOTT — Mission rules and techniques. We got an early handle on mission techniques by having the data priority meetings about 6 months before flight and got all those out of the way, and the techniques documents established as to what we were going to do. We proceeded along in that direction while the paper work caught up with us. I thought that gave us a good position for understanding the mission rules as they were developed during the simulations. We always had a fairly good handle on the techniques of the mission rules. I recommend an early start on those things. The flight directors kept us well abreast of changes in the philosophy in the mission rules, which I thought was very good. I don't think we had any disagreement on mission rules as we went. Everybody was in agreement on exactly what we were to do.

26.0 MISSION CONTROL

SCOTT — Mission control. Go no-go's were timely. We were never behind on understanding those. Jim did you ever have any problems with flight updates?

IRWIN — No, they were all very timely.

SCOTT — Those late PIPA bias updates were surprising on the descent as was the 3000-foot call. We'll get that worked out. The consumables seemed to work out well, passing back and forth the data. Our inflight gauges were somewhat different from the ground, particularly in the RCS. It was quite different. We were all abreast of the situation throughout the mission. Anything in consumables, problems anywhere? It was a good thing somebody found the 25-pound water delta, that day we had the leak in the LM. That was a sharp bite because we could have ended up with an unexplained 25-pound water loss, and a bunch of water in the back of the LM that might have frozen up and broken a line or a wire or something.

RCS fuel. I didn't look at the command module's when we got back down, but I'm sure we had plenty. We had both rings and it was a nominal entry. Service module: we ended up with 35 percent. In the orbital operations, you never got close to the red line, did you?

IRWIN — The last number I recall hearing was 15 percent above the red line on one quad. There was some concern about the second day of lunar orbit operation that we were expending fuel more rapidly than we should have been. We were going to have a problem with the red line. They called up a weight change for the CSM to try and take the DAP into firing fewer times, to conserve some of that fuel. That didn't work. So we went back to the actual weight and nothing else was said. We never compromised any of the operation and we ended up not even close to the red line. I never did hear a number after the 15 percent, but I assume that we were comfortably above it.

SCOTT — The LM RCS was great. We ended up with something like 80 per cent when we came in there, in the final braking. DPS fuel. We had plenty. We had a minute and 51 seconds of hover time remaining. SPS fuel. I think that was all planned and utilized, with a couple percent left. Real-time changes from mission control. No comments. Communications. The system worked well all the time.

<table>
<tr><td>27.0 HUMAN
FACTORS</td><td>SCOTT — The health stabilization and control program is a good idea. There was some concern about us catching a bug just before launch and one of the reasons we didn't catch it probably was because of the isolation.</td></tr>
<tr><td>27.1 PREFLIGHT</td><td>IRWIN — Yes, I thought it was a good idea. It gave us a chance to rest up a little bit, too.</td></tr>
</table>

WORDEN — I felt quite rested before we went.

SCOTT — I did too. I think we all got a lot of rest; everybody was in good physical shape, healthy, and ready to go. Medical care. We had no problems in that area. Time for exercise, rest, and sleep was adequate. There was no pressure at all during the last 3 weeks. As a matter of fact, we were trying to find things to do. Medical briefing and exams. My impression is that the medical protocol seems to grow and grow, and everybody wants a little bit more. It's an awful lot. Because of the requirements of the mission and the greater demands on the crew during a mission, it would be nice if we could keep the same level of medical, activity instead of increasing it. In each test, each group of people seems to have one or two more little things they want, and it just adds up to one big, big step. I think we're reaching the limit.

IRWIN — I'm glad we went on that low residue diet before flight.

SCOTT — You think it helped.

IRWIN — I might have had to go the first day, if I hadn't been on that diet.

SCOTT — Eating habits and amount of food consumed at F minus 5 to F minus 0. We went on the low residue, and I guess it helped.

27.2 FLIGHT   SCOTT — Appetite and food preference. Before we went we decided we were going to try to eat everything on board as prescribed on the meals. That's exactly what we did. We ate everything there was to eat except the bacon squares and a few other things. The more you eat, the more your system works and the more waste you have. That created a time problem in that everybody had to use the waste management procedures at least every other day. That's time you have to allow for. It's still a good idea to eat because I think we all felt very good throughout the flight. Once we got on the curve of eating all the meals, it was no problem. In fact, I think we all got pretty hungry when it came to be meal time.

IRWIN — In that connection, I just wish the packages in the pantry had been labeled per meal without having to search for items. We wasted a little time there.

SCOTT — In other words, have meals all the way through, and have a pantry with extras. I think that's a good idea. Changes in food preferences as the flight progressed, were not noticeable, except we all wanted scrambled eggs for breakfast.

WORDEN — That's right, and we all got off the bacon squares.

IRWIN — I sure could have gone for some more of that chalais soup.

SCOTT — That's the kind of thing that's an individual preference, and I don't think it's too meaningful for us to go through our preferences on the foods. In general, the food was very good and there was enough variety that we were all happy and everybody ate good.

WORDEN — The wetpacks were great for giving you something to chew, giving you the bulky kind of food that you wanted, something that's already prepared. However, the flavor in the wetpacks left something to be desired. It discouraged me, upon first opening up a wetpack, to first see an amount of gray grease that the food had been cooked in. That didn't make it very tasteful.

SCOTT — Catsup and mustard helped. Chili sauce, next time, might even be better. No deviations. We stuck to the eat periods and and the programed menus. Food preparation.

IRWIN — Well, you noticed gas.

SCOTT — Yes, I thought there was too much gas. The cans worked and once we learned how to use the soup packs, they worked okay.

WORDEN — Yes, there's a certain amount of readjustment you have to undergo to get used to opening those things. A comment on the canned food. Those things that are packaged in metal cans in a liquid, caused us some problems, and I don't think we ever really solved the secret of opening the cans without some of the liquid out.

IRWIN — Some of them had almost too much liquid in them, such as the peaches.

WORDEN — They were very difficult to open without spilling some of it; without getting some drops of liquid.

SCOTT — We used the germicidal tablets until we ran out of them. Some of the packages we brought back were without the tablets because we just ran out. Odors were okay. We ate everything on the surface. I thought the water tasted good, except for the gas we mentioned.

WORDEN — I was expecting more of a chlorine taste in the water than I actually found. That was quite a pleasant surprise.

IRWIN — I frankly couldn't tell the difference in LM and command module water.

SCOTT — We talked about the sleep. We recorded all that. Restraints were okay. Everybody sleeps a little different. Everybody will find out the way they want to sleep. Exercise. We discussed that. Everybody exercised when they were supposed to, and we talked about the ergometer. Anybody get any muscle soreness?

WORDEN — The first couple of nights, I had a very sore back. In talking to the doctors, after the flight, it was readily explainable in terms of the one-g conditioning that your back muscles have. It's a normal thing. It's just something to be aware of.

SCOTT — Inflight oral hygiene. We only had one tube of toothpaste - and Al happened to get it. We always had to borrow Al's toothpaste, and we didn't have any on the lunar surface.

IRWIN — You took your tooth brush to the lunar surface. I didn't I just gave up brushing my teeth after reaching the lunar surface and from then on I just forgot it.

SCOTT — Sunglasses were okay. Unusual or unexpected visual. I don't remember any problems there. Distance judgement versus aerial perspective during EVA. Everybody knows it's hard to judge distances on the surface. Everything appears closer than it really is.

WORDEN — We've already commented on the Lexan shield. During those periods when we were exposed to the ultraviolet, I never noticed any discomfort; never noticed any effects on me at all, during the times when the Lexan shield was on.

SCOTT — Medical kits. We didn't use them, except for the biosensors and these were discussed.

IRWIN — I'd like to suggest that they put the right size...

SCOTT — Housekeeping. We discussed that. Shaving. We didn't. Dust. We discussed that, relative to the vacuum cleaner and the cabin fans. Radiation dosimetry. The PRDs were a bit of a problem because they just kept having to be searched for. We were being asked about PRDs at odd times. We got out of sync the first day and never got back in. I didn't realize they were quite that important, especially the low numbers we were getting. The crew ought to pay attention to having those things around because the doctors will ask you. We never touched the radiation survey meter. Personal hygiene. I rarely used those wipes.

WORDEN — I found the wipes a little bit disagreeable, in a way. I didn't like the odor of the wipes.

IRWIN — We commented that they ought to be scented.

WORDEN — The one bar of soap that we had, that Jim took, was a very good idea and I would suggest that in the future. You get almost as much a refreshing feeling from the scent of that soap as you do from using it. I think that is a good idea and I think we ought to consider scenting the wet wipes.

SCOTT — I thought the towels worked great. If you had a bar of soap and towels, you'd be in good shape.

WORDEN — I don't think we used but about 10 percent of those wet wipes.

IRWIN — I threw most of mine away.

WORDEN — There was a large package in the food locker when we finished.

IRWIN — You guys used the combs. Somewhere, I lost, mine.

SCOTT — Your light-weight headset never worked.

WORDEN — That's right. The microphone in my light weight headset never did work.

SCOTT — Jim left it on the surface somewhere.

IRWIN — The last time we saw it was when we were on the surface.

WORDEN — A spare light-weight headset should be considered.

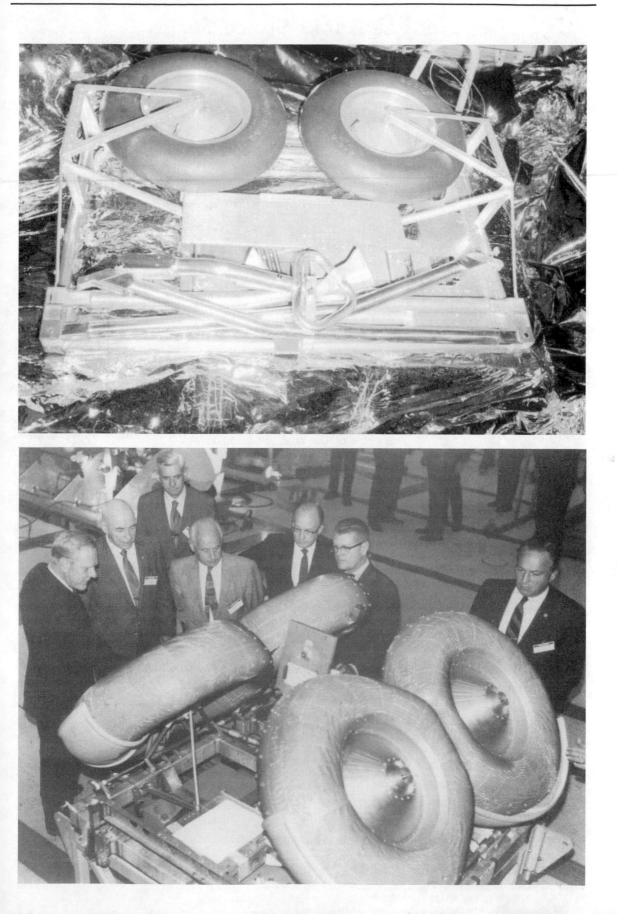

Apogee Books Space Series